Functional Polymer Foams

Green Fabrication Methods, Performance and Applications

Haoyang Mi

WILEY VCH

Author

Professor Haoyang Mi
Zhengzhou University
Zhengzhou 450003
China

Cover Design: Wiley

Cover Image:
© Polygraphus/Shutterstock

Library of Congress Card No.: applied for

British Library Cataloguing-in-Publication Data
A catalogue record for this book is available from the British Library.

Bibliographic information published by the Deutsche Nationalbibliothek
The Deutsche Nationalbibliothek lists this publication in the Deutsche Nationalbibliografie; detailed bibliographic data are available on the Internet at <http://dnb.d-nb.de>.

Print ISBN: 978-3-527-35295-1
ePDF ISBN: 978-3-527-84458-6
ePub ISBN: 978-3-527-84459-3
oBook ISBN: 978-3-527-84460-9

Typesetting Straive, Chennai, India

Printed and bound by
CPI Group (UK) Ltd, Croydon, CR0 4YY

C9783527352951_061124

Contents

Author Biography

Haoyang Mi received his BS and PhD degrees from South China University of Technology in 2010 and 2015, respectively. He served as a visiting scholar at the University of Wisconsin-Madison from 2011 to 2014. Subsequently, he worked as a postdoctoral researcher at the Wisconsin Institutes for Discovery, University of Wisconsin-Madison, from 2016 to 2018, and as a research fellow at Hong Kong Polytechnic University from 2018 to 2019. In 2019, he joined the National Engineering Research Center for Advanced Polymer Processing Technology, Zhengzhou University, as an associate professor.

Dr. Mi has amassed over 15 years of experience in the field of polymer composite foams, specializing in the fabrication and application of multifunctional porous polymer composites. His work has contributed to advancements in various domains, including energy absorption, self-cleaning surfaces, oil–water separation, thermal insulation, solar steam generation, electromagnetic interference shielding, tissue engineering scaffolds, flexible sensors, and energy harvesting devices. Notably, he has been an advocate for the environmentally friendly technique of supercritical carbon dioxide foaming in advanced applications.

Dr. Mi directs three grants funded by the National Natural Science Foundation of China and many regional and corporate projects within the realm of polymer processing and polymer foams. He also holds the role of associate editor for international journals such as *Bio-design and Manufacturing* and *Scientific Reports*. His academic contributions include over 100 SCI journal papers, where he served as either the first author or corresponding author. His work has received significant recognition with a total of over 8500 citations and an impressive h-index of 52 according to Google Scholar. Furthermore, he holds 20 invention patents, serving as the first inventor for 16 of them.

Preface

This book delves into the world of advanced polymer foams, which have evolved into functional materials and integral components within cutting-edge fields. Three-dimensional porous foams are at the forefront of advanced functional materials, driving innovations in diverse cutting-edge applications. The microstructure and material composition of these polymer foams play pivotal roles in defining their unique functional properties. Crafting polymer and polymer composite foams with precise and desirable structures poses a formidable challenge, yet is essential for unlocking their full potential across a range of applications. This book seeks to bridge the existing gap by introducing readers to the manifold applications of functional porous polymer foams, elucidating the underlying fundamentals that drive their functionality, and providing insights into their fabrication processes.

Furthermore, this book will mainly focus on the polymer foams that are prepared by physical foaming where the technique is applied, especially the foaming process using supercritical carbon dioxide ($scCO_2$) as the foaming agent, since the $scCO_2$ foaming is an environmentally friendly, cost-effective, and scalable technique that is promising for large-scale production of multifunctional polymer foams for advanced applications. The content of this book transcends the conventional, showcasing innovative applications of polymer foams in energy absorption, acoustic absorption, superhydrophobicity, electromagnetic interference shielding, tissue engineering scaffolds, flexible sensors, triboelectric nanogeneration, and solar steam generation.

Comprising 10 comprehensive chapters, this book embarks on a journey through the multifaceted realm of functional polymer foams and their advanced applications in recent years. Our journey begins in Chapter 1, where we establish the foundational principles of $scCO_2$ foaming, its classifications, and its significance. We then venture into a spectrum of applications in subsequent chapters. Chapter 2 offers an in-depth examination of functional polymer foams in the realm of energy absorption. We uncover the fundamental theories, the impact of foam structure, methods for fabricating energy-absorbing foams, and applications of energy-absorbing foams. In Chapter 3, the focus shifts to thermal insulation, elucidating the mechanism behind foams' insulation properties, their structural influence, fabrication techniques, and pertinent advanced research. Chapter 4 brings us to the world of acoustic absorption, where we unravel the intricacies of basic theories, influential factors, fabrication approaches, and advanced sound-absorbing foams reported in the literature. Chapter 5 introduces us to superhydrophobic polymer foams, indispensable in self-cleaning and oil absorption applications. We delve into superwetting theories, influential factors, fabrication methods, and the relevant literature and applications. In Chapter 6, we explore composite polymer foams designed for electromagnetic interference (EMI) shielding, deciphering the shielding mechanism, influential factors, and advanced EMI shielding foam fabrication methods. Biomedical foams for tissue engineering are introduced in Chapter 7, outlining the requirements for tissue engineering scaffolds, the primary biodegradable polymers, scaffold fabrication methods, and applications of the polymer-based scaffolds. Chapter 8 steers us toward composite polymer foams for flexible piezoresistive pressure sensors, unveiling their sensing mechanisms, factors affecting sensing properties, common materials, fabrication methods, and relevant literature. The journey continues in Chapter 9, as we investigate polymer foams tailored for triboelectric nanogenerators (TENG) in energy generation. We probe into TENG working mechanisms, factors influencing output performance, innovative fabrication methods, and advanced TENGs developed using polymer foams. Chapter 10 ushers us into the world of polymer foams for solar steam generation, shedding light on solar steam generation mechanisms, the impact of material and porous structure on performance, fabrication methods, and pertinent literature.

We believe that this book will serve as a bridge between environmentally friendly polymer foam fabrication methods and their limitless potential in advanced functional applications. Readers will not only expand their knowledge of foam development but also gain insights into enhancing their performance and extending the applicable areas of polymer foams to potential cutting-edge fields.

Welcome to a journey through the world of advanced polymer foams.

10 Aug 2024

Haoyang Mi
National Engineering Research Center for Advanced Polymer Processing Technology
Zhengzhou University
Zhengzhou, China

Acknowledgment and Dedications

I would like to express my heartfelt gratitude to my advisors, Professor Xiang-Fang Peng, Professor Lih-Sheng Turng, Professor Han-Xiong Huang, Professor Shaoqin Gong, Professor Heng Li, and Professor Xiaoming Tao, for their unwavering support, invaluable guidance, and enduring patience throughout the journey of my research career. Your wisdom, expertise, and mentorship have been instrumental in shaping the content and direction of this work.

I am also indebted to my dedicated graduate students for their diligent efforts in collecting data, conducting experiments, and providing critical insights. Your contributions have been indispensable in ensuring the accuracy and depth of the information presented within these pages. Specifically, Chengzhi Yuan helped with Chapter 1, Qingli Tian helped with Chapter 2, Zemian Zuo helped with Chapter 3, Miaomiao Zhang helped with Chapter 4, Lin Wang and Xiao Li helped with Chapter 5, Jinghao Qian helped with Chapter 6, Yi Yang helped with Chapter 7, Xiaoyue Ren helped with Chapter 8, Ruixue Li helped with Chapter 9, and Shuangjie Sun helped with Chapter 10.

Furthermore, I extend my gratitude to all the colleagues, collaborators, and peers who have shared their knowledge and experiences, enriching the scope of this book. This endeavor would not have been possible without the support and encouragement of my family and friends. Your unwavering belief in my pursuits has been a constant source of inspiration.

Last but not least, I extend my heartfelt appreciation to the readers of this book. It is my hope that the knowledge and insights shared within the book will be of value to the scientific community and beyond.

Haoyang Mi

1

Introduction

1.1 Overview of Polymer Foams

Foam materials, characterized by highly porous structures, are prevalent in both natural and synthetic forms [1]. Examples in nature include natural sponges with open cellular structures and wood, which humans have used for millennia. Polymer foams are defined as a kind of polymer material formed by a large number of microcellular cells containing a gas medium uniformly dispersed in the polymer matrix. Almost all polymers can be made into polymer foams. In the twentieth century, with the growth of the polymer industry, various types of polymer foam products emerged. The introduction of the gas phase makes polymer foams possess excellent performance in material weight reduction, heat insulation, sound absorption and noise reduction, shock absorption, etc. As a gas phase and solid phase mixed material, polymer foams possess special properties. It is widely used to prepare various packaging materials, automotive and aircraft parts, sports equipment, building materials, etc.

Traditional polymer foams are mainly prepared by direct mixing of molten polymer and gas. The foams usually have large cell diameters and low cell density, and the average cell size is generally larger than 100 μm. These large cells often become the starting point of cracks, which become one of the factors restricting the performance improvement of traditional polymer foams.

With the continuous development of human society and the enhancement of people's pursuit of a higher quality of life, the performance of traditional polymer foams has gradually entered a bottleneck that cannot meet the new requirements in the fields of automobile, aircraft, aerospace, electronic devices, and medical devices. To address this problem, Professor N P Suh from MIT and colleagues first proposed the microcellular foaming technology and defined the cell size and density range. The average cell size and cell density in modern microcellular foam are in the range of 0.1–100 μm and within 10^9–10^{15} cells cm^{-3}. The microsized cells in the foamed part could blunt the crack tip and block the crack propagation when the cell size is smaller than the crack size, which greatly improves the mechanical properties of the polymer foams compared with traditional large cell foams [2].

Microcellular foaming enhances mechanical properties compared to traditional foams, offering more than fivefold improvements in impact strength, toughness, and fatigue life, with density reductions ranging from 5% to 95% [3].

Functional Polymer Foams: Green Fabrication Methods, Performance and Applications,
First Edition. Haoyang Mi.

In addition, microcellular foaming leads to lower dielectric constants and thermal conductivities [4] expanding its range of applications. Owing to their low density and high toughness, robust shock strength, and fatigue resistance, these materials are suitable for packaging, as well as shock-absorbing buffers.

In applications requiring lightweight and high-strength soundproofing, such as aircraft and automobiles, microcellular foams are preferred due to their high specific strength and effective soundproofing. Their low dielectric constant, thermal conductivity, and excellent electromagnetic shielding/absorption properties, arising from their unique vesicular structure, make microcellular foams desirable for aerospace and electronics industries. Moreover, adjusting foaming process conditions, such as temperature and pressure, allows for control over the final material structure and properties.

Beyond these properties, microcellular foams with high open porosity find applications in various fields, including biological tissue scaffolds, filtration adsorption, catalyst carriers, and sustained drug release. Traditional chemical blowing agents, like chlorofluorocarbons (CFCs) and hydrochlorofluorocarbons (HCFCs), have environmental limitations due to the emission of chlorinated pollutants. In contrast, microcellular foaming with supercritical CO_2 as the foaming agent is environmentally friendly and offers excellent performance making it highly promising for various applications. To delve into the advanced applications of polymer foams, it is crucial to first grasp the fundamental principles of polymer foaming technology.

1.2 Polymer Foaming Methods

The basic process in polymer foaming contains the blending of a polymer matrix with a foaming agent, separating gas from the polymer, and fixing the polymer matrix to form a unique uniform cellular structure. According to the gas introduction methods, polymer foaming can be roughly divided into three categories, namely, mechanical foaming, physical foaming, and chemical foaming.

1.2.1 Mechanical Foaming

Mechanical foaming is a method in which air is sucked into the polymer matrix by intense mechanical agitation to form a uniform foam body. Air and emulsifier or surfactant can be added to shorten the molding cycle. It is not necessary to add a foaming agent, but it has the disadvantage that the bubbles generated by this process are easy to disappear, and it is difficult to meet the production requirements of microcellular foam.

1.2.2 Physical Foaming

Physical foaming refers to the process in which gas is directly injected into a polymer melt to create a cellular structure. There are typically three approaches utilized:

(1) Inert gas is dissolved into a plastic melt or paste under pressure, then the pressure is released quickly, resulting in the creation of cells within the polymer matrix.
(2) Low-boiling temperature liquids that are dissolved in the polymer melt and are then evaporated at elevated temperature to form gas bubbles in the polymer matrix;
(3) Hollow microspheres or expandable microspheres are directly added to the polymer matrix via melt blending to form a porous foam structure.

In the process of physical foaming, the physical foaming agent undergoes only a change in state, such as from a saturated solution to liquid/gas, or a supercritical fluid (SCF) state to a gas, and the composition of the gas does not change. Physical foaming agents typically can be divided into two categories, namely, inorganic foaming agents and organic foaming agents. Inorganic foaming agents include CO_2, N_2, air, etc., while organic foaming agents include hydrocarbons, chlorinated hydrocarbons, fluorinated and chlorinated hydrocarbons, etc. However, due to potential environmental pollution and safety issues associated with the chemical foaming agent, such as flammability, explosiveness, and ozone depletion, most of them have been phased out or restricted for use, as people have become more aware of safety and environmental protection.

The physical foaming method using physical foaming agents has a relatively low cost, especially for CO_2 and N_2, which are inexpensive and nonpolluting. They are also flame retardant, making them highly valuable. However, it usually requires special molding machines and auxiliary equipment, and inert gases such as CO_2 and N_2 require high pressures to achieve foaming, making the process highly technically challenging and equipment demanding.

SCF foaming technology is a physical foaming method using SCF as the foaming agent. SCF refers to fluids in their supercritical state where both the temperature and pressure exceed their critical points (Figure 1.1) [6]. SCF is in the state between liquid and gas, with viscosities and diffusion coefficients close to gases and densities and solvation capabilities close to liquids. It has the advantages of both liquid and gas, such as good fluidity, large mass transfer coefficient, easy adjustment of fluid density, good diffusivity, and solubility [7]. The most commonly used SCFs are CO_2, N_2, water, ethane, etc. As listed in Table 1.1, the critical point of CO_2 is the temperature above 31.1 °C and pressure above 7.38 MPa, which is relatively easy to reach

Table 1.1 Critical points of common supercritical fluids.

Supercritical fluid	Critical temperature (°C)	Critical pressure (MPa)
$scCO_2$	31.1	7.38
ScN_2	−147.1	3.39
scH_2O	374.2	21.83
ScC_2H_6	32.3	4.82

Source: Adapted from Cha [8].

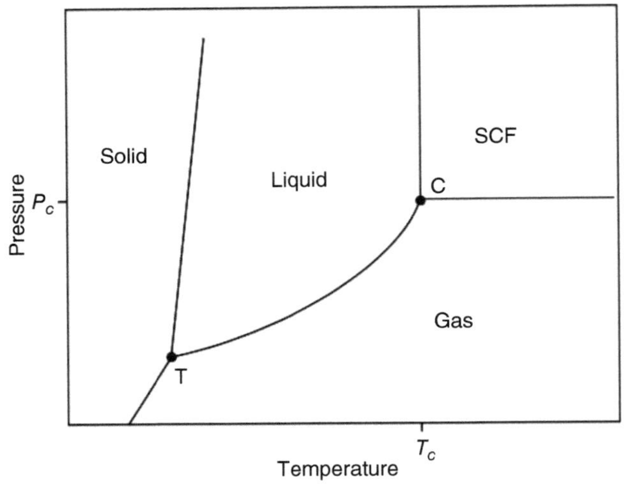

Figure 1.1 Phase diagram of fluid: T, triple point; C critical point. Source: Nalawade et al. [5]/with permission of Elsevier.

compared with other gas/liquid [5]. In addition, attributing to the high solubility and high diffusion rate of supercritical CO_2 ($scCO_2$), physical foaming using $scCO_2$ as the foaming agent could achieve a higher gas content, greater expansion ratio, and higher cell density. Considering the advantages of SCF foaming using $scCO_2$ and scN_2, it has been widely researched in recent years and employed in the production of various thermoplastic polymer products.

1.2.3 Chemical Foaming

The chemical foaming method relies on chemical reactions to generate gases to form cells in the polymer matrix. Gases are typically generated by the decomposition reaction via the heating of a chemical foaming agent added to a polymer matrix. At present, common chemical foaming agents mainly include sodium bicarbonate, ammonium nitrate, azo compounds, sulfonyl hydrazides, and nitroso compounds, etc. In the chemical foaming of materials like polyurethane, gas is generated by the cross-linking reaction between the isocyanate functional groups and water molecules. For chemical foaming, the temperature and gas content of the reaction is the key to determining the foaming quality, the reaction temperature at the time of gas generation should match the processing temperature of the polymer, the rate of gas generation needs to be controllable, and the gas content needs to be adequate.

The main advantage of the chemical foaming agent is that it does not require any modification of existing plastic processing equipment, and the process of injection molding/extrusion of polymer foams using a chemical foaming agent is essentially the same as the general injection molding/extrusion process. The heating, mixing, plasticization, and most of the foaming expansion of the plastics were done in an injection molding machine/extruder. Compared to physical foaming agents, the disadvantages of chemical foaming are mainly that they are more demanding for the reaction conditions, the foaming agents usually cost more, the potential environmental pollution, and the volatile organic compound (VOC) emission.

1.3 Fundamentals of SCF Foaming

In the polymer foaming process, a homogeneous system of polymer/gas solution is obtained by blending dissolution or saturation of a two-phase system consisting of polymer and gas, and then the equilibrium state of the system is broken by changing the external conditions (usually pressure and temperature) to trigger the nucleation of the cells due to the separation of the gas phase from the polymer phase owing to the transfer of the gas from the supercritical state to the gas state. Thus, the cells are the product of this phase separation process. The formed cells can, therefore, be seen as a way to counteract changes in external conditions [9], while the first half of the foaming process can be seen as a transition from a stable (homogeneous) state to a metastable or unstable (multiphase) state.

However, it is crucial to recognize that the cells formed within this metastable state lack stability due to the high molecular mobility inherent to the polymer matrix. Consequently, a reinforcement treatment, such as cooling, applied to the polymer-dense phase surrounding the cells is imperative. This treatment serves to immobilize the advantageous structure acquired during the foaming process, ultimately leading to the attainment of a stable foaming product. The latter half of the foaming process can thus be construed as a transformation of unstable foaming products into stable ones.

From the foregoing exposition, it becomes evident that the polymer foaming process is an intricate thermodynamic and kinetic undertaking. The fundamental process can be broadly delineated into five stages, as depicted in Figure 1.2 [10].

1.3.1 Preparation of Homogeneous Solution

A homogenous solution needs to be constructed as the basis for the physical foaming process. During the formation of gas polymer homogeneous systems, the solubility

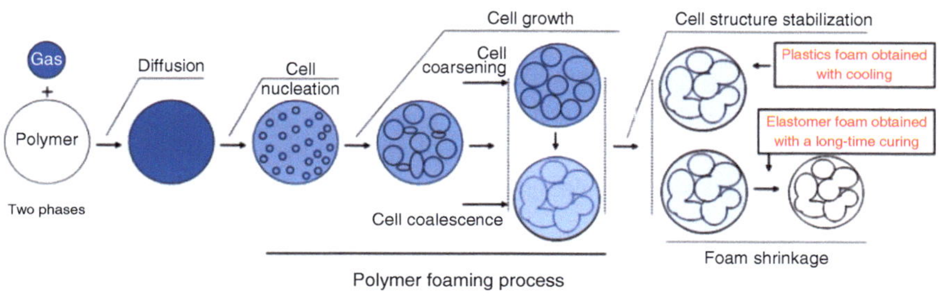

Figure 1.2 Schematic illustration of the five stages in the SCF foaming process. Source: Zhai et al. [6]/with permission of Taylor & Francis.

of the gas is one of the most important parameters deciding the content of the foaming agent that can be introduced to the system, which in turn affects the final foam density and cell size. Determining the solubility of the foaming agent in the polymer is fundamental to the overall subsequent foaming process, so understanding the interaction between gases and the polymer melt and the influencing factors on solubility is critical.

Binary systems of gases/polymers are usually described using Henry's law, and the solubility of gases in polymers can be derived from Henry's law [11]

$$C = K_H P \tag{1.1}$$

where C is the gas solubility, K_H is Henry's constant, and P is the gas pressure.

Thus, increasing pressure is an effective mean to enhance gas solubility, and of course, a similar effect is obtained by lowering temperature. Henry's law combined with van't Hoff equation can explain the relationship between temperature and gas solubility [12].

$$K_H = K_0 \exp\left(\frac{-\Delta H_S}{RT}\right) \tag{1.2}$$

where K_H is Henry's constant, K_0 is pre-exponential factor, ΔH_S is the heat of the solution, R is the gas constant and T is temperature.

Henry's equation is based on ideal dilute solution solvent conditions without considering the interaction between gas and polymer melt, especially under high-pressure conditions, where the interaction between SCF and the polymer is more complicated. When a large amount of SCF is dissolved in the polymer, the plasticizing effect would affect the surface tension and rheological properties of the system.

The Flory Huggins equation [9] is a good guideline to determine the amount of gas used in a system, and it is expressed as:

$$\Delta F_m = kT(n_g \ln \varphi_g + n_p \ln \varphi_p + \chi n_g \varphi_g) \tag{1.3}$$

$$\chi = 0.3 + V_g/RT(\delta_p - \delta_g)^2 \tag{1.4}$$

The left-hand side of Eq. (1.3) ΔF_m refers to the mixing of free energy after the mixing of polymer and gas. On the right-hand side of the equation, n_g and φ_g refer to the moles and volume fraction of the gas, while n_p and Φ_p refer to the moles and volume fractions of the macromolecules, χ is a parameter to describe the action of macromolecules and gases (consists of both entropic and enthalpic components, with an average value of about 0.3 for an entropic component of most polymer/gas systems, and the enthalpic component is determined by the solubility parameter [10]), K is the Boltzmann constant, and T is the thermodynamic temperature. V_g in Eq. (1.4) represents the molar volume of the gas, R is the ideal gas constant, and δ is the solubility parameter.

In practice, solubility is greatly affected by the temperature, pressure, and crystallization behavior of polymers [13]. Normally, the solubility of liquid/gas in a polymer matrix would increase with the increase of pressure, while decrease as the increase of temperature [14].

1.3.2 Cell Nucleation

The basic theory of cell nucleation in polymer foaming originates from the classical nucleation theory established by Gibbs in the twentieth century. Colton and Suh [15] described the nucleation process during foaming. The formation of cells is actually a phase separation process initiated by the rapid pressure change or temperature change. The cell nucleation step is driven by the thermodynamic instability of gas/polymer homogeneous systems under high-temperature and high-pressure conditions. The physical foaming method caused a sharp decrease in the gas solubility in the polymer by a rapid pressure drop to form very high supersaturations. When the unstable high-energy state gas molecules cross the free energy barrier, high-energy state molecules would aggregate with each other through the activation transition, and then stable nuclei are formed on these aggregation sites.

In the phase separation process, the nucleation of gas cells needs to overcome the phase transition energy barrier (i.e., the phase transition activation energy), the dynamic source of the phase transition being the difference between the free energy of the system's initial state and the end state. There are typically two types of cell nucleation mechanisms according to the difference in the initial state of the nucleation system, namely, homogeneous nucleation and heterogeneous nucleation (Figure 1.3).

1.3.2.1 Homogeneous Foam Nucleation

When the system is composed of a single homogeneous phase of gas/polymer mixture, it is assumed that there are no impurities within the system that can induce nucleation. At this point, each gas molecule is a theoretical nucleation point. In a thermodynamic system composed of a polymer melt and dissolved gas, the total Gibbs free energy of the system changes according to the law of thermodynamic energy conservation ΔG consists of three parts: the change in bulk free energy, the change in chemical potential, and the change in interfacial free energy, as shown in Eq. (1.5).

$$\Delta G = -\Delta P \cdot V_G + n_g(\mu_G - \mu_L) + \gamma_{LG} \cdot A_b \tag{1.5}$$

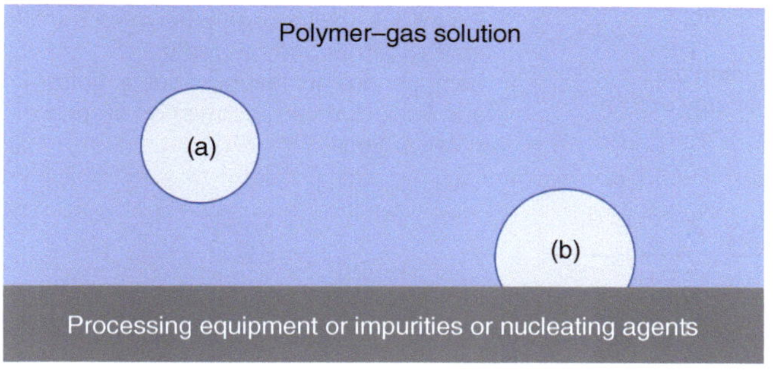

Figure 1.3 Two nucleation types in a polymer–gas system (a) homogeneous nucleation, and (b) heterogeneous nucleation.

where ΔP is the pressure difference between inside and outside bubbles, μ_L is the chemical potential of gas molecules in polymer melts, μ_G is the chemical potential of gas molecules in bubbles, n_g is the number of gas molecules in the gas phase, A_b is the surface area of the bubble, and γ_{LG} is the gas/liquid interfacial tension.

The first term is the work done by the volume expansion of the gas inside the cells, the second term is the difference in chemical potentials before and after nucleation, and the third term is the work required to create the liquid gas interface. Since the chemical potential difference is zero for the nucleus in chemical thermal equilibrium [15a], and the cell is assumed spherical for homogeneous nucleation, Eq. (1.5) can be written as:

$$\Delta G = -\frac{4}{3}\pi r^3 \cdot \Delta P + 4\pi r^2 \cdot \sigma \tag{1.6}$$

where r is the cell radius and σ is the surface tension of the polymer matrix.

The function curve of the free energy barrier ΔG and the radius r of the cell nucleus during homogeneous nucleation can be obtained, and the maximum value of a ΔG corresponding to the radius size of R_c can be obtained from Figure 1.4.

Since the thermodynamic system tends to maintain a low-energy state, the cell nucleus tends to collapse when the radius is smaller than R_c, while it can spontaneously grow up when the radius is larger than R_c seeking to balance Eq. (1.7) to 0.

$$\frac{d\Delta G}{dr} = 0 \tag{1.7}$$

The formula for the critical radius R_c can be derived:

$$R_c = \frac{2\sigma}{\Delta P} \tag{1.8}$$

Substituting Eq. (1.8) into Eq. (1.7) yields a uniform nucleation free energy barrier of:

$$\Delta G^*_{hom} = W_{hom} = \frac{16\pi\sigma^3}{3\Delta P^2} \tag{1.9}$$

According to the classical nucleation theory, the formula of homogeneous cell nucleation rate can be derived [15c]:

$$N_{hom} = f_0 C_0 \exp\left(\frac{-\Delta G^*_{hom}}{kT}\right) \tag{1.10}$$

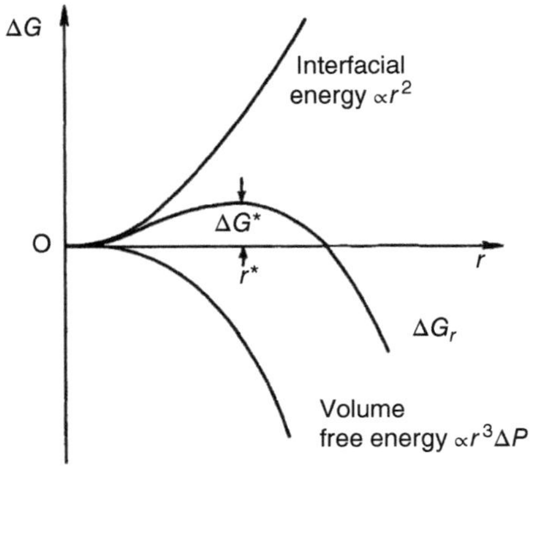

Figure 1.4 Relationship between the free energy barrier and cell radius in homogenous nucleation. Source: Colton and Suh [15a]/with permission of John Wiley & Sons.

where f_0 is the frequency factor of gas molecules entering the cell nucleus, expressed in 1/s, C_0 is the unit volume gas concentration, k is the Boltzmann constant, and T is the absolute temperature.

From Eqs. (1.8) and (1.9), the critical radius R_c and the critical free energy barrier W_{hom} have a strong dependence on the surface tension of the melt and the pressure difference inside and outside of the cell nucleus. Therefore, increasing the supersaturation degree, decreasing the polymer melt surface tension, and increasing the internal and external pressure difference are approaches to reduce the critical radius and the critical independent energy barrier, which would enhance the cell nucleation rate and the number of cells that can be formed. In addition, increasing the depressurization rate, the gas concentration, and the system temperature would also increase the cell nucleation rate. However, some conditions are contradicted naturally such as the gas concentration would be reduced when elevating the temperature. Hence, saturation pressure is usually recognized as the most influential factor in the cell nucleation rate.

1.3.2.2 Heterogeneous Foam Nucleation

If there is a second phase in the system, such as foreign impurities or nucleating agents, the energy barrier for cell nucleation would be lower with the coexistence of the gas–liquid–solid three phases. The composite interface can serve as a starting point for cell nucleation in heterogeneous nucleation. The critical radius R_c and free energy barriers ΔG in heterogeneous nucleation were calculated similarly to homogeneous nucleation, as shown in Eq. (1.11).

$$\Delta G = -\Delta P \cdot V_G + (\gamma_{sg} - \gamma_{sl})A_{sg} + \gamma_{lg}A_{lg} \tag{1.11}$$

where γ_{sg} is the surface tension of the solid–gas interface, γ_{sl} is the surface tension of the solid–liquid interface, A_{sg} is the solid–gas interface surface area, and A_{lg} is the liquid–gas interface surface area.

The first term in Eq. (1.11) is the work done by the volume expansion of the gas inside the cell, the second term is the energy required to replace the solid–liquid interface with the solid–gas interface, and the third term is the work required to create the liquid–gas interface that constitutes the cell. The triple-phase composite interfacial energy follows the relationship described by Young's equation (1.12).

$$\gamma_{sg} = \gamma_{sl} + \gamma_{lg}\cos\theta \tag{1.12}$$

The shape of the heterogeneous nucleating cell nucleus takes a deficient spherical shape as shown in Figure 1.3(b). The free energy barriers ΔG can be expressed as Eq. (1.13):

$$\Delta G = -\Delta P \cdot V_G + \gamma_{lg} \cdot A_{lg} - \pi r^2 \cdot \gamma_{lg} \cdot \cos\theta \tag{1.13}$$

The volume V_g of the cell nucleus is:

$$V_G = \pi R^3 \frac{2 - 3\cos\theta + \cos^3\theta}{3} \tag{1.14}$$

The surface area of the cell nucleus A_{lg} is:

$$A_{lg} = 2\pi R^2(1 - \cos\theta) \tag{1.15}$$

The contact surface radius is:

$$r = R \sin \theta \tag{1.16}$$

By substituting the above equations into Eq. (1.13), deriving it and making it equal to 0, the critical radius of the cell nucleus of heterogeneous nucleation can be obtained as:

$$R_c = \frac{2\sigma}{\Delta P} \tag{1.17}$$

$$\Delta G^*_{het} = W_{het} = \frac{16\pi\sigma^3_{bp}}{3\Delta p^2} S(\theta) \tag{1.18}$$

$$S(\theta) = \frac{(2 + \cos\theta)(1 - \cos\theta)^2}{4} \tag{1.19}$$

where θ is the wetting angle, $S(\theta)$ is the out-of-phase factor and a function of the contact angle, and σ_{bp} is the interfacial tension of polymer–gas cells.

Colton and Suh [15] found that when the wetting angle is 20°, the free energy barrier between homogeneous nucleation and heterogeneous nucleation could be on the order of 10^{-3}. The heterogeneous nucleation rates can be derived as follows.

$$N_{het} = f_1 C_1 \exp\left(\frac{-\Delta G^*_{het}}{kT}\right) \tag{1.20}$$

where f_1 is the frequency factor of heterogeneous nucleation, N_{het} is the heterogeneous nucleation rate, and C_1 is the concentration of heterogeneous nucleation points.

Therefore, the activation energy required for heterogeneous nucleation is much lower than that required for heterogeneous nucleation, as compared in Figure 1.5.

1.3.2.3 Mixed Nucleation Theory

Cell nucleation during the foaming process may take homogeneous and heterogeneous nucleation modes, but they are not mutually exclusive. Due to the relatively low activation energy required, heterogeneous nucleation is performed first in a system contains cell nucleation agent. Whereas homogeneous nucleation would also

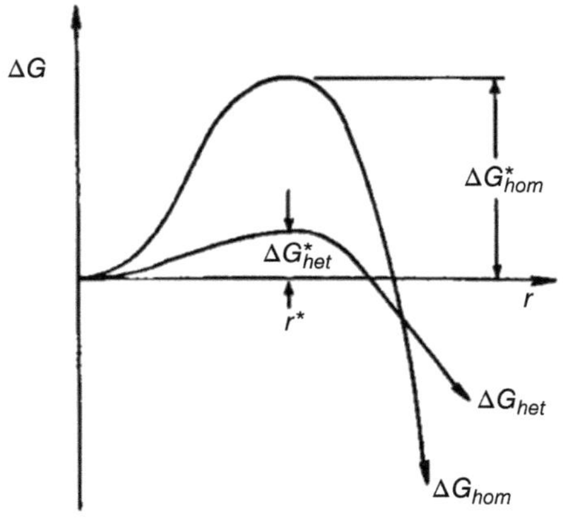

Figure 1.5 Comparison of the free energy barrier between homogeneous nucleation and heterogeneous nucleation. Source: Colton and Suh [15a]/with permission of John Wiley & Sons.

occur since the nucleation time is short, gas diffusion in the polymer melt system is hindered by the melt viscoelastic resistance during the heterogeneous nucleation, which would form a local supersaturation. Thus, heterogeneous nucleation followed by homogeneous nucleation would take place. The two-nucleation processes are not simply additive but rather interact with each other.

On the one hand, the first occurring heterogeneous nucleation consumes part of the gas, which makes the system supersaturated, and then the subsequent homogeneous nucleation rate drops and decreases in quantity [15a, b]; On the other hand, the internal pressure was relatively high in the cells with smaller size. When physical contact is established among these cells, the cells tend to merge, which will result in a decrease in cell density and uneven cell size. The gas concentration after heterogeneous nucleation started for time t can be expressed by:

$$C' = C_0 - N_{het} \cdot t \cdot n_b \tag{1.21}$$

where t is the time calculated from the occurrence of the first homogeneous nucleation, and n_b is the number of gas molecules in cell nuclei.

Substituting Eq. (1.21) into Eq. (1.10) yields the rate expression for homogeneous nucleation in heterogeneous systems:

$$N'_{hom} = f_0 C' \exp\left(\frac{-\Delta G'_{hom}}{kT}\right) \tag{1.22}$$

Then, the total nucleation rate of the heterogeneous system is:

$$N = N'_{hom} + N_{het} \tag{1.23}$$

In the foaming system with a large amount of nucleation agents, the homogenous nucleation is usually ignored due to the high heterogenous nucleation rate. Moreover, besides the classical nucleation theory, new nucleation theories have been established on the basis of classical nucleation theory, such as interface nucleation theory [16], free volume theory [17], shear nucleation [18], and hot spot nucleation [19].

1.3.3 Cell Growth

Right after the cell nucleation, the nucleus with a radius over the critical value would start to grow into cells. The vesicular structure is mainly formed at this stage, and the size, geometric type, density, and distribution situation of the cells all have an important influence on the performance of the foam. To study the process of cell growth, the dynamics and resistance during cell growth need to be analyzed. Due to the complexity of polymer melts and the complex transfer process of mass, momentum, and heat between cells and melt components, it is difficult to accurately describe the growth process of each cell.

Cell growth kinetics can be studied in the way of theoretical simulations and experimental observations. Ramesh [20] summarized the theoretical and experimental analysis of the bubble growth model since 1917 using the prevalent single bubble growth model [21] and the cell model [22]. The single bubble

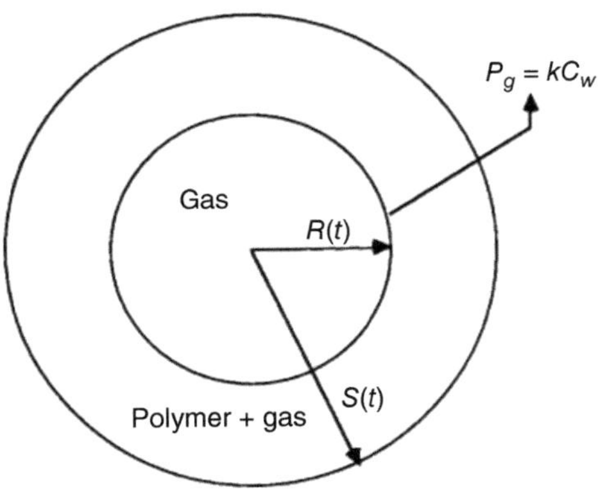

Figure 1.6 The single bubble growth model in the foaming process. Source: Han and Yoo [21]/with permission of John Wiley & Sons.

growth model (Figure 1.6) describes the process of bubble length growth behavior of a single bubble in an infinite melt, but it has limitations in practical applications.

Cell models are roughly divided into two categories: one is a closed cell model without a blowing agent and gas loss, and the other is an extrusion foaming modified cell model with a blowing agent and gas loss, which is suitable for different application scenarios – the former is suitable for injection molding, and the latter is more suitable for extrusion molding. Based on the bubble growth kinetics of the cell model, the following equation describing the isothermal growth of cells is obtained. To analyze the cell growth process, a set of control equations needs to be solved simultaneously: the continuity of the polymer–gas solution around the cell interface, the momentum equilibrium and gas diffusion equations, the constitutive equations describing the viscoelastic properties of the polymer–gas solution, and the mass conservation equation of the gas molecules [22, 23].

$$\ddot{R}R + \frac{3}{2}\dot{R}^2 + \frac{4K(2\sqrt{3})^{n-1}}{n\rho_l}\left(\frac{\dot{R}}{R}\right)^n = \rho_l\left(P_g - P_\infty - \frac{2\gamma_{pb}}{R}\right) \tag{1.24}$$

$$\frac{\partial C}{\partial t} + V_r\frac{\partial C}{\partial r} = \frac{D}{r^2}\frac{\partial}{\partial r}\left(r^2\frac{\partial C}{\partial r}\right) r \geq R \tag{1.25}$$

$$\frac{d}{dt}(\rho_g R^3) = 3R^2\rho_l D\left(\frac{\partial C}{\partial r}\right)_{r=R} \tag{1.26}$$

where C is the gas concentration, D is the diffusion rate, P_g is the internal pressure of cells, K, n is the viscoelastic characteristic parameter obtained from the power law equation $\mu = K\dot{\gamma}^{n-1}$, P_∞ is the pressure at the outer boundary of polymer cells, V_r is the radial velocity component = $\dot{R}R^2/r^2$, $\dot{\gamma}$ is the shear rate, and γ_{pb} is the surface energy of the polymer–cell interface. Equation (1.24) is a continuity equation that assumes that the polymer is a non-Newtonian fluid, which can be described by the power law equation. Equation (1.25) is the gas diffusion process equation, mainly resulting from the gas concentration gradient around the gas cell. Equation (1.26) describes the gas consumption equation, mainly with the process of cell expansion at the gas melt interface. By modeling the cell growth process using numerical or finite element methods, it is possible to understand the effect of different

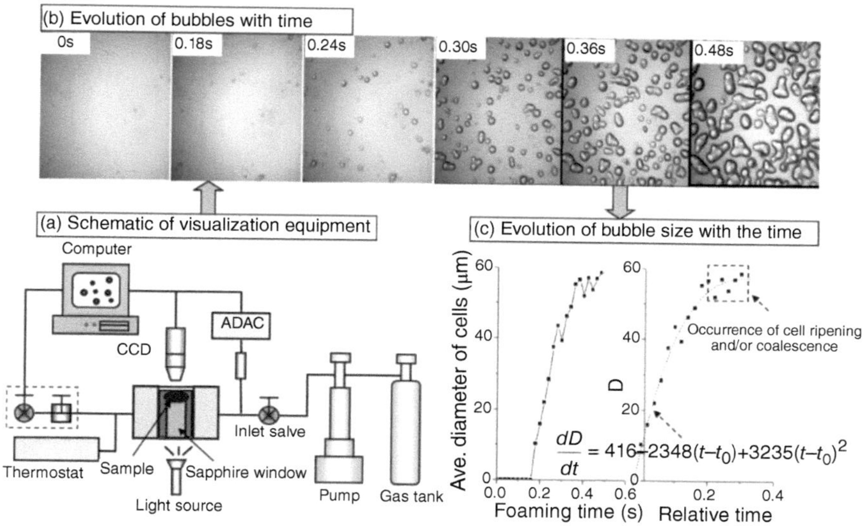

Figure 1.7 *In situ* visualization investigating the evolution of nucleated cells. Source: Zhai et al. [24]/with permission of John Wiley & Sons.

parameters in the cell growth process and give guidance on the actual production of polymer foams.

The experimental observational study is another avenue to investigate the growth of gas cells. Using visualization techniques, it is possible to directly observe the evolution of the cellular morphology during foaming so as to verify the theoretical model and study the actual foaming process. Park et al. have made significant contributions in this field. They [24] used an *in situ* visualization technique to study the cell growth kinetics during PEO foaming, and the actual process of cell growth is shown in the following Figure 1.7. They also developed a visualization system for observing the plastic foaming process under extensional stress [25] and shear stress [26].

1.3.4 Cell Coalescence and Rupture

Cell coalescence is extremely easy to occur during the cell growth phase, when the cells grow into contact, the expansion tends to move toward the "weak" side, which is usually the larger cell, so coalescence and rupture usually occur successively. From the viewpoint of cell growth dynamics, cell rupture is due to the joint force between contacting cells to promote the growth of cells under the combined effect of internal and external pressure differences, surface tension, and normal stress. Thermodynamically, the surface area after cell merging is smaller, and the total free energy of the system is lower, so connecting cells tend to merge. The merging of cells can lead to uneven distribution of cells, reduced density, and poor mechanical properties of the foam. However, when the cell rupture behavior is accurately controlled, it is possible to increase the cell opening rate and alter the microstructure of the cells. Thus, the bull rupture mechanism deserves an in-depth analysis.

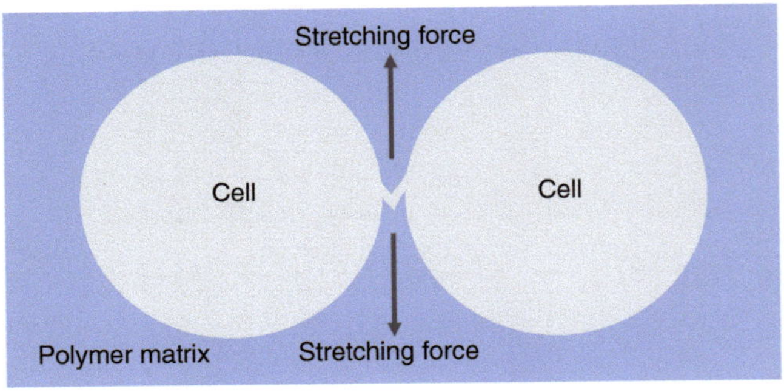

Figure 1.8 Cell merging of adjacent cells due to the cell wall rupture.

1.3.4.1 The Mechanism of Cell Rupture

During cell growth, the cell wall would be subject to a stretching force along the direction of cell growth, as illustrated in Figure 1.8. The cell wall would rupture if the stretching force is much greater than the surface tension of the melt [27]. During the foaming process, the cell size is not very uniform due to the curvature radius difference between the small and large cells and the merge of cells. The larger the size difference between adjacent cells, the greater the gas pressure difference inside them, and the more likely the small cells to merge into the large cells [28].

Cell coalescence reduces the number of cells and increases the cell size. Even if a large number of cell nuclei are formed during the nucleation stage, the cell coalescence during the cell growth process often causes a reduction in cell density. Therefore, to obtain foamed materials with high cell density and small size, it is necessary to strictly control the cell coalescence during cell growth. Cell coalescence can be suppressed by improving the melt strength of the matrix polymer and controlling process conditions. The melt strength of the polymer can be enhanced by introducing branching molecular chains, cross-linking, and blending with other polymers with higher melt strength. In terms of process control, within the temperature range at which the polymer can be foamed lowering the foaming temperature of the polymer, and increasing the cooling rate can all play a role in preventing cell coalescence [28, 29].

1.3.4.2 Mechanism of Cell Opening

Foam coalescence and rupture will cause the mechanical properties of microcellular materials to decline, while in producing open-cell structure foaming materials, controllable obtaining open cell structure is sometimes necessary. Open-cell foam materials are materials in which both phases of the gas–solid phase are continuous and have a unique three-dimensional morphology, which can be further divided into fully open and partially open cell types. Due to the characteristics of reciprocal flow between the pores of the open-cell foams, such foams usually have excellent absorption and penetration properties. Open-cell foams are widely used in filtration, sound reduction, electromagnetic shielding, and biomedicine fields, especially in tissue engineering scaffolds, biological dialysis materials, etc.

So far, most open-cell foams are produced by the phase separation method or chemical foaming during material synthesis. It is challenging to produce open-cell foams via physical foaming or SCF foaming. In the phase separation (phase inversion) method, a two-phase or multiphase mixed system containing a polymer matrix and solvents is prepared, followed by triggering the phase separation process under the action of temperature or solvent to produce a cell structure within the polymer collective. It suffers from hazardous disposal of chemical solvents and residual solvents, as well as complicated procedures that usually use large amounts of water. In addition, adding water-soluble porogens into the polymer matrix is another approach to fabricating open-cell foams. Mosanenzadeh et al. [30] prepared polylactic acid (PLA) foams with high sound-absorbing properties by using salt granules as porogen, and this method was called the particle leaching technique.

Aiming at the preparation of open cell foams by SCF foaming, a large number of scholars have proposed different mechanisms and technical practices. Rodeheaver and Colton [31] proposed a theoretical model of open-cell foaming on the basis of classical foaming theory. He divided the process of cell opening into two stages, one was the collision stage, which contained the process of cell nucleation, growing up to the contact of cell walls, and the other was the stage in which the cell wall became thinner and broken into a hole. The theoretical model for the first stage is similar to the closed cell foaming process in which there must be sufficient nucleation points to guarantee that the cells grow to final contact, assuming that the number of nucleation points is equal to the number of cells eventually formed, the critical number of nucleation points required to form open cell foams can be deduced, and then the pressure versus temperature range required for foaming can be deduced. In the second stage, when the walls of the cells are in contact, by studying the kinetics of the rupture of the film, according to the Young Laplace equation and the equation of surface wave initiated by thermodynamic instability, the unstable conditions of the walls of the cells are deduced. Therefore, the time of the rupture of the cells, which is the time when the cell walls reach a critical thickness, can be estimated. According to the above model, the formula to estimate the rupture time of the cells can be derived:

$$\Delta t = \left(\frac{2\sqrt{3}A}{V}\right)\ln\left(\frac{h_0}{2\sqrt{6}A}\right) \tag{1.27}$$

$$A = \sqrt{\frac{kT}{\sigma}} \tag{1.28}$$

where Δt is the time of cell wall rupture, V is the rate of cell wall thinning, A is the size of surface waves, h_0 is the original wall thickness, k is Boltzmann constant, and T is the cell wall temperature.

From the equations, it can be seen that the rupture time Δt is a function of the cell wall thinning rate V, Boltzmann constant k, foaming temperature T. Rodeheaver and Colton [31] conducted a batch foaming experimental study on PS using nitrogen gas as foaming agent to verify this model, and the experimental results led to the following process conditions that enabled PS to produce open cell structures, that is, foaming pressure was more than 17.2 MPa, foaming temperature was more than 200 °C, and foaming time was controlled between one and two seconds. Enayati et al. [32], based on this model, investigated the foaming process parameters of the PS batch foaming system by using $scCO_2$ as the foaming agent, and the experimental results proved that open-cell foams can be obtained with a minimum foaming time of six seconds at a minimum pressure of 13 MPa and a minimum temperature of 150 °C, which in general agree with the theoretical predictions.

Derrick [33] first proposed the immiscible polymer blending theory model in 1994, which is based on the principle of using immiscible two polymers to form a two-phase structure. The interface between the two phases makes it easier to form nucleation points due to the lower free energy required for nucleation, similar to the heterogeneous nucleation in classical nucleation theory. Due to the weak interaction between the two incompatible phases at the interface, the bonding between the two phases begins to detach as the cells grow, ultimately forming an open-cell structure. Yu et al. [34] prepared biodegradable polybutylene succinate (PBS)/PLA foam with open-cell structure by batch foaming owing to the small dispersion PBS phase and the relatively low melt strength of PBS compared with PLA as shown in Figure 1.9.

Lee and Park et al. [35] also prepared open-cell microcellular foam materials using an immiscible polymer blending system. They proposed two mechanisms for cell opening (Figure 1.10). One is to use polymers with higher melt strength as the continuous phase and to add a small amount of polymers with lower melt strength as the dispersed phase, which is similar to the PBS/PLA blends system. During the cell growth process, the low melt strength dispersion phase would be easily broken by the stretching force of the cell wall. Another method is to use polymers with low melt strength as the continuous phase and polymers with higher melt strength as the dispersed phase. As the cell size of the cell continues to grow, the stretching of the soft phase polymer matrix is transmitted to the hard phase polymer, and the peeling of the cell wall occurs at the junction of the two phases due to different degrees of deformation of the two phases. There are also some scholars using ultrasound-assisted techniques for the preparation of open-cell foam [36], and found that ultrasound can induce the rupture of the cell walls to form an open-cell structure.

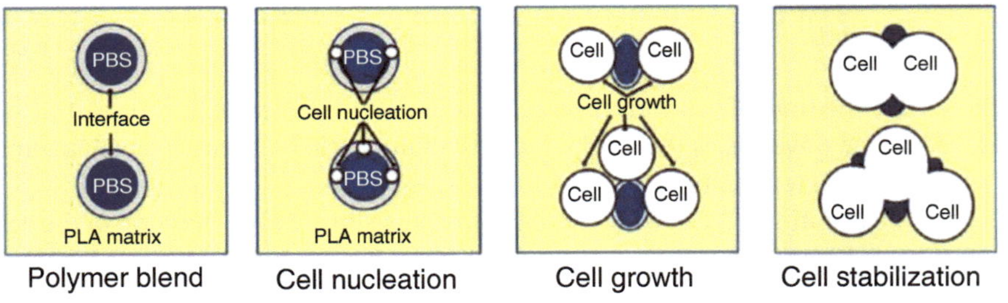

Figure 1.9 Cell opening mechanism during the $scCO_2$ foaming of PBS/PLA blends. Source: Yu et al. [34]/with permission of American Chemical Society.

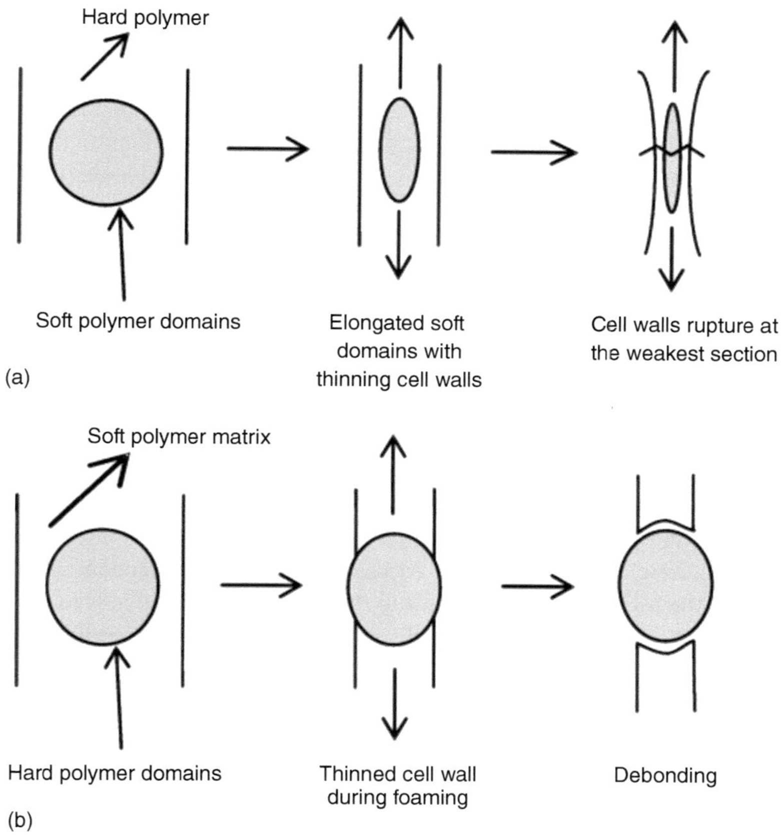

Figure 1.10 Two cell-opening mechanisms for the system composed of (a) a high melt strength matrix phase and a low melt strength dispersion phase, and (b) a low melt strength matrix phase and a high melt strength dispersion phase. Source: Lee et al. [35b]/with permission of American Chemical Society.

1.3.5 Solidification and Curing

Solidification and curing, the final step in the foaming process, have an important impact on the final structure and properties of polymer foams. Since the foaming process is inherently unstable, both cell nucleation and cell growth are essential energy-dissipating processes to reestablish equilibrium. The transformation of polymers from the molten state to the solid state is a process of molecular chain mobility reduction, which enables the maintenance of the cell structure. In the case of amorphous polymers, sample cooling below the vitrification temperature allows the solidification of the cell morphology. For crystalline polymers, solidification is achieved by a sharp rise in viscosity with the crystallization of the polymer chains. For some polyurethane foams, curing is achieved by cross-linking reactions. In general, the heat is taken away by cooling the foam directly or indirectly through air, water, or other cooling media. However, polymer foams are poor conductors of heat, which

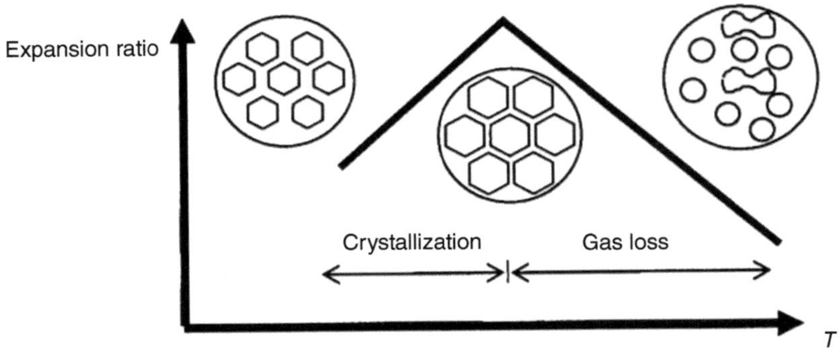

Figure 1.11 Effect of cooling temperature on the volume expansion and cell structure of PP foam. Source: Naguib et al. [37]/with permission of John Wiley & Sons.

often causes that the cells close to the surface have been stereotyped by cooling, while the interior cells are still at a high temperature. Therefore, the outward heat transfer from the interior may cause the surface temperature to rise and destroying the cellular structure close to the surface. At the same time, the temperature difference between the inside and outside leads to the tendency of inner cells to have expansion relative to the surface layer, which will lead to stress concentration and even deformation and rupture of the cells. So, the solidification and curing process is important for the foams to maintain a stable and uniform porous structure.

Naguib et al. [37] studied the expansion behavior of PP foams and found that the final volume expansion ratio of butane as a foaming agent for extruded PP foams was related to the cooling temperature, exhibiting a trend as shown in Figure 1.11. The PP volume expansion ratio was jointly determined by the loss of the foaming agent and the crystallization of the polymer matrix. When the temperature was too high, the high diffusion rate of the foaming agent would cause the rapid loss of gas and reduce the expansion degree of the cells. When the temperature was too low, the polymer crystallized too fast, and the cells were fixed without full expansion.

1.4 Influencing Factors of Cell Structure in the Foaming Process

The structure of cells has a decisive influence on the performance of polymers, and the cell structure is directly related to the foaming material and the foaming parameters. The influential factors for cell structure can be roughly divided into two categories. One is the influence of physical properties of substances in the system, such as rheological properties of polymers, solubility and diffusion rate of the foaming agent in the polymer matrix, the interaction between foaming agent and polymer matrix, the influence of added nucleating agent and nanoparticles, etc. The other is the influence of foaming process parameters, such as foaming temperature, saturation pressure, depressurization rate, and depressurization method, etc.

1.4.1 Effects of Polymer Properties

1.4.1.1 Rheological Properties of Polymers

Rheological properties of polymer matrix have an important impact on the overall foaming process. During foaming, the melt around the cells undergoes tensile action, and the rheological properties of polymers directly affect cell growth. The cell would be difficult to grow within a polymer matrix with high melt strength. When the melt strength is too low, the cells tend to rupture and collapse, resulting in a low expansion rate.

The molecular structure has a huge influence on the tensile viscosity of polymers, and the viscosity of long-chain branched polymers is higher than that of linear-chain polymers. Park and Cheung [38] found that the closed cell rate of branched PP foams in extrusion foaming was much higher than that of linear PP foams, suggesting that the cells are more easily ruptured for the low melt strength linear PP. Fang et al. [39] used γ radiation to prepare a series of long-branched PLA materials and produced a bimodal foam morphology, which verified that the introduction of long-branched structures enhanced the complex viscosity, shear thinning, and storage modulus of the matrix and improved the foaming properties of the PLA materials.

The influence of the polymer matrix crystallization behavior on the rheological properties is not negligible during the foaming process. For example, PP is a crystalline polymer that barely flows below the melting temperature due to the restriction of the crystalline regions. However, as soon as the melting temperature is reached, the viscosity and melt strength decreases sharply, which induces fast gas diffusion and cell merge during foaming, making it difficult to form a well-established cellular structure. Therefore, the melt strength of polymers is often improved from the perspective of polymer modification, and the current common methods of polymer modification mainly include crosslinking modification [40], grafting modification [41], blending modification [42], and filling with nanoparticles [43].

1.4.1.2 Solubility and Diffusion of Blowing Agent

In general, the solubility of gases in molten polymers increases with increasing pressure and decreases with increasing temperature, with pressure and temperature showing opposite trends in promoting the solubility of gases in polymers (Figure 1.12). Sato et al. [44] measured the solubility of CO_2 and N_2 in molten PP and high-density polyethylene (HDPE), and the experimental data coincided with the trend of solubility change represented in Figure 1.12.

The diffusion rate of gas in a polymer matrix represents the rate of movement of gas molecules through the molten polymer, which is the mass transport capacity of gas molecules through the matrix. For large-scale production of industrial foaming products, the diffusion rate is usually a more important influencing factor. The general trend of the gas diffusion rate in the molten polymer increases with increasing temperature, quite the opposite trend of the solubility change with temperature. The decrease of surface tension is beneficial to the diffusion and mixing of substances, and the adoption of supercritical gas can reduce the surface tension of gas and polymer melts, thus, elevating the diffusion rate. Sota and Suh et al. [44, 45] studied the

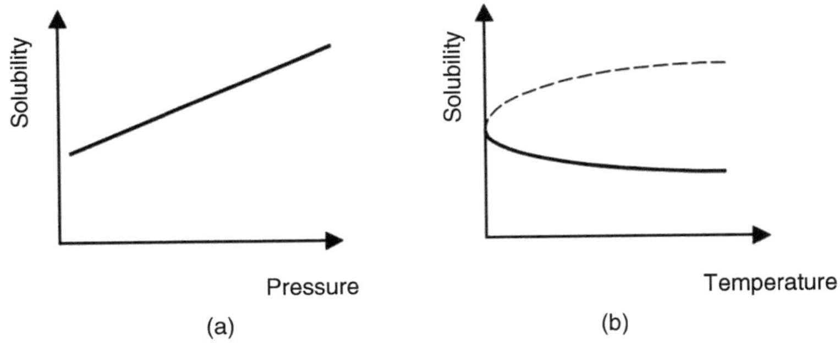

Figure 1.12 Solubility of gas in most molten polymers: (a) solubility changes with increasing pressure, and (b) solubility changes with increasing temperature. Source: Adapted from Xu [3a] and Cha [8].

gas diffusion theory in molten polymers to provide a basis for estimating the diffusion rates of gases. Peng and Qu [46] found that introducing a vibrational force field could promote the diffusion of small molecule gases in polymers, demonstrating that changing the external force field could elevate the diffusion rate of gases.

Controlling the diffusion direction of gas in the melt also affects the cell structure of foams, since the gas diffuses into the cell voids also diffuses to the surface and external environment. Siripurupu et al. [47] found that reducing the diffusion speed of gas to the melt surface resulted in a more dense cell structure with a smaller cell size. Mi and coworkers [48] successfully produced fine cells on the surface of the thermoplastic polyurethane (TPU) and polyvinylidene fluoride (PVDF) foamed sheets by introducing various gas barrier films to reduce the diffusion speed of CO_2.

1.4.1.3 Interaction Between Foaming Agent and Polymer Matrix

As mentioned above, the addition of foaming agents may interact with the polymer matrix thus affecting the properties of the polymer as well as the foaming behavior. For example, $scCO_2$ has a plasticizing effect on the melt when saturated within the polymer matrix, which would improve the activity capacity of molecular chains, thus affecting the properties of the polymer matrix. For amorphous polymers, it decreases the glass transition temperature and viscosity of the polymer. For crystalline polymers, it usually affects the crystallinity, crystallization temperature, crystallization kinetics, and crystal structure. Garg et al. [49] proposed two mechanisms for the plasticizing effect of SCF addition. One is the alleviation of entanglement of polymer molecular chains after the addition of gas, and the other is the generation of additional free volume, which increases the mobility of molecular chains.

Chiou et al. [50] investigated the effect of CO_2 on the glass transition temperature (T_g) of polymers such as PMMA, PS, and polyvinyl chloride (PVC) using a high-pressure differential scanning calorimetry, and showed that CO_2 can significantly decrease the T_g of glassy polymers with high CO_2 solubility at moderate pressures. The effect of $scCO_2$ on the T_g of polymers can be divided into three regions, as shown in Figure 1.13. When the pressure is low (region I), the absorption of CO_2

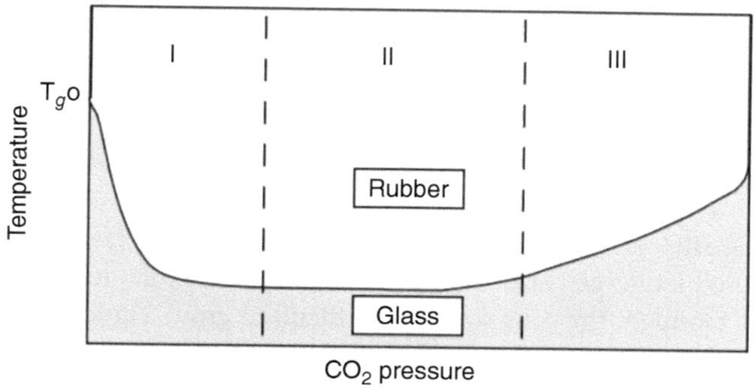

Figure 1.13 Proposed model of glass–liquid transition for a polymer–CO_2 mixture. Source: Royee [51]/with permission of NC State University Libraries.

causes swelling, and the T_g significantly decreases; With the increasing pressure (region II), the T_g tends to stabilize and is less affected by the pressure; As the pressure further increases (region III), the free volume of the system decreases, and the T_g begins to rise [51].

Zhang et al. [52] found that the presence of $scCO_2$ could significantly improve the crystallinity of PP, and the crystallinity was increased with the increase of CO_2 pressure. Liao et al. [53] found that the content of γ crystals increases continuously with the increase of CO_2 pressure in PP. Some studies found that under the influence of CO_2, the crystallization kinetics, and morphology of polymers such as PVDF [54], polyethylene terephthalate [54, 55], and polycarbonate [56] have also changed.

1.4.1.4 Nucleating Agent and Nanoparticles

The introduction of fillers into the melt is an effective method to control the cell structure mainly due to the heterogeneous nucleation effect of the nanofillers and the effect of nanofillers on the melt strength of the matrix polymer. According to the classical nucleation theory, introducing heterogeneous nucleation can greatly reduce the nucleation barrier, thus obtaining more cell nuclei during the nucleation stage and greatly increasing the cell density. Turng [57] and Park [58] studied the effect of nanoclay on the foaming of PE, and verified the heterogeneous nucleation effect of nanoclay in improving the cell density and the performance of the foams. Antunes et al. [59] studied the effects of changing foaming parameters and introducing nano montmorillonite (MMT nanoparticles) on the PP foaming system. They found that MMT nanoparticles played a dual role in heterogeneous nucleation and improving melt strength, reducing the overall cell size and increasing the cell density.

1.4.2 Effects of Foaming Process Parameters

1.4.2.1 Foaming Temperature

The processing temperature is an important parameter for polymer processing since the properties of a material are highly related to its temperature. In the foaming process, temperature affects the viscosity of the polymer matrix, the solubility of the foaming agent in the molten polymer, the gas diffusion rate, and even the time needed for cooling. Foaming temperature affects the final cell morphology by influencing these physical properties.

In general, the higher the foaming temperature during the process, the lower the polymer viscosity, the lower the solubility of the foaming agent, and the faster the diffusion rate, which usually corresponds to lower cell nucleation rate, faster cell growth, lower density, and larger cell size. When the temperature is too low, and the polymer viscosity is too high, the cells would be difficult to grow. Therefore, a suitable foaming temperature window needs to be decided in the first place.

Yokoyama et al. [60] studied the foaming behavior of polystyrene (PS)/polystyrene perfluoro methyl acrylate copolymer (PFMA) blend system and obtained different cellular structures based on the change of viscosity of PS as a function of temperature. When the foaming temperature was set above the T_g of PS, cells in micrometer scale were obtained; When the foaming temperature was below T_g of PS, the cell size could be reduced to nano scale. Goel and Beckman [61] investigated the plasticization effect of $scCO_2$ on PMMA, and found that the solubility of CO_2 decreases with the increase in temperature, while the surface tension of the melt and the free energy barrier to nucleation rises with temperature increase, which leads to a decrease in the nucleation rate and the cell density.

1.4.2.2 Saturation Pressure or Foaming Pressure

According to the classical nucleation theory, the pressure of the saturated gas has a great influence on cell nucleation. The higher the foaming pressure is the more gas will be absorbed by the polymer, which is responsible for a greater plasticizing effect. In addition, a high foaming pressure is usually associated with a high-pressure drop rate. All these properties are favorable for enhancing the cell nucleation rate and the cell density of the foam product, while it usually causes a decrease in cell size. Goel et al. [62] found that the size of the cells decreased with increasing pressure and the density of the cells increased with increasing pressure in the foaming of PMMA using $scCO_2$ as the foaming agent. Baldwin et al. [63] investigated the foaming process of amorphous and crystalline PET depending on the foaming pressure and found that the increasing pressure had a more significant effect on the amorphous PET than the crystalline PET. This was because high pressure could activate more homogeneous nucleation points in the amorphous PET, while the crystalline region in the crystalline PET could induce heterogeneous nucleation which will cause a high cell density even without the increase in foaming pressure.

1.4.2.3 Depressurization Rate

In physical foaming with high-pressure gases, the cell nucleation and growth are generally achieved by rapid depressurization. Since cell nucleation and growth occur almost simultaneously in the system, there is clearly a competitive relationship between them. The final cell morphology is largely determined by the two processes. In the moment of pressure release, the cell nucleus started to appear in the region with a lower free energy barrier, and subsequently, more gas molecules diffused into the cell nucleus to cause the cell growth. At a low depressurization rate, the sudden pressure drop was small, which is adverse for cell nucleation, and the time for cell growth is prolonged. Thus, more gas could diffuse into the cell nucleus forming a structure with a large cell size. On the contrary, a high depressurization rate usually refers to high cell density and relatively small cell size. Han et al. [64] studied the continuous extrusion foaming behavior of PS and adjusted the depressurization rate by changing the shape of the die. They basically found a similar tendency and prepared PS foams with different cell structures by varying the depressurization rate. Park et al. [65] designed three nozzles that can generate different depressurization rates in injection foaming to study the effect of depressurization rate on the gas competition between cell nucleation and cell growth processes. They predicted that there would be a maximum depressurization rate to enable high cell nucleation and balance the cell growth processes.

1.4.2.4 Multistage Saturation and Depressurization

In most cases, the high pressure is released in one step in the foaming process. However, the pressure could also be released step-by-step in batch foaming, which is capable of controlling the cell morphology of the final foams. Step-by-step depressurization refers to the step-by-step removal of saturated pressure after supersaturation. Similarly, the foaming temperature and supersaturation pressure could also be programed stepwise to influence the final cellular structure of the foam. Typically, a bimodal cellular morphology with distinct different cell sizes in a single foam can be obtained by these multistage foaming processes since the conventional continuous nucleation and growth process is divided into two or more stages.

The foams with a bimodal cell structure that combines large cells and small cells may have advantages in their properties. The large cells could contribute to the weight reduction and the small cells could provide particular performance. Compared with the conventional singular cell structure, the bimodal cell structure may render the foams better performance in sound absorption, heat insulation, bio-tissue engineering, and other fields. Arora et al. [66] fabricated PS foams with a bimodal cell structure using a two-step depressurization method. Bao et al. [67] studied the factors affecting the bimodal cell structure in the two-step pressure drop method. They proposed that "Holding time," the duration after the first pressure drop, is important for the control of the bimodal cell structure since it controls the time for the growth of the large cells. Mi and coworkers [68] developed a multistage soaking approach in which the saturation temperature was elevated during the soaking stage to cause the precipitation of $scCO_2$ in the PLA matrix due to the lower gas solubility at higher temperatures. The precipitated $scCO_2$ formed the large cells in the subsequent depressurization step, and the prepared bimodal foams have a dramatic difference between the large cells and the small cells.

1.5 Previlant Foaming Methods for Microcellular Foams

Foams with cell size in the micrometer level are called microcellular foams. Compared to conventional microporous foams, microcellular foams have superior mechanical, absorption, and insulation properties. Thus, they are been heavily investigated in recent years. Microcellular foams prepared by SCF-based foaming methods are highly recognized since the SCF are environmentally friendly and will cause no chemical residual in the polymer matrix and no VOCs emission. It also has high productivity and fabrication reliability. In a typical SCF-foaming process, SCF is first injected into the polymer to form a gas-saturated system, then the system is supersaturated by means of rapid pressure drop, which induces cell nucleation and growth, and finally, the foam with microcellular structure is cooled and solidified. SCF foaming process is mainly divided into batch foaming, continuous extrusion foaming, and injection foaming.

1.5.1 Batch Foaming

Batch foaming was the earliest method used to manufacture microcellular foams. The first patent for microcellular foaming was disclosed in the United States in 1984 [69]. Two batch foaming methods are widely used in practice. The first is temperature-induced foaming, in which the cell nucleation and growth are triggered by the fast temperature rising as illustrated in Figure 1.14. In this method, solid polymer preform is put in a high-pressure vessel charged with $scCO_2$ or scN_2 at a low temperature to form a saturated homogeneous system. Then, the saturated preform is taken out from the vessel and rapidly transferred to a hot media (i.e. hot oven and hot bath) to quickly increase the temperature above the T_g of the polymer while below the melting point (T_m) of it. Thus, cell nucleation and growth will be triggered due to the increase in molecular chain

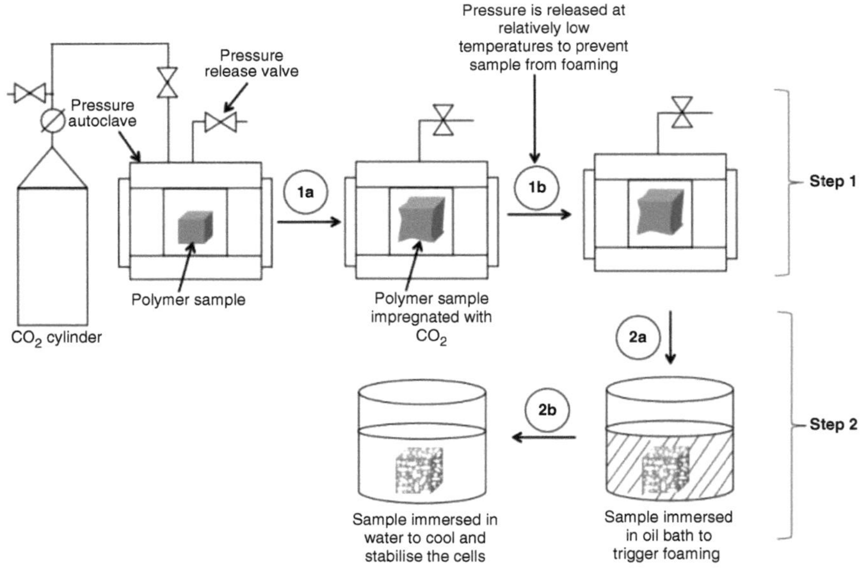

Figure 1.14 Temperature-induced batch foaming process. Source: Okolieocha et al. [70]/with permission of Elsevier.

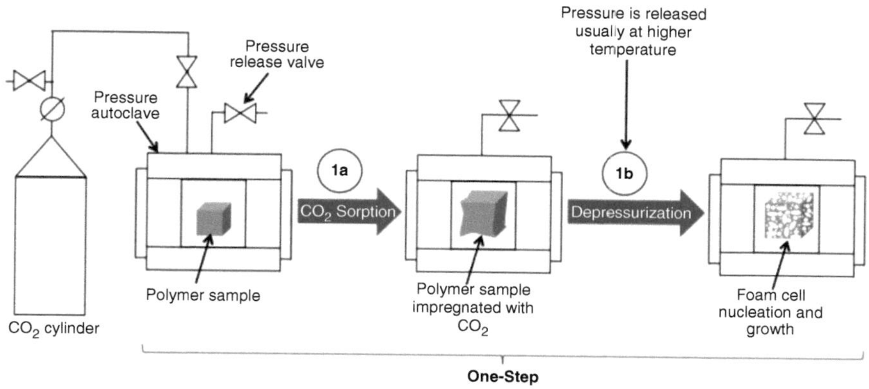

Figure 1.15 Pressure-induced batch foaming process. Source: Okolieocha et al. [70]/with permission of Elsevier.

mobility. Finally, after the expansion process, the temperature will be reduced and the foams will be solidified by quenching. Since the samples need to be transferred from a high-pressure environment to a high-temperature environment to allow the foaming process, this method is also referred to as a two-step foaming method.

The second batch foaming method, which is used more often in practice, is called one-step foaming approach. As illustrated in Figure 1.15, in the pressure-induced foaming process, the preform samples are saturated at a higher temperature that is usually the foaming temperature with the SCF. The cell nucleation and cell growth are triggered by the rapid depressurization. Then, the expanded samples would be subject to a cooling process to allow the maintaining of the foamed cellular structures [62a].

Batch foaming method has the advantages of simplicity in foaming device, easy control of the foaming process, and easy regulation of product cell structure. However, the batch foaming method is difficult to achieve continuous production in large quantities. Batch foaming has been mainly used in laboratory research on a small scale for decades, while it is now more and more used in the industrial production of foam products attributing to the invention of novel large-scale batch foaming instruments.

1.5.2 Continuous Extrusion Foaming

Continuous extrusion foaming of microcellular plastics was issued in 1990s [71], which realized the continuous foaming process and promoted the industrial application of microcellular foamed products. In industry, foam extrusion lines are usually in tandem (as shown in Figure 1.16), and simple foam extrusion single line is also used in practice. In a typical extrusion foaming process, the polymer is added through the feed hopper into the extruder, it is compressed and plasticized by the screw in the first barrel and mixed with $scCO_2$ injected into the barrel to form a homogeneous system under the action of high temperature, high pressure,

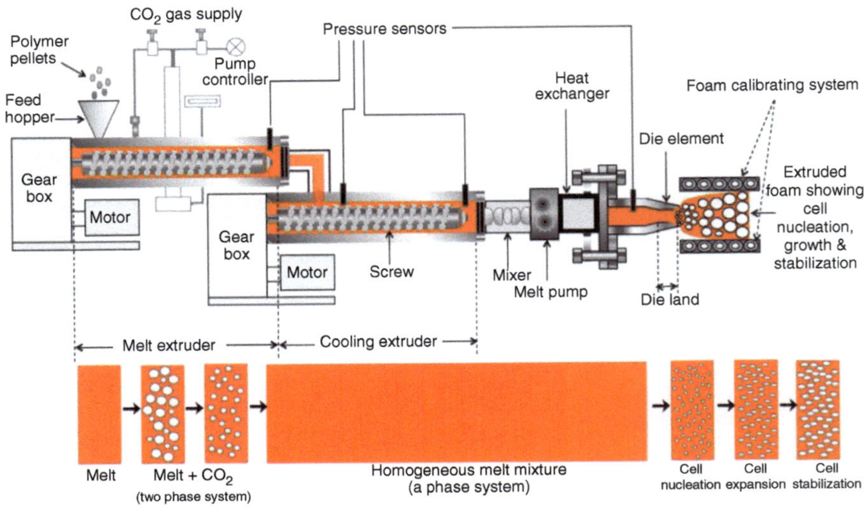

Figure 1.16 Schematic representation of foam extrusion on a tandem line.
Source: Okolieocha et al. [70]/with permission of Elsevier.

and screw shear. Subsequently, the mixed system is transferred to the second barrel where preliminary cooling takes place and higher pressure continues to increase gas concentration and saturation pressures. A static mixer, a melt pump, and a heat exchanger are sometimes used between the barrel and the die to further increase the homogeneousness and the pressure of the polymer/gas system before the extruding. Finally, as the melt passes through the die, the cell nucleation and cell growth take place rapidly due to the rapid pressure drop. Continuous extrusion foaming greatly improves production efficiency and is a key step toward the commercialization of microcellular foamed products. It has been widely used in the production of foamed sheets, plates, pipes, and beads. However, due to the restriction of die geometry, continuous extrusion foaming has significant limitations in the production of foamed products with complex shapes.

1.5.3 Injection Foaming Technique

In 1995, Axiomatics (later Trexel) began to develop microcellular injection molding technology based on extrusion technology and registered MuCell® [3a], the trademark most commonly used for microcellular molding technology, which is also called injection foaming or foam injection molding. Figure 1.17 shows the basic process of the injection foaming method. Similar to extrusion foaming, the polymer was added into the barrel through the funnel, and it becomes a melt under the compression and plasticization of the screw. To the melt, a defined amount of SCF was injected via a precisely controlled valve and mixed with the polymer melt by the rotation of the screw. When the polymer/gas system is plasticized and metered in front of the screw, a high-pressure state is still maintained to prevent early foaming. Subsequently, the melt is injected into the cavity of the mold, and

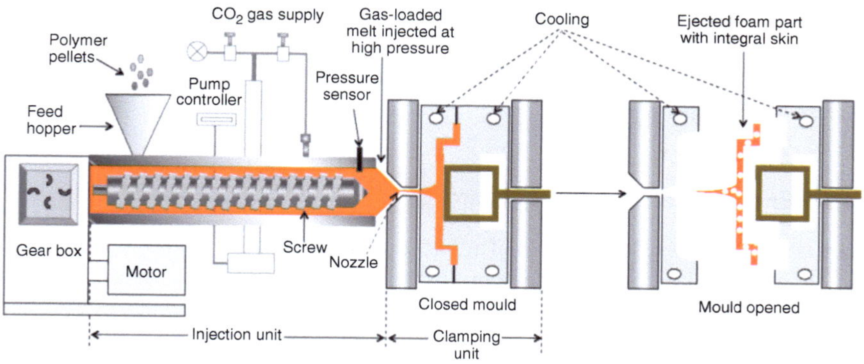

Figure 1.17 Schematic representation of foam injection molding process. Source: Okolieocha et al. [70]/with permission of Elsevier.

the melt starts to foam as the melt fills the mold cavity due to the rapid pressure drop. For injection foaming, the holding stage is removed to allow sufficient expansion of the foamed part during the cooling cycle to achieve high weight reduction. Since the injection foaming contains plasticizing, storage, injection, and cooling stages, the cell structure of injection foamed parts is affected by various parameters, including processing temperature, injection speed, injection pressure, injection volume, mold cavity geometry, back pressure, cooling temperature, cooling time, gas content, gas injection time, L/D ratio of the screw, etc. In general, injection foaming improves the productivity of foamed parts and allows the production of products with complex shapes. Compared with traditional injection molding, foam injection molding has the advantages of more stable dimensions, shorter foaming cycles, and material savings. However, the high cost of the SCF supply system, the flow marks left on the surface of the foamed parts, and the quality fluctuation in production are the main concerns for this technique.

1.6 Advanced Applications of Functionalized Polymer Foams

With the fast development of polymer-foaming techniques and foamed products in recent years, more and more attention has been attracted. Researchers and pioneers have been devoted to exploring the advanced applications of polymer foams and trying to develop various functionalized polymer foams in the advanced fields in the new era as cutting-edge techniques. This book will focus on the advanced polymer foams that are used in cutting-edge fields. The approach to developing such foams will be mainly focused on microcellular foaming using SCF as the physical foaming agent so as to promote the potential applicability of this green technique. The content of this book will cover novel applications of the foams, including energy absorption, acoustic absorption, superhydrophobic, electromagnetic interference (EMI) shielding, tissue engineering scaffold, flexible sensors, triboelectric nanogenerator (TENG), and solar steam generation. A brief introduction to these advanced fields and the role or potential role of polymer foams in these fields is given below.

1.6.1 Energy Absorbing Buffer Foam

Foam materials have a long history of applications in energy absorption and cushioning. They have found use in various industries, ranging from sports equipment like exercise pads for gymnastics, high jump, and combat, to safety products like cycling helmets and bulletproof vests. Polymer foams offer several advantages over metallic or inorganic foams. They provide greater design flexibility, are easier to process and recycle, and offer a wide range of compressive strengths, from highly elastic to semi-rigid and rigid. In addition, polymer foams are lightweight, cost-effective, easy to shape, and offer numerous advantages in commercial applications. The mechanism of energy absorption, the factors influencing the energy absorption property of polymer foams, as well as the advanced fabrication approaches of energy-absorbing polymer foams will be covered in Chapter 2.

1.6.2 Thermal Insulation Polymer Foams

Heat transfer occurs through three fundamental mechanisms, namely, conduction, convection, and radiation. Insulation materials are designed to impede heat transfer by reducing its occurrence through these channels. The effectiveness of insulation is determined by the material's thermal conductivity, with lower values indicating better insulation properties. Polymer foams utilize a unique cellular structure to simultaneously obstruct thermal conduction and convection processes. Their porous structure creates multiple internal interfaces, significantly enhancing their ability to block thermal radiation. Attributing to the lightweight, high surface area, and low thermal conductivity of the polymer foams, they have become essential in various applications. In Chapter 3, the thermal insulation mechanism of polymer foams, the structural design, influential factors, and the methods of fabricating advanced thermal insulation foams will be covered.

1.6.3 Acoustic Absorption Polymer Foams

Noise pollution is a major concern in both industrial settings and daily life, with long-term exposure negatively affecting mental and physical health. Porous materials, featuring interconnected pores that allow sound waves to enter and reflect within the material, excel at absorbing sound energy. Polymer foams, with their intricate microporous structures, provide effective sound absorption due to their ability to distort and scatter incident and reflected sound waves. Consequently, various polymer foams, such as melamine foam, PS foam, polyurethane foam, and their composites, find wide application in buildings, automobiles, ships, and aircraft for sound absorption. The sound absorption mechanism of polymer foams, the critical influencing factors, the measurement approach, and the fabrication methods of sound absorption foams will be covered in Chapter 4.

1.6.4 Superhydrophobic Polymer Foams

Superhydrophobicity, observed in nature on surfaces like lotus leaves and water striders, relies on low surface energy and multilevel surface roughness. By mimicking these features through bionanotechnology, polymer foams can be transformed into superhydrophobic materials suitable for various applications, including anti-fouling, self-cleaning, selective oil absorption, and oil–water separation. The three-dimensional porous structure of polymer foams, both internally and on the surface, enhances their roughness, robustness, and stability compared to surface-engineered materials. In Chapter 5, the surperwetting theory, the chemical and structural design for superhydrophobic surfaces and foams, the methods to fabricate superhydrophobic polymer foams, and their advanced applications will be introduced.

1.6.5 Electromagnetic Shielding Conductive Polymer Foam

Electromagnetic shielding technology is vital for countering EMI and radiation and ensuring information security in our technology-driven world. CPFs possess multiple conductive interfaces within the micro-cell structure, which increases electromagnetic wave reflection and scattering, extending their transmission path and enhancing electromagnetic wave attenuation. Thus, the CPFs have natural advantages in EMI shielding and excellent EM wave absorption properties. The EMI shielding mechanisms for CPFs, the effect of CPF material composition and porous structure, the performance characterization methods, and the advanced approaches to fabricate CPFs will be introduced in Chapter 6.

1.6.6 Medical Tissue Engineering Repair

Organ and tissue transplantation faces a significant supply–demand gap, prompting the exploration of the tissue engineering field. Tissue engineering involves creating three-dimensional structures with cells and biomaterials. Due to their adjustable highly porous structure, tunable mechanical properties and surface chemistry, and biocompatible and biodegradable properties, polymer foams have been recognized as promising candidate materials for tissue repair and artificial organs. Porous polymer scaffolds have been used in the repair of various tissues, including bone, cartilage, skin, vascular, and neural tissues. In Chapter 7, the tissue engineering process, basic requirements for tissue engineering scaffolds, the property evaluation methods, and fabrication methods of tissue engineering scaffolds will be covered.

1.6.7 Flexible Sensors Based on Porous Polymer Foams

In the era of the Fourth Industrial Revolution, there is a growing demand for lightweight, flexible, highly sensitive sensors with fast response times. Porous polymer foams, with a lightweight nature and adjustable mechanical properties, are promising candidates for flexible pressure sensors. These materials can be customized to control piezoresistance and sensitivity by regulating cell microstructure, surface microstructure, and conductive nanofiller networks. Their versatility enables applications in health monitoring, mechanical analysis, robotic tactile perception, and more. The piezoresistive sensing mechanism, the microstructure design, the criteria of pressure sensors, and the advanced fabrication approaches will be introduced in Chapter 8.

1.6.8 Triboelectric Nanogenerator Based on Polymer Foams

Micro/nano energy has gained prominence with the rise of mobile electronics and microdevices. TENGs offer a novel approach to converting mechanical energy into electrical energy by making use of the electrostatic charges generated on different triboelectric materials. Polymer foams have shown promising potential to be used as a new class of friction materials for TENGs due to their rough surface, high surface area, flexibility, abrasion resistance, and low cost. In addition, their internal porous structure enhances charge generation capacity and charge density during TENG operation, making them an attractive choice for sustainable and scalable energy generation. In Chapter 9, the classification and mechanism of TENGs, the prominent factors influencing TENG performance, the advantages of polymer foams in TENG, and advanced achievements in polymer foam-based TENGs will be covered.

1.6.9 Porous Polymers for Solar Steam Generation

Solar steam generation is a crucial technology for addressing water scarcity and seawater desalination. Porous polymers play a vital role in enhancing solar-to-steam conversion efficiency because their highly porous structure facilitates water transport, their lightweight makes them able to float on water, and their light scattering and absorption properties render them high energy conversion efficiency. The fundamental basis of solar steam generation, the effects of porous structure and the surface properties on the solar-thermal energy conversion efficiency, the methods to characterize the solar steam generation performance, and the different types of porous polymers based solar steam generators will be covered in Chapter 10.

References

1 Gibson, L.J. and Ashby, M.F. (1997). *Cellular Solids: Structure and Properties*. Cambridge University Press.
2 (a) Martini, J., Waldman, F., and Suh, N.P. (1982), The production and analysis of microcellular thermoplastic foams. SPE Technical Papers, Vol. 28, p. 674;

(b) Martini, J.E. (1981). The production and analysis of microcellular foam. MS Thesis, Massachusetts Institute of Technology.

3 (a) Xu, J. (2011). *Microcellular Injection Molding*. John Wiley & Sons. (b) Park, C.B., Behravesh, A.H., and Venter, R.D. (1998). Low density microcellular foam processing in extrusion using CO_2. *Polymer Engineering & Science* 38 (11): 1812–1823.

4 Sun, X., Liu, H., Li, G. et al. (2004). Investigation on the cell nucleation and cell growth in microcellular foaming by means of temperature quenching. *Journal of Applied Polymer Science* 93 (1): 163–171.

5 Nalawade, S.P., Picchioni, F., and Janssen, L. (2006). Supercritical carbon dioxide as a green solvent for processing polymer melts: processing aspects and applications. *Progress in Polymer Science* 31 (1): 19–43.

6 Zhai, W., Jiang, J., and Park, C.B. (2022). A review on physical foaming of thermoplastic and vulcanized elastomers. *Polymer Reviews* 62 (1): 95–141.

7 (a) Wood, C., Butler, R., Hebb, A. et al. (2002). Synthesis and processing of porous polymers using supercritical carbon dioxide. *Progress in Rubber, Plastics and Recycling Technology* 18 (4): 247–258. (b) Tomasko, D.L., Li, H., Liu, D. et al. (2003). A review of CO_2 applications in the processing of polymers. *Industrial & Engineering Chemistry Research* 42 (25): 6431–6456.

8 Cha, S.W. (1994). *A Microcellular Foaming/Forming Process Performed at Ambient Temperature and a Super-Microcelluar Foaming Process*. Massachusetts Institute of Technology.

9 Lee, S.-T. and Ramesh, N.S. (2004). *Polymeric Foams: Mechanisms and Materials*. CRC Press.

10 Merrill, E.W. (1996). I. States of polymers II. *Polymer Devolatilization* 33: 13.

11 (a) Suh, N.P. (2003). Impact of microcellular plastics on industrial practice and academic research. *Macromolecular Symposia*, Wiley Online Library 201: 187–202. (b) Shieh, Y.T., Su, J.H., Manivannan, G. et al. (1996). Interaction of supercritical carbon dioxide with polymers. I. Crystalline polymers. *Journal of Applied Polymer Science* 59 (4): 695–705.

12 (a) Colton, J. (1989). The nucleation of microcellular foams in semi crystalline thermoplastics*. *Materials and Manufacturing Processes* 4 (2): 253–262. (b) Durrill, P.L. and Griskey, R.G. (1966). Diffusion and solution of gases in thermally softened or molten polymers: Part I. Development of technique and determination of data. *AIChE Journal* 12 (6): 1147–1151.

13 Naguib, H.E., Park, C.B., and Song, S.-W. (2005). Effect of supercritical gas on crystallization of linear and branched polypropylene resins with foaming additives. *Industrial & Engineering Chemistry Research* 44 (17): 6685–6691.

14 Park, C.B. (2006). *Polymeric Foams: Science and Technology*. CRC Press.

15 (a) Colton, J. and Suh, N. (1987). The nucleation of microcellular thermoplastic foam with additives: Part I: Theoretical considerations. *Polymer Engineering & Science* 27 (7): 485–492. (b) Colton, J. and Suh, N. (1987). The nucleation of microcellular thermoplastic foam with additives: Part II: Experimental results and discussion. *Polymer Engineering & Science* 27 (7): 493–499. (c) Colton, J.S.

and Suh, N.P. (1987). Nucleation of microcellular foam: theory and practice. *Polymer Engineering & Science* 27 (7): 500–503.

16 (a) Matuana, L.M., Park, C.B., and Balatinecz, J.J. (1997). Processing and cell morphology relationships for microcellular foamed PVC/wood-fiber composites. *Polymer Engineering & Science* 37 (7): 1137–1147. (b) Matuana, L.M., Park, C.B., and Balatinecz, J.J. (1998). Cell morphology and property relationships of microcellular foamed pvc/wood-fiber composites. *Polymer Engineering & Science* 38 (11): 1862–1872.

17 Hall, P. and Stoeckli, H. (1969). Adsorption of nitrogen, n-butane and neo-pentane on chloro-hydrocarbon polymers. *Transactions of the Faraday Society* 65: 3334–3340.

18 (a) Deshpande, N.S. and Barigou, M. (1999). Performance characteristics of novel mechanical foam breakers in a stirred tank reactor. *Journal of Chemical Technology & Biotechnology* 74 (10): 979–987. (b) Djelveh, G., Gros, J., and Cornet, J. (1998). Foaming process analysis for a stirred column with a narrow annular region. *Chemical Engineering Science* 53 (17): 3157–3160.

19 Hansen, R.H. and Martin, W.M. (1964). Novel methods for the production of foamed polymers. Nucleation of dissolved gas by localized hot spots. *Industrial & Engineering Chemistry Product Research and Development* 3 (2): 137–141.

20 Lee, S.-T. and Park, C.B. (2014). *Foam Extrusion: Principles and Practice*. CRC Press.

21 Han, C.D. and Yoo, H.J. (1981). Studies on structural foam processing. IV. Bubble growth during mold filling. *Polymer Engineering & Science* 21 (9): 518–533.

22 Ramesh, N., Rasmussen, D.H., and Campbell, G.A. (1991). Numerical and experimental studies of bubble growth during the microcellular foaming process. *Polymer Engineering & Science* 31 (23): 1657–1664.

23 Koopmans, R.J., den Doelder, J.C., and Paquet, A.N. (2000). Modeling foam growth in thermoplastics. *Advanced Materials* 12 (23): 1873–1880.

24 Zhai, W., Wang, J., Chen, N. et al. (2012). The orientation of carbon nanotubes in poly (ethylene-co-octene) microcellular foaming and its suppression effect on cell coalescence. *Polymer Engineering & Science* 52 (10): 2078–2089.

25 Wong, A., Chu, R.K., Leung, S.N. et al. (2011). A batch foaming visualization system with extensional stress-inducing ability. *Chemical Engineering Science* 66 (1): 55–63.

26 Wong, A. and Park, C. (2012). A visualization system for observing plastic foaming processes under shear stress. *Polymer Testing* 31 (3): 417–424.

27 Baldwin, D.F., Park, C.B., and Suh, N.P. (1996). An extrusion system for the processing of microcellular polymer sheets: shaping and cell growth control. *Polymer Engineering & Science* 36 (10): 1425–1435.

28 (a) Taki, K., Tabata, K., Kihara, S.-i., and Ohshima, M. (2006). Bubble coalescence in foaming process of polymers. *Polymer Engineering & Science* 46 (5): 680–690. (b) Zhai, W., Wang, H., Yu, J. et al. (2008). Cell coalescence suppressed by crosslinking structure in polypropylene microcellular foaming. *Polymer Engineering & Science* 48 (7): 1312–1321.

29 Park, C.B., Behravesh, A.H., and Venter, R.D. (1997). *A Strategy for the Suppression of Cell Coalescence in the Extrusion of Microcellular High-Impact Polystyrene Foams*. ACS Publications.

30 Mosanenzadeh, S.G., Naguib, H.E., Park, C.B., and Atalla, N. (2014). Development of polylactide open-cell foams with bimodal structure for high-acoustic absorption. *Journal of Applied Polymer Science* 131 (7): 39518.

31 Rodeheaver, B.A. and Colton, J. (2001). Open-celled microcellular thermoplastic foam. *Polymer Engineering & Science* 41 (3): 380–400.

32 Enayati, M., Famili, M.H.N., and Janani, H. (2013). Open-celled microcellular foaming and the formation of cellular structure by a theoretical pattern in polystyrene. *Iranian Polymer Journal* 22: 417–428.

33 Derrick, T. (Derrick Edward)(1994). *Processing and Analysis of Microcellular Open-Cell Foams*. Massachusetts Institute of Technology.

34 Yu, P., Mi, H.-Y., Huang, A. et al. (2015). Effect of poly (butylenes succinate) on poly (lactic acid) foaming behavior: formation of open cell structure. *Industrial & Engineering Chemistry Research* 54 (23): 6199–6207.

35 (a) Park, C.B., Padareva, V., Lee, P.C., and Naguib, H.E. (2005). Extruded open-celled LDPE-based foams using non-homogeneous melt structure. *Journal of Polymer Engineering* 25 (3): 239–260. (b) Lee, P.C., Wang, J., and Park, C.B. (2006). Extruded open-cell foams using two semicrystalline polymers with different crystallization temperatures. *Industrial & Engineering Chemistry Research* 45 (1): 175–181.

36 (a) Gandhi, A., Asija, N., Gaur, K.K. et al. (2013). Ultrasound assisted cyclic solid-state foaming for fabricating ultra-low density porous acrylonitrile–butadiene–styrene foams. *Materials Letters* 94: 76–78. (b) Guo, G., Ma, Q., Zhao, B., and Zhang, D. (2013). Ultrasound-assisted permeability improvement and acoustic characterization for solid-state fabricated PLA foams. *Ultrasonics Sonochemistry* 20 (1): 137–143.

37 Naguib, H.E., Park, C.B., and Reichelt, N. (2004). Fundamental foaming mechanisms governing the volume expansion of extruded polypropylene foams. *Journal of Applied Polymer Science* 91 (4): 2661–2668.

38 Park, C.B. and Cheung, L.K. (1997). A study of cell nucleation in the extrusion of polypropylene foams. *Polymer Engineering & Science* 37 (1): 1–10.

39 Fang, H., Zhang, Y., Bai, J. et al. (2013). Bimodal architecture and rheological and foaming properties for gamma-irradiated long-chain branched polylactides. *RSC Advances* 3 (23): 8783–8795.

40 (a) Schulze, D., Trinkle, S., Mülhaupt, R., and Friedrich, C. (2003). Rheological evidence of modifications of polypropylene by β-irradiation. *Rheologica Acta* 42 (3): 251–258. (b) Nam, G., Yoo, J., and Lee, J. (2005). Effect of long-chain branches of polypropylene on rheological properties and foam-extrusion performances. *Journal of Applied Polymer Science* 96 (5): 1793–1800.

41 (a) Bahreini, E., Aghamiri, S.F., Wilhelm, M., and Abbasi, M. (2018). Influence of molecular structure on the foamability of polypropylene: linear and extensional rheological fingerprint. *Journal of Cellular Plastics* 54 (3): 515–543.

(b) Xu, Z., Zhang, Z., Guan, Y. et al. (2013). Investigation of extensional rheological behaviors of polypropylene for foaming. *Journal of Cellular Plastics* 49 (4): 317–334.

42 (a) Doroudiani, S., Park, C.B., and Kortschot, M.T. (1996). Effect of the crystallinity and morphology on the microcellular foam structure of semicrystalline polymers. *Polymer Engineering & Science* 36 (21): 2645–2662. (b) Huang, H.-X., Wang, J.-K., and Sun, X.-H. (2008). Improving of cell structure of microcellular foams based on polypropylene/high-density polyethylene blends. *Journal of Cellular Plastics* 44 (1): 69–85.

43 (a) Guo, M.C., Heuzey, M.C., and Carreau, P.J. (2007). Cell structure and dynamic properties of injection molded polypropylene foams. *Polymer Engineering & Science* 47 (7): 1070–1081. (b) Jiang, X.-L., Bao, J.-B., Liu, T. et al. (2009). Microcellular foaming of polypropylene/clay nanocomposites with supercritical carbon dioxide. *Journal of Cellular Plastics* 45 (6): 515–538.

44 Sato, Y., Fujiwara, K., Takikawa, T. et al. (1999). Solubilities and diffusion coefficients of carbon dioxide and nitrogen in polypropylene, high-density polyethylene, and polystyrene under high pressures and temperatures. *Fluid Phase Equilibria* 162 (1–2): 261–276.

45 Suh, N. (1996). *Innovation in Polymer Processing* (ed. J.F. Stevenson). Cincinnati, OH: Hanser Gardner Publications.

46 Peng, X. and Qu, J. (1999). Extrusion mixing characteristics of filled polymer system on vibration force field. *China Plastic Industry* 2: 41–43.

47 Siripurapu, S., Coughlan, J.A., Spontak, R.J., and Khan, S.A. (2004). Surface-constrained foaming of polymer thin films with supercritical carbon dioxide. *Macromolecules* 37 (26): 9872–9879.

48 (a) Ni, G.-L., Zhu, X., Mi, H.-Y. et al. (2021). Skinless porous films generated by supercritical CO_2 foaming for high-performance complementary shaped triboelectric nanogenerators and self-powered sensors. *Nano Energy* 87: 106148. (b) Wang, L., Cui, W., Mi, H.-Y. et al. (2022). Fabrication of skinless cellular poly (vinylidene fluoride) films by surface-constrained supercritical CO_2 foaming using elastic gas barrier layers. *The Journal of Supercritical Fluids* 184: 105562.

49 Garg, A., Gulari, E., and Manke, C.W. (1994). Thermodynamics of polymer melts swollen with supercritical gases. *Macromolecules* 27 (20): 5643–5653.

50 Chiou, J., Barlow, J.W., and Paul, D.R. (1985). Plasticization of glassy polymers by CO_2. *Journal of Applied Polymer Science* 30 (6): 2633–2642.

51 Royee, J.R. (2000). *Supercritical Fluid Assisted Polymer Processing: Plasticization, Swelling and Rheology*. North Carolina State University.

52 Zhang, R.-H., Li, X.-K., Cao, G.-P. et al. (2011). Improved kinetic model of crystallization for isotactic polypropylene induced by supercritical CO_2: introducing pressure and temperature dependence into the Avrami equation. *Industrial & Engineering Chemistry Research* 50 (18): 10509–10515.

53 Liao, R., Yu, W., Zhou, C. et al. (2008). The formation of γ-crystal in long-chain branched polypropylene under supercritical carbon dioxide. *Journal of Polymer Science Part B: Polymer Physics* 46 (5): 441–451.

54 Chiou, J., Barlow, J., and Paul, D. (1985). Polymer crystallization induced by sorption of CO_2 gas. *Journal of Applied Polymer Science* 30 (9): 3911–3924.

55 Lambert, S. and Paulaitis, M. (1991). Crystallization of poly (ethylene terephthalate) induced by carbon dioxide sorption at elevated pressures. *The Journal of Supercritical Fluids* 4 (1): 15–23.

56 Beckman, E. and Porter, R.S. (1987). Crystallization of bisphenol a polycarbonate induced by supercritical carbon dioxide. *Journal of Polymer Science Part B: Polymer Physics* 25 (7): 1511–1517.

57 Lee, J., Turng, L.-S., and Kramschuster, A. (2010). The microcellular injection molding of low-density polyethylene (LDPE) composites. *Polymer-Plastics Technology and Engineering* 49 (13): 1339–1346.

58 Lee, Y.H., Kuboki, T., Park, C.B., and Sain, M. (2011). The effects of nanoclay on the extrusion foaming of wood fiber/polyethylene nanocomposites. *Polymer Engineering & Science* 51 (5): 1014–1022.

59 Antunes, M., Velasco, J.I., Realinho, V., and Solorzano, E. (2009). Study of the cellular structure heterogeneity and anisotropy of polypropylene and polypropylene nanocomposite foams. *Polymer Engineering & Science* 49 (12): 2400–2413.

60 Yokoyama, B.H., Li, L., Nemoto, T., and Sugiyama, K. (2004). Tunable nanocellular polymeric monoliths using fluorinated block copolymer templates and supercritical carbon dioxide. *Advanced Materials* 16 (17): 1542–1546.

61 Goel, S. and Beckman, E. (1993). Plasticization of poly(methyl methacrylate) (PMMA) networks by supercritical carbon dioxide. *Polymer* 34 (7): 1410–1417.

62 (a) Goel, S.K. and Beckman, E.J. (1994). Generation of microcellular polymeric foams using supercritical carbon-dioxide. II: cell-growth and skin formation. *Polymer Engineering & Science* 34 (14): 1137–1147. (b) Goel, S.K. and Beckman, E.J. (1994). Generation of microcellular polymeric foams using supercritical carbon dioxide. II: cell growth and skin formation. *Polymer Engineering & Science* 34 (14): 1148–1156.

63 (a) Baldwin, D.F., Park, C.B., and Suh, N.P. (1996). A microcellular processing study of poly (ethylene terephthalate) in the amorphous and semicrystalline states. Part I: microcell nucleation. *Polymer Engineering & Science* 36 (11): 1437–1445. (b) Baldwin, D.F., Park, C.B., and Suh, N.P. (1996). A microcellular processing study of poly(ethylene terephthalate) in the amorphous and semicrystalline states. Part II: cell growth and process design. *Polymer Engineering & Science* 36 (11): 1446–1453.

64 Han, X., Koelling, K.W., Tomasko, D.L., and Lee, L.J. (2002). Continuous microcellular polystyrene foam extrusion with supercritical CO_2. *Polymer Engineering & Science* 42 (11): 2094–2106.

65 Park, C.B., Baldwin, D.F., and Suh, N.P. (1995). Effect of the pressure drop rate on cell nucleation in continuous processing of microcellular polymers. *Polymer Engineering & Science* 35 (5): 432–440.

66 Arora, K.A., Lesser, A.J., and McCarthy, T.J. (1998). Preparation and characterization of microcellular polystyrene foams processed in supercritical carbon dioxide. *Macromolecules* 31 (14): 4614–4620.

67 (a) Bao, J.-B., Liu, T., Zhao, L., and Hu, G.-H. (2011). A two-step depressurization batch process for the formation of bi-modal cell structure polystyrene foams using $scCO_2$. *The Journal of Supercritical Fluids* 55 (3): 1104–1114. (b) Bao, J.-B., Weng, G.-S., Zhao, L. et al. (2014). Tensile and impact behavior of polystyrene microcellular foams with bi-modal cell morphology. *Journal of Cellular Plastics* 50 (4): 381–393.

68 Huang, J.-N., Jing, X., Geng, L.-H. et al. (2015). A novel multiple soaking temperature (MST) method to prepare polylactic acid foams with bi-modal open-pore structure and their potential in tissue engineering applications. *The Journal of Supercritical Fluids* 103: 28–37.

69 Martini-Vvedensky, J.E., Suh, N.P., and Waldman, F.A. (1984). Microcellular closed cell foams and their method of manufacture. Gloogle Patents: 1984.

70 Okolieocha, C., Raps, D., Subramaniam, K., and Altstädt, V. (2015). Microcellular to nanocellular polymer foams: progress (2004–2015) and future directions–a review. *European Polymer Journal* 73: 500–519.

71 Baldwin, D.F., Park, C.B., and Suh, N.P. (1998). Microcellular sheet extrusion system process design models for shaping and cell growth control. *Polymer Engineering & Science* 38 (4): 674–688.

2

Energy-Absorbing Polymer Foams

2.1 Overview of Energy-Absorbing Foam

Polymer foams represent a class of polymer materials with numerous tiny cellular voids dispersed within the polymer matrix [1]. Due to their highly compressible nature, they exhibit exceptional energy absorption and buffering properties, rendering them greatly advantageous for applications in numerous industrial sectors, such as automotive, aerospace, construction, and packaging industries [2–4]. Compared to metallic or inorganic foams, polymer foams provide superior design flexibility, ease of processing and recycling, and a wider range of compressive strengths ranging from highly elastic to semirigid and rigid.

The prominent energy absorption capacity of polymer foams is owing to the special deformation mechanism of the elastic or plastic cells within the foams. During compression, the cell walls undergo bending deformation, followed by subsequent collapse of the cells, thereby, dissipating energy through the frictional effects between the cell walls and the gas within the cells. Therefore, the porous structure is a critical factor that governs the mechanical properties of polymer foams as well as their energy absorption performance. Furthermore, dynamic mechanical considerations must be taken into account along with the static compressive mechanical properties, as different strain rates can significantly influence the buffering behavior of foam materials [5, 6].

2.1.1 Classification of Energy-Absorbing Polymer Foams

Polymer foams can be classified according to the structure of the cells. Open-cell foams consist of interconnected cell units connected by opening channels. In this foam, both the matrix and the gas phase are continuous phases, and the gas phase can flow freely through the foam. During compression and deformation, the edges of the cells play a prominent role in the mechanical response.

Conversely, the cells of closed-cell foams are entirely isolated from each other, with the gas-filled cavities being encased by the cell walls. During compression, the gas pressure within the cells should be considered since the gas is difficult to flow out from the cells in the process. Thus, the mechanical strength of closed-cell foams is

Functional Polymer Foams: Green Fabrication Methods, Performance and Applications,
First Edition. Haoyang Mi.

typically higher than that of open-cell foams, and their energy absorption capability is superior.

However, polymer foams can possess both open-cell and closed-cell structures with varying degrees of prevalence. Consequently, foams containing more than 90% closed-cell structures are generally categorized as closed-cell foam, while those containing less than this threshold are classified as open-cellular foams.

Auxetic foams are characterized by a concave morphology within their cell struts, which confers negative Poisson's ratio (NPR) behavior. This unique property is manifested by transverse expansion upon uniaxial stretching and transverse contraction under uniaxial compression. In contrast to conventional foams, the NPR effect in auxetic foams arises from the inward bending of the cell walls, rather than the outward bending of the cell walls upon compression or impaction.

Wrinkled foams are a type of cellular material with an intricate surface structure consisting of regularly spaced micro-sized folds or wrinkles, which could resemble the wrinkled surfaces of desiccated fruits and vegetables. The formation of folds or wrinkles usually originates from the release of internal stress and the fast contraction of the cells right after the cell expansion stage. The unique structure of wrinkled foams presents opportunities for effective energy absorption and impact mitigation applications, such as buffering and shock attenuation.

2.1.2 Factors Affecting Mechanical Properties of Polymer Foams

2.1.2.1 Relative Density

The mechanical behavior of polymer foams is governed by their internal architecture, which serves as the primary determinant of their macroscopic mechanical properties. Specifically, when subjected to external loading, low-density foams typically undergo a fast deformation process owing to rapid densification and low plateau stress, while high-density foams exhibit relatively low deformation when subjected to the same energy input [7, 8]. Therefore, Young's modulus and energy dissipation capacity of polymer foams are highly sensitive to their relative density or bulk density. Therefore, precise control of foam density is essential for tuning its mechanical properties, which in turn can be tailored for specific engineering applications. The relative density of foam materials is defined as the density of foam material over the density of matrix material:

$$\rho_0 = \rho_f / \rho_s \tag{2.1}$$

ρ_f is the relative density of the foam material, ρ_s is the relative density of the foam base material, it can be seen that $\rho_0 \in (0,1]$.

Since the relative density will be more or less affected by the shape, size, and opening rate of the cells, the relative density is also related to the mechanical properties of polymer foams. To study the mechanical properties of foams, it is essential to establish the relationship between the relative density and the mechanical property.

Gibson and Ashby proposed an equation to describe the relationship between relative density and modulus in 1982 [9, 10]. The equation includes three terms, reflecting the influence of pillar bending, membrane (cell surface) stretching, and internal gas pressure of the obturator. The stiffness of closed-cell foam can be attributed to these three contributions as follows [11]:

$$\frac{E^*}{E_S} \approx \phi^2 \left\{ \frac{\rho^*}{\rho_S} \right\}^2 + (1 - \phi)\frac{\rho^*}{\rho_S} + \left(\frac{\rho_0}{E_S} \right) \frac{(1 - 2v^*)}{1 - \left(\frac{\rho^*}{\rho_S} \right)} \tag{2.2}$$

where E^* and E_S are the modulus of foam and solid polymer, respectively; ρ^* and ρ_S are the relative density of foam and density of solid polymer, respectively; v^* is the Poisson ratio of the polymer foams. According to formula (2.2), foam materials with higher relative density have a larger modulus than those with lower relative density and will absorb more energy during impact.

2.1.2.2 Morphological Features of the Cells

In the case of polymer foams, the term "cells" is usually used to indicate the "pores" or "bubbles" inside of the polymer matrix. Different types of cellular structures can be distinguished based on their morphological features, which greatly affects the macroscopic properties and application fields of the final foam products. In the case of a closed-cell foam structure, the cells are completely isolated from each other, and the gas-filled cavity is surrounded by closed-cell walls. While the air in open-cell foam can flow freely through the cells, resulting in a soft, more sponge-like appearance [12]. Thus, closed-cell foam usually has a higher modulus than open-cell foams for a given matrix material.

The cells in closed-cell foam could present in different morphologies, such as spherical shape, polygonal shape, and ellipsoidal shape (Figure 2.1). The cell shapes are highly related to the physical properties of the polymer and the parameters in the foaming process, like foaming pressure and temperature. Special multilevel cellular morphology could be formed by manipulating the foaming process. For example, wrinkled cellular structures can be created by introducing a flow field of supercritical CO_2 during the gas soaking stage, and foams with concave corner cellular structures can be formed by a water-assisted foaming process. Foams with special cellular structures sometimes exhibit unique properties such as NPR properties.

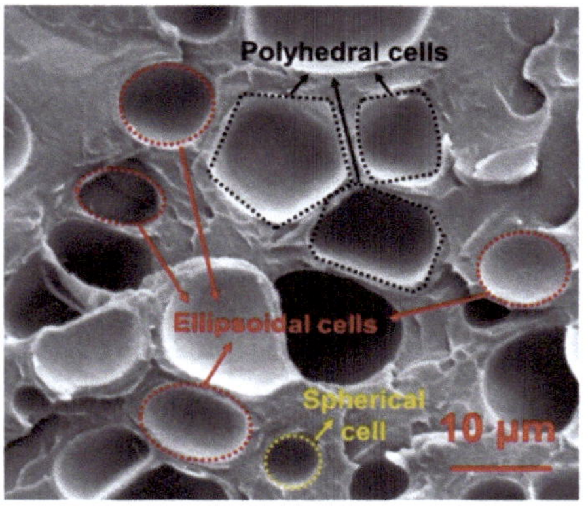

Figure 2.1 Different cell morphologies in polymer foams. Source: Zhu et al. [13]/with permission of Elsevier.

To establish a mathematical model, the cell morphology can be characterized by the shape factor and the pore shape coefficient. The isoparametric quotient expressed as (2.3) describes the degree to which the shape of a bubble is nearly spherical [13].

$$\text{isoparametric-quotient} = 36\pi V^2/A^3 \tag{2.3}$$

where V is the volume of the cells and A is the surface area of the cells. According to the formula, the shape factor of the sphere is equal to 1, and that of other shapes is less than 1. When the average value of all the shape factors of the cells in the foam is close to 1, it indicates that the foam is mainly composed of spherical cells; otherwise, the cells would be mostly polyhedron or ellipsoid.

The cell shape coefficient, or the anisotropy of foam cell structure, refers to the differences in pore size in different directions. It can be represented by the average height-to-diameter (H/D) ratio of cells. As the increase of the H/D ratio, the cellular structure of the foam transfers to anisotropic gradually, and the mechanical properties including the compression, bending strength, and Young's modulus of polymer foams in the cell alignment direction would increase with the increase of the cell shape coefficient, showing a basically proportional relationship. Under quasi-static compression, as the strain increases, the stress in the cells is mainly concentrated at the edge of the cells.

2.1.2.3 Cell Size Influence on Mechanical Properties

The size of the cell is represented by its aperture and its aperture distribution. Aperture refers to the average cell diameter of the polymer foams, which can be defined in the following ways:

$$d_n = \sum n_i d / \sum n_i \tag{2.4}$$

$$d_w = \sum n_i d_i^2 / \sum n_i d_i \tag{2.5}$$

$$d_v = \sum n_i d_i^3 / \sum n_i d_i^2 \tag{2.6}$$

where d_n, d_w, and d_v refer to the number average, weight average, and volume average diameter of cells, respectively.

Cell size distribution refers to the size distribution of cells in the foam body, and is expressed by the standard deviation of cell size:

$$\sigma_d = \sqrt{\frac{1}{N-1}\sum_{i=1}^{N}(d_i - d)^2} \tag{2.7}$$

According to the average cell size, foams can be divided into the following groups:

1) Bulbous cell structure, cell size > 100 μm
2) Microcellular structure, cell size = 1–100 μm
3) Ultrafine cellular structure, cell size = 0.1–1 μm
4) Nanocellular structure, cell size < 0.1 μm.

For most polymer foams, Young's modulus usually increases with the decrease of cell size [14]. Since the edge polymer fraction φ would be larger with the increase of cell size and the contribution of polymer foams to mechanical properties mainly comes from the edge, the mechanical properties would decrease with the increase of the cell diameter, resulting in the decline of the compressive strength of polymer foams, when the relative density and wall thickness maintains the same. However, Smits studied the polyurethane (PU) foam and found that the compression strength first decreased and then increased with the increase of the cell size when the relative density remained constant. This was because when the cell size increases to a certain extent, the polymer fraction $(1 - \varphi)$ of the cell wall is significantly increased. At this time, the compressive strength of the foam mainly comes from the contribution of the cell walls, so the compressive strength increases with the increase of the cell size.

2.1.2.4 Matrix Influence on Mechanical Properties

For different buffering polymer foams, the requirements for the materials vary according to different applications. For example, shock absorption is very important in the high jump and pole vault, so the density of the buffer cushion is small, and the foam thickness is large. Combat sports, on the other hand, require a safe position to move in, so the MATS used in these sports (e.g. karate and taekwondo) are dense and thin [15]. To meet different requirements, the polymer matrix material selection plays a pivotal role since the property of the foam is largely determined by the matrix material, which will affect the stress–strain relationship and capacity absorption of foams.

Several commonly used matrix materials for polymer foams are described below.

Ethylene vinyl acetate (EVA) foam. Ethylene vinyl–vinyl acetate is a copolymer of vinyl and vinyl acetate materials, in which the vinyl acetate content is usually between 10 and 40 mol%. Due to its transparency, flexibility, and low production costs, it is increasingly used in the packaging and pharmaceutical industries. Attributing to its high energy absorption capacity, it can also be applied in the automotive and sports equipment industries. For example, the sole subjected to a dynamic load of 200 kPa is mainly supported by a density of 150–250 kg m^{-3} closed-cell EVA foam [16]. The core layer of protective equipment used in cricket is also made of EVA foam, and its sides are surrounded by a polycarbonate shell layer. This sandwich structure is capable of preventing impact speeds of up to 45 m s^{-1}.

PU Foam. PU series include PU connection —NH—C=O—O— containing materials. In the production of PU, polyisocyanates are produced by the reaction between diols and polyols. PU foam accounts for the majority of the PU market, with a market share of about 68% [15]. By changing the amount of raw material composition, foam properties can be changed from soft foam to monolithic foam to hard foam on a large scale, with foam density ranging from 100 to 250 kg m^{-3} and elastic modulus ranging from 10 to 300 kPa. Thermoplastic PU (TPU) elastomer is a linear polymer formed by the interblock copolymerization of hard and soft chain segments. TPU has tensile, wear, and heat resistance and other physical properties, similar to rubber elasticity. TPU particles after foaming, the volume expansion of 5–8 times, the diameter of the bubble is only 30–300 microns, the density of 0.15–0.25 g cm^{-3} makes people almost feel its weight, the industry image called "popcorn." TPU foam material overcomes the shortcomings of TPU raw materials such as large weight, high hardness, and poor shock absorption performance, and is transformed into a performance material with all the advantages of TPU and foam materials, which has a lighter weight, stronger permeability, high rebound, compression resistance, and excellent low-temperature resistance. Environmental protection, wear resistance, high elasticity, yellowing resistance, and many other advantages of performance materials.

PVC/nitrile Butadiene Rubber Foam. PVC/nitrile butadiene rubber materials belong to the category of thermoplastic elastomers. Most of these materials are linear block copolymer (–AAAA–BBBB–AAAA–) structures, in which each block forms a soft segment and a hard chain segment. Similarly, the mechanical properties of this structural material can be easily changed by changing the proportion of hard and soft segments.

Crosslinked polyethylene (PE) foams usually have a closed-cell structure. Compared to non-crosslinked PE foams, they have better heat resistance. Due to the modification of their molecular structure, they also have better resilience after static loads, which is a basic requirement for cushion pads.

2.1.2.5 Effect of Open-Cells on Mechanical Properties

The fraction of open-cell refers to the percentage of the air-accessible volume of the total foam volume. The fraction of close-cell, contrary to the open-cell rate, refers to the percentage of the air inaccessible volume of the total foam volume. The effective fraction of open-cells in foam is estimated by the following equation:

$$v = \frac{N_{open} + \frac{1}{2}N_{part}}{N_{open} + N_{part} + N_{pin} + N_{closed}} \tag{2.8}$$

where N_{open} is the open-cell foam component, N_{closed} is the closed-cell foam component, N_{part} is a half-cell foam component.

When subjected to compressive stress, the surface, and edge of the closed-cell foam bear the load together, and the rupture of the cell wall will not cause damage to the adjacent edge, so the compression strength is higher than that of the open-cell foam under the same condition. Under the action of tensile stress, the cell wall of closed-cell foam is easy to rupture, and this rupture will accelerate the damage of adjacent edges, resulting in relatively lower tensile strength and strain than the open-cell foam under the same condition. Therefore, the compressive strength decreases with the increase of the open-cell ratio, and the tensile strength usually increases with the increase of open-cell ratio when the relative density is the same. Under impact and static compression, closed-cell foam has better bearing capacity than open-cell foam.

2.2 Energy Absorption Mechanism of Polymer Foams

Polymer foams, as excellent buffer packaging materials, absorb, store, or dissipate energy generated during shock and vibration, mainly through the buckling, plastic deformation, fracture, and friction of cell walls and the compression of air preserved inside the cells. Gibson and Ashby proposed a material energy absorption diagram to evaluate the energy absorption properties of foam materials [9]. The energy absorption and buffering properties can be analyzed by the stress–strain curve, and the energy absorption of foam materials can be represented by the integrated area from the stress–strain curve. It can be expressed by the following formula:

$$W = \int \sigma d\varepsilon \tag{2.9}$$

2.2.1 Open-Cell Structure Model

The special deformation and energy-absorbing mechanism of polymer foams usually include cell wall buckling, cell collapse, and compression of solidified foam. The deformation of polymer foam under compressive loads can be divided into three zones (Figure 2.2). In a small strain range (Zone I), the deformation property is almost linearly elastic, and the slope of the curve is the elastic modulus of the foam material. For strain less than 0.05, the bending of the bubble wall is dominant, and it is found that the foam deformation is mainly in the linear elastic region [17]. As the load increases (Zone II), the cells begin to collapse, and the stress changes very little at this stage, which leads to the emergence of the so-called stress platform. As the load continues to increase (Zone III), the opposite cell walls contact and enter the densification zone, at which time the pressure level increases sharply again. In this case, the foam behaves as a solid polymer material. The increased energy absorption capacity of the polymer foam is provided by the second region (the platform stage). At this stage, foam products can absorb large amounts of energy without significantly increasing stress levels. If the deformation during the loading process is too high and reaches the densification zone, the whole foam will be densified, and the energy absorption will be mainly attributed to the deformation of the matrix

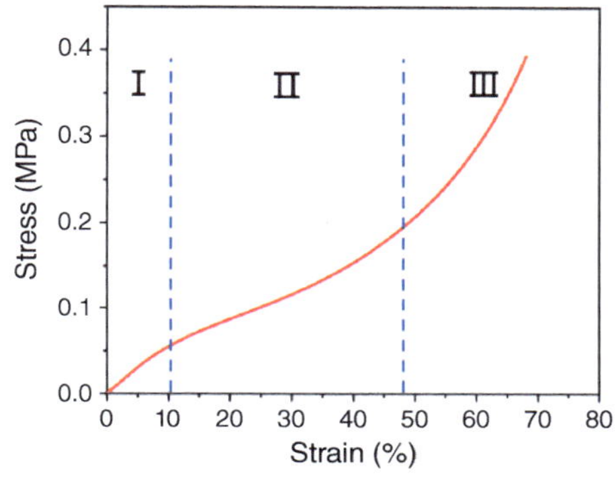

Figure 2.2 Typical compression strain–stress curve of polymer foams showing the three compression states.

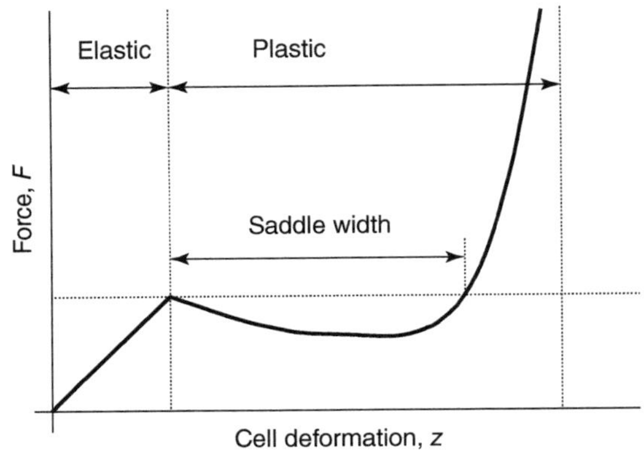

Figure 2.3 Typical compression strain–stress curve of viscoelastic foams. Source: Li and Mines [18]/with permission of John Wiley & Sons.

material. Therefore, the foams with higher elastic modulus and higher density usually have sharper slopes in the stress–strain curves, which represents greater energy absorption under the same strain.

The deformation mechanism of foams would be different when they have different rigidity. For foams with relatively high rigidity, the initial collapse stress is usually used to represent the yield strength in the initial compression stage, which can be obtained through the first peak stress (as shown in Figure 2.3), and plastic deformation begins to occur at the same time. However, the typical viscoelastic foam dissipates the compressive stress through a local viscoelastic process through the bending, twisting, and buckling of the struts forming its structural skeleton. The stress plateau region is owing to the nonlinear fully recoverable bubble wall buckling, resulting in continuous collapse for the viscoelastic foam.

In closed-cell foam, in addition to the contribution of the matrix structure of the cell, the compression of the gas in the bubble also absorbs part of the energy. The cell surface, air, and air–polymer interaction are crucial aspects of foam mechanics. When the gas and polymer act simultaneously during compression, they experience the same strain as the bubble and gas were compressed. In the uniaxial compression process of the closed-cell foam, the contribution of gas within the cell can be described as the equation [19]:

$$\sigma_G = \frac{p_a \varepsilon f}{1 - \varepsilon - R} \tag{2.10}$$

where p_a is one atmosphere (assumed gas in undeformed foam cell), ε is the compressive strain applied, f is the proportion of closed pores in the foam, and R is the relative density of the foam.

The phenomenological constitutive model is a mathematical expression between stress and strain obtained by fitting experimental data. It can be used to describe physical phenomena such as linear elastic region, plateau region, and densification region. Rusch first proposed a general model [20]:

$$\sigma(\varepsilon) = \varepsilon E_f \psi(\varepsilon)$$

where $\psi(\varepsilon) = A\varepsilon^m + B\varepsilon^n$ is a dimensionless function of compressive strain ε. The parameters A, B, m, and n are curve-fitting constants of the logarithm graph of $\psi(\varepsilon)$

versus $\psi(\varepsilon)$, but the model does not accurately describe the densified region. Avalle [21] improved connectivity in the elastic and platform areas and proposed a new model:

$$\sigma(\varepsilon) = A(1 - e^{-(E/A)\varepsilon(1-\varepsilon)^m}) + B\left(\frac{\varepsilon}{1-\varepsilon}\right)^n$$

where A and E are the stress and elastic modulus of the platform, and the index m represents the connection between the elastic region and the platform region. The parameters B and n describe the densification region. Jeong [22, 23] added a modulus function (strain rate correlation) to the Avalle model [21] and proposed a constitutive model of the stress–strain relationship in foam medium at a low strain rate:

$$\sigma(\varepsilon) = \left\{A(1 - e^{-(E/A)\varepsilon(1-\varepsilon)^m}) + B\left(\frac{\varepsilon}{1-\varepsilon}\right)^n\right\} M(\varepsilon, \dot{\varepsilon}) \tag{2.11}$$

where $M(\varepsilon, \dot{\varepsilon})$ was introduced as shape and scale factors:

$$M(\varepsilon, \dot{\varepsilon}) = 1 + n(\varepsilon) \ln\left(\frac{\dot{\varepsilon}}{\dot{\varepsilon}_0}\right) \tag{2.12}$$

$$n(\varepsilon) = a_1 + b_1\varepsilon \quad \varepsilon \leq \varepsilon_1 \tag{2.13}$$

$$n(\varepsilon) = c_1 + c_2\varepsilon + c_3\varepsilon^2 + c_4\varepsilon^3 \quad \varepsilon_1 \leq \varepsilon \leq \varepsilon_2 \tag{2.14}$$

$$n(\varepsilon) = a_2 + b_2\varepsilon \quad \varepsilon_2 \leq \varepsilon \tag{2.15}$$

Initial linear region ($\varepsilon_1 \leq \varepsilon \leq \varepsilon_2$) and compact region ($\varepsilon_2 \leq \varepsilon$) adopt $n(\varepsilon)$ linear function. The tight strain range is $\varepsilon_1 \leq \varepsilon \leq \varepsilon_2$. c_1, c_2, c_3, and c_4 can be obtained from the $\varepsilon = \varepsilon_1$, and $\varepsilon = \varepsilon_2$ boundary conditions of the sum.

In a study in 1997, Gibson et al. described the mechanical behavior of closed-cell and open-cell foam based on the cubic cell model [10]. Using this simplified structure, the properties of real foam can be estimated, and the model is similar to the deformation and failure behavior of real foam structures. Figure 2.4 shows a model

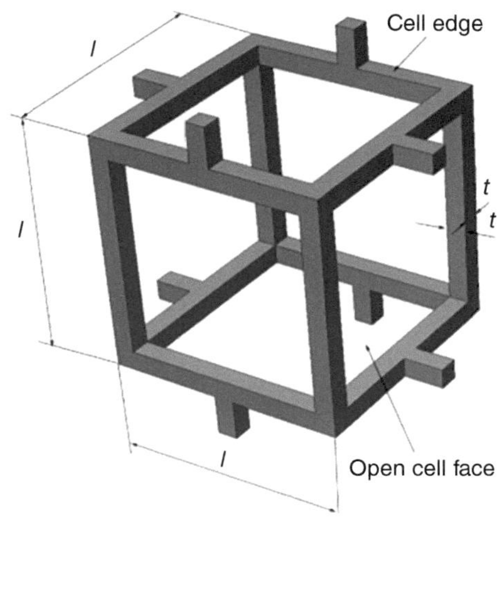

Figure 2.4 Illustration of cubic cell model for an open-cell foam. Source: Hössinger-Kalteis et al. [24]/with Sage Publications, Inc, CC BY 4.0.

of an open-cell foam consisting of an image edge of length l and thickness t. There is a relationship between the relative density and the pore size $\frac{\rho}{\rho_0} \propto \frac{V}{V_0} \propto \left(\frac{t}{l}\right)^2$. The elastic modulus of the bubble structure can be calculated by the elastic deformation of the beam under the load acting on the point in the middle. The deformation required by the deformation can be obtained by the beam theory $\delta = \frac{Fl^3}{E_0 I}$. Where the polar moment of inertia of the edge is $I \propto t^4$.

Therefore, when stress is applied to the whole foam, each edge is subjected to a force of magnitude F, thus obtaining the Young's modulus of the foam as follows:

$$E = \frac{\sigma}{\varepsilon} = \frac{C_1 E_0 I}{l^4} \tag{2.16}$$

Among them, C_1, as a proportional coefficient, contains all the geometric constants that can be obtained through the fitting of experimental data.

2.2.2 Closed-Cell Foam

In 1997, Gibson modeled and analyzed the closed-cell foam [10]. The model consisted of an image edge of length (l) and thickness (t_e) and an image plane of thickness (t_f) (Figure 2.5). It is deduced that the contribution of Young's modulus of closed-cell foam is as follows: (i) The bending of the cell wall; (ii) The gas phase in the cell during compression. For the stress of the cell mask caused by the bending of the cell edge, the formula of the contribution of three parts to the foam modulus can be derived as follows:

$$\frac{E}{E_0} \approx \phi^2 \left(\frac{\rho}{\rho_0}\right)^2 + (1-\phi)\frac{\rho}{\rho_0} + \frac{p(1-2v)}{E_0(1-\rho/\rho_0)} \tag{2.17}$$

Compared to open-cell foam, there are some additional parameters in the equation for closed-cell foam. The initial pressure p of a gas in the cell is usually atmospheric pressure. The ϕ is the solid part contained in the edge of the cell, and the v is the Poisson's ratio of the foam.

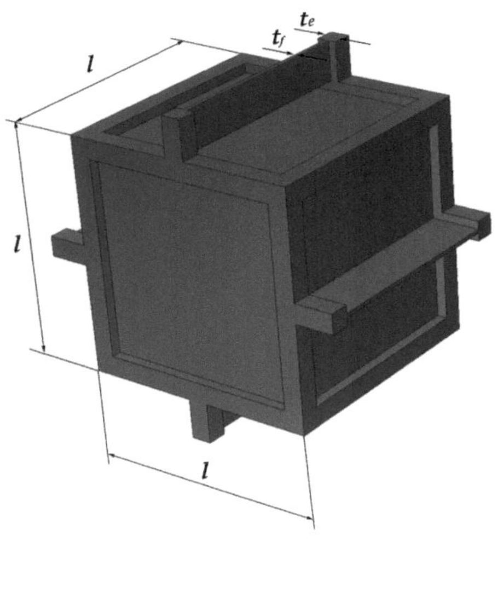

Figure 2.5 Illustration of cubic cell model for a closed-cell foam. Source: Hössinger-Kalteis et al. [24]/with Sage Publications, Inc, CC BY 4.0.

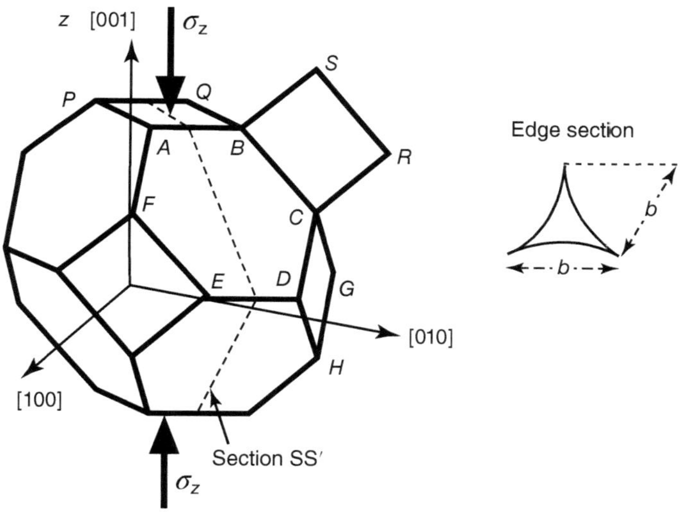

Figure 2.6 Illustration of Kelvin model for a closed-cell foam. Source: Almanza et al. [26]/with permission of ELSEVIER.

The Kelvin foam model is a spatially filled arrangement of equal volume cells whose final arrangement has the smallest surface area. The polyhedron used as the base unit of Kelvin's foam is a tetrahedral dodecahedron, consisting of 14 slightly curved faces and 24 vertices. Simone [25] gave an analytical description of closed-cell tetradodecahedral foam used to calculate Young's modulus and peak stress in 1998. It is assumed that the thickness of the entire cell wall is constant (Figure 2.6). Finite element analysis (FEA) is carried out on the model, and its Young's modulus is obtained by the polynomial fitting [26]:

$$\frac{E^*}{E_s} \approx A_E\left(\frac{\rho^*}{\rho_s}\right) + B_E\left(\frac{\rho^*}{\rho_s}\right)^2 \tag{2.18}$$

Mills et al. [27] established a finite element micromechanical model of closed-cell polymer foams under impact compression (Figure 2.7). Using a model compressed in the direction of Kelvin closed-cell foam [1 1 1], the dynamic (explicit) FEA model diagram represents a sample of infinite layers of foam, with a thickness of two or four holes in diameter in the direction of compression. The limited thickness reduces the number of cells and the task of defining cell surfaces in ABAQUS CAE. Each unit remains sealed, and the contained air mass remains the same.

ABAQUS calculates the air mass m_c of the cavity from the gas molecular mass M and cavity volume V_c by using $M_c = \frac{V_c M p_a}{R_0 T_0}$. Where R_0 is the gas constant, and T_0 is the absolute reference temperature. By setting boundary conditions and material characteristics, the compression impact was modeled. The average axial compression engineering strain in the foam was e_c and it is related to displacement U_1: $e_c = \frac{U_1}{2D}$. The stress σ_{TA} at the top of the barometric contribution model is calculated from the relative pressure in cells 1–3: $\sigma_{TA} = 0.4062(p_1 + p_2) + 0.1875p_3$.

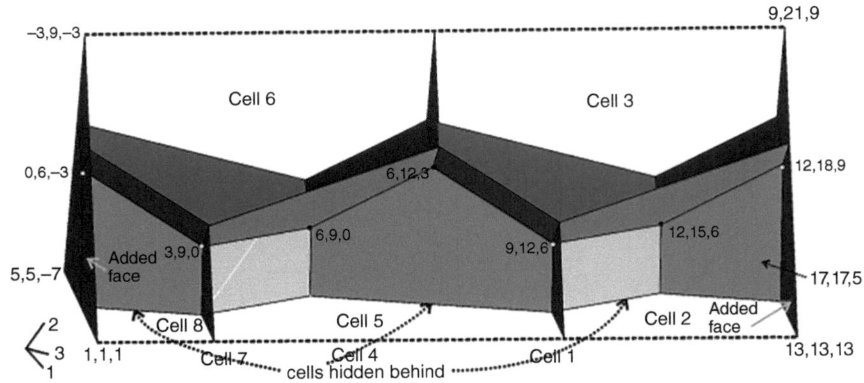

Figure 2.7 A finite element micromechanical model of closed-cell polymer foams by Mills. Source: Mills et al. [27]/with permission of Elsevier.

2.2.3 Other Special Structural Models

Auxetic foams with NPA are a kind of foam with volume expansion under tensile strain and volume contraction under compressive strain. Since Lakes prepared polymer pull-up foams in 1987, studies on the relationship between concave structure and foam properties have been enriched [28]. The concept of auxetic foams was first proposed by Evans in 1991 [29]. Different from traditional foam materials, auxetic foams will undergo lateral contraction under a compression load, resulting in increased material density and improved compressive capacity and energy absorption. The bending of the cell rib is the main deformation mechanism during auxetic transformation, so the stretching effect in the foam is achieved by converting the convex polyhedral-shaped cells of the traditional foam into a concave and more curly-shaped porous structure. Figure 2.8 compares the compressive force response of a conventional foam and an auxetic foam. Materials would be pushed away from the force for conventional foam, while the materials are contracted towards the applied force for an auxetic foam [30]. Due to the complexity and

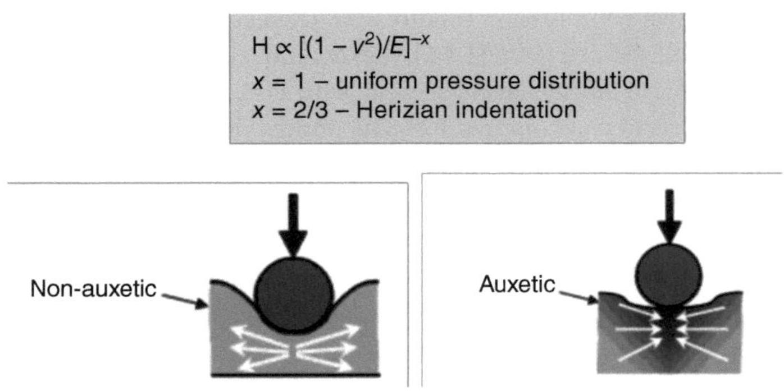

Figure 2.8 Compressive force response of a conventional foam and an auxetic foam. Source: Evans and Alderson [30]/with permission of John Wiley & Sons.

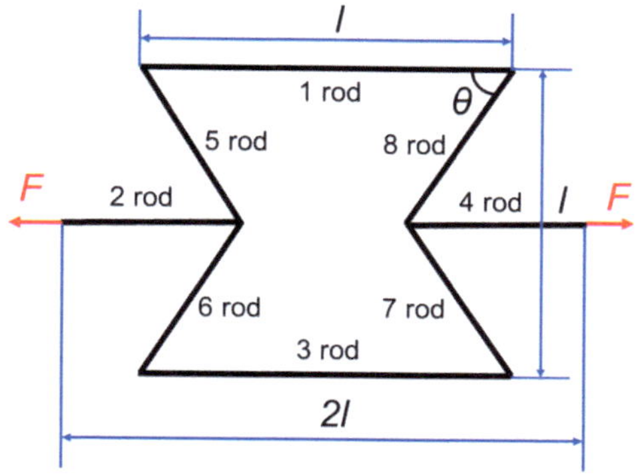

Figure 2.9 An auxetic concave hexagon monocell design.

poor regularity of NPR foam structure, the current research mainly focuses on the performance of a single structure or regular structure.

The concave angle structure is typically used to represent auxetic concave hexagon monocell design (Figure 2.9). The cross-section of all the rods in the structure is a circular section with radius R, and the angle between the transverse and oblique rods is θ. The length and width of the structural cell are $2l$ and l. The elastic modulus of the material is E, and the shear elastic modulus is G. When subjected to axial load, through the overall analysis of the structure, the 1, 2, 3, and 4 rods are stretched, and the 5, 6, 7, and 8 rods are compressed, bent, and sheared.

When the axial unit force load is applied, the internal force of each bar, the bending moment equation of bar three, and the shear equation are as follows:

$$\overline{F}_{Ni} = 1 \quad (i = 2, 4) \tag{2.19}$$

$$\overline{F}_{Ni} = \frac{1}{2}(i = 1, 3) \tag{2.20}$$

$$\overline{F}_{Ni} = -\frac{1}{2}\cos\theta \quad (i = 5, 6, 7, 8) \tag{2.21}$$

$$\overline{M}_i(x) = \frac{1}{2}x\sin\theta - \frac{1}{8}l \quad (i = 5, 6, 7, 8) \tag{2.22}$$

$$\overline{F}_{si} = \frac{1}{2}\sin\theta \quad (i = 5, 6, 7, 8) \tag{2.23}$$

According to Mohr's theorem formula:

$$\Delta = \int \frac{\overline{F}_N F_N}{EA}dx + \int \frac{\overline{M}(x)M(x)}{EI}dx + \int \frac{k\overline{F}_s F_s}{GA}dx \tag{2.24}$$

The Poisson's ratio numerical model is obtained

$$\nu = -\frac{\frac{4l^2}{3R^2E\sin\theta\tan\theta} + \frac{32\cos\theta}{27G} - \frac{\cos\theta}{E}}{\frac{1}{E}\left(2 + \frac{1}{\tan\theta} + \frac{\cos\theta}{2\tan\theta}\right) + \frac{l^2}{96R^2E\sin\theta} + \frac{16\sin\theta}{27G}} \tag{2.25}$$

Ma et al. [31] proposed a novel three-dimensional-folded structure (3D-FS) with an NPR. The deformation mechanism and energy absorption properties are investigated by the experiment and simulation. An analytical model describing the two

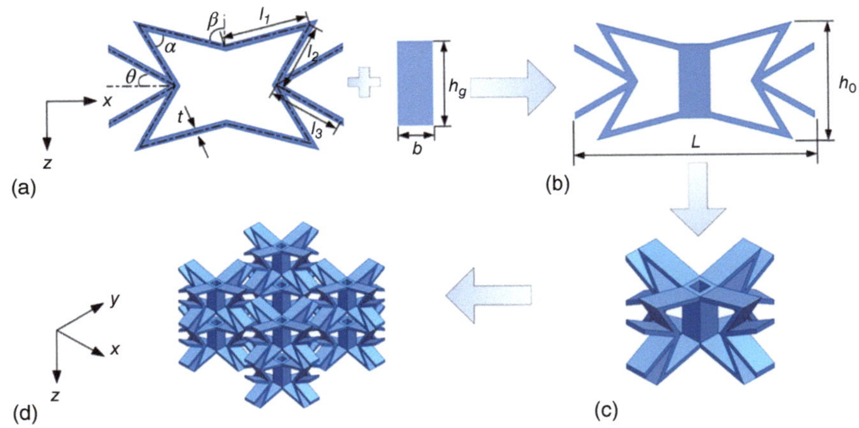

Figure 2.10 3D-FS and its unit cell structure: (a) 2D star structure and thin-walled square tube, (b) 2D structure of 3D-FS unit cell, (c) 3D-FS unit cell, and (d) 3D-FS. Source: Ma et al. [31]/with permission of John Wiley & Sons.

plateau stress phases is presented based on the experimental data. Numerical simulations are performed to analyze the influence of the structural geometry parameters on the plateau stress, energy absorption properties, and NPR, and to establish the relationship between specific energy absorption and wall thickness ratio for the two plateau stress phases. The unit cell of the illustrated 3D-FS consists of two parts, and Figure 2.10a–d shows the construction process of the 3D-FS. One is a star-shaped structure consisting of four upper inclined walls, four lower inclined walls, and two oblique crossing walls. The two side plates can form a star structure between the two unit cells and enhance the NPR effect. The other is a thin-walled hollow square tube. The reason for using a hollow tube is to achieve a second plateau of the structure and to reduce the mass of the 3D-FS.

They found that both the deformation of the 3D-FS and its unit cell contain double-plateau stress phases, and the 3D-FS has an apparent NPR effect. In the present study, based on a rigid-plastic beam with a perfectly plastic foundation model, the deformation mechanism of the 3D-FS cell is analyzed using the energy dissipation theory of plastic hinges. The deformation process of the 3D-FS cell is shown in Figure 2.11a–c.

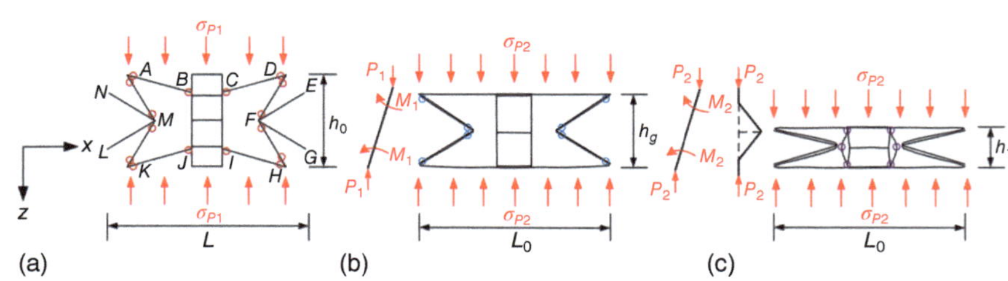

Figure 2.11 The deformation phases of the unit cell: (a) the initial phase of deformation, (b) the initial configuration of the second-plateau stress phase, and (c) the densification phase. Source: Ma et al. [31]/with permission of John Wiley & Sons.

The relationship between stress and structure of the two platforms is deduced by formula as:

$$\sigma_{P1} = \frac{4\sigma_s bt^2(\pi - 2\beta + \alpha - \theta)}{2l_1 \cos\beta\{4b[l_3 \cos\theta - l_2 \sin(\beta - \alpha) + l_1 \sin\beta] + b^2\}}$$

$$\sigma_{P2} = \frac{4\sigma_s t^2\theta}{(4l_1 + b)\{2[l_2 \cos(\beta - \alpha) - l_1 \cos\beta] - 6t\}} + \frac{9.56\sigma_0 b^{-\frac{2}{3}} t^{\frac{5}{3}}}{4l_1 + b} \tag{2.26}$$

where α is a re-entrant angle, β is an angle between the upper inclined wall and the thin-walled square tube, θ is an angle between the double-inclined wall and the horizontal line, b is a side length of the thin-walled square tube and t is a thickness of the cell wall.

2.2.4 Energy Dissipation Mechanism of Polymer Foams

As a cushioning material, the main function of polymer foams is to reduce the force applied to the product, thereby, effectively protecting the objects. When a compressive load is applied to the material, the material structure undergoes deformation, and at this stage, some of the energy is stored, and some of the energy is dissipated through friction and other means. By applying cyclic stress to polymer foams, the stress–strain curves during loading and unloading exhibit a certain phase shift, thus obtaining hysteresis curves as shown in Figure 2.12. The lag between the loading and unloading curves can be interpreted as a measure of mechanical energy dissipation. The hysteresis for each curve could be normalized by the work done during the loading period and marked as the defined energy dissipation ratio:

$$R = E_1/(E_1 + E_2) \tag{2.27}$$

The lag between the loading and unloading curves is represented as E_1, and the area under the unloading curve is represented as E_2. During the compression process, the system always tends to find ways to dissipate energy to achieve stability. Therefore, energy dissipation during this process mainly occurs through the following pathways:

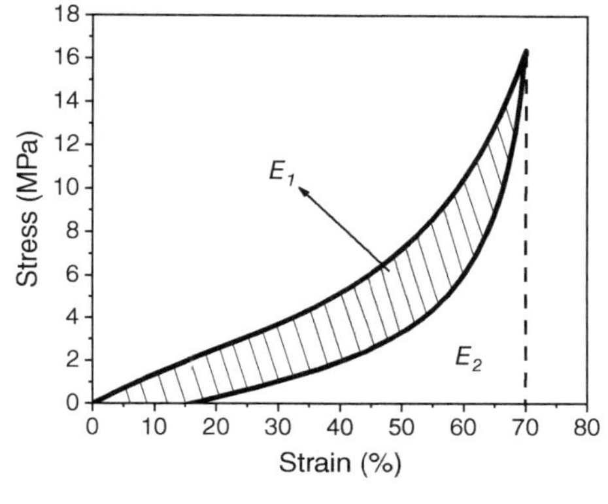

Figure 2.12 Typical cyclic compression curves of polymer foams with different densities.

(1) Energy dissipation caused by polymer chains. This includes the contribution of polymer chains relaxation and mesoscopic rotation to dissipate energy [15]. Although existing research indicates that this is related to loading rate and temperature, the direct relationship between deformation rate and temperature and energy dissipation in polymer foams is still elusive and deserves further investigation.
(2) In the loading and unloading processes, part of the kinetic energy will be converted into thermal energy. Even in purely elastic deformation without energy dissipation, heat will inevitably be generated during compression (friction generated by structural deformation, compression of gas inside pores, etc.). Fundamentally, this is a reversible process, and if the load is removed, the temperature will return to its original value. However, there is a stress field gradient and a corresponding temperature distribution gradient, which will lead to heat flux between the hotter and colder regions. When the stress is eliminated in the later stage of the negative load cycle, the temperature distribution is no longer the same as the temperature distribution caused by loading. Therefore, it is impossible for the compressed foam to return locally to its original state, which becomes a source of energy dissipation.
(3) The influence of shape recovery of foam materials. For example, compared with rigid polymer foams, the advantage of using soft foam is that it can withstand repeated mechanical damage caused by shape recovery and the viscoelastic relaxation mechanism helps to enhance energy dissipation. However, these mechanisms may also impair their ability to quickly recover shape under environmental conditions, thereby reducing the dissipation ratio during subsequent impact.
(4) Energy taken away by sound vibrations. When the foam material is subjected to a force load, a certain surface vibration will be generated, which will displace the surrounding air or other media, thereby generating sound waves. These sound waves will take away part of the energy to achieve the purpose of dissipation.

When a foam material is subjected to an impact load, the kinetic and elastic potential energies generated by the structure are transformed into other forms of energy through various physical processes, resulting in energy loss during vibration, which is known as damping. The damping magnitude is typically expressed by the damping factor or damping ratio. The motion equation of a system with damping and no external load is

$$m\ddot{u} + c\dot{u} + ku = 0 \tag{2.28}$$

Dividing by the mass m to normalize the equation.

$$\ddot{u} + 2\zeta\omega_0\dot{u} + \omega_0^2 u = 0 \tag{2.29}$$

where ω_0 is the natural frequency of the system in the absence of damping, ζ is the damping ratio, with a range of $0 \leq \zeta < 1$. The amplitude of free vibration will decay accordingly:

$$e^{-\zeta\omega_0 t} = e^{\frac{-2\pi\zeta t}{T_0}} \tag{2.30}$$

T_0 represents the period of an undamped oscillation.

To better represent the energy dissipation of materials during loading and unloading processes, it is assumed that the material properties are complex valued. Therefore, when the stress–strain relationship is linearly elastic, the complex-valued stress–strain relationship can be expressed as:

$$\dot{\sigma} = \dot{E}\dot{\varepsilon} = (E' + iE'')\dot{\varepsilon} \tag{2.31}$$

Here, the real part of Young's modulus is called the storage modulus, and the imaginary part is called the loss modulus. The loss modulus is typically described by the loss factor η, which is defined as:

$$\dot{E} = E(1 + i\eta) \tag{2.32}$$

To further characterize the damping behavior, the damping factor can also be expressed in terms of energy. Assuming that the energy dissipated during one cycle is given by:

$$D = \pi E'' \varepsilon_a^2 \tag{2.33}$$

The strain amplitude is denoted by ε, and the maximum elastic potential energy during one cycle of loading and unloading is given by:

$$W_s = \frac{1}{2} E' \varepsilon_a^2 \tag{2.34}$$

The damping factor can be expressed in terms of energy as follows:

$$\eta = \frac{D}{2\pi W_s} \tag{2.35}$$

The description based on dissipated energy can be used as long as the values of two energies can be determined, regardless of whether the hysteresis curve is elliptical.

2.2.5 Effect of the Matrix Material

The matrix material of polymer foams is also a critical factor affecting the energy absorption performance. Due to differences in crystallinity, atactic polypropylene (APP) exhibits a viscous or elastic state since it cannot crystallize, while isotactic polypropylene (iPP) has a certain toughness and hardness due to its higher crystallinity. The mechanical properties of the polymer amorphous region differ significantly under different states, with a smaller difference in modulus in the glassy state. When the amorphous region is in the rubber state, the modulus of the polymer will increase with the increase of crystallinity so that the modulus of the polymer with high crystallinity can be higher by orders of magnitude. In general, above the glass transition temperature (T_g), the intermolecular forces increase, and the tensile strength increases as the crystallinity increases. At the same time, the microcrystals play a role in physical cross-linking, reducing chain slip, so that the increase in crystallinity can reduce creep and stress relaxation phenomena. Below the T_g, the polymer becomes brittle as the crystallinity increases, and the tensile strength usually decreases. The change in matrix modulus with the change of crystallinity would render the same trends to the polymer foams fabricated using different matrix materials [32, 33].

Soft elastic materials, such as gels and rubbers, are typically referred to as elastomers. Elastomers undergo large deformations at low stresses and recover their original shape upon release of the stress. To exhibit such a response, the elastomer must have sufficient molecular mobility to allow conformational deformation of the polymer under the forces and cross-linking between the polymers to prevent flow and provide restoring forces simultaneously. The elongation-at-break of cross-linked elastomers is proportional to half the molecular weight between cross-link points of the polymer chains constituting the elastomer. The elongation-at-break is rapidly compromised when increasing cross-link density to enhance elastic modulus. Typically, the ductility and elasticity of conventional elastomers with crosslinked structures are in a trade-off relationship, and it is difficult to improve the toughness of elastomers, resulting in poor energy dissipation. Therefore, polymer foams fabricated using elastic polymers usually have high elasticity and recoverability, and their stretchability is also highly related to the elongation-at-break of the matrix material.

Meanwhile, energy dissipation also occurs due to viscosity between polymers, and the polymer chains of the elastomer are tightly packed together, so viscosity is generated by the frictional force between polymer chains. Therefore, in addition to the elastic behavior exhibited by the chain component with hooked elasticity of the cross-linked structure, elastomers also have viscous properties. Elastomers composed of amorphous polymers with viscous properties exhibit different stress behaviors during deformation loading and show different stress paths during unloading, which is known as hysteresis. Typical elastomers store energy through the application of strain, but because the stored energy is released without strain, no hysteresis is exhibited. In other words, it is well known that the origin of hysteresis is energy loss due to the properties of viscous materials, and the presence of hysteresis increases the fracture energy of elastomers.

For highly tough elastic materials, in addition to traditional cross-linking structures, a 3D cross-linked polymer that contains reversible bonds can be constructed, which can effectively dissipate energy by cleavage of the reversible chemical bonds. Reversible bonds such as hydrogen bonds, coordination bonds, and ionic bonds are often employed to create these supramolecular structures (Figure 2.13). Traditional stable cross-linking structures have the role of maintaining the elasticity and shape of the elastic material, and the cross-linking structure does not change even when the elastic material is deformed. On the other hand, reversible bonds

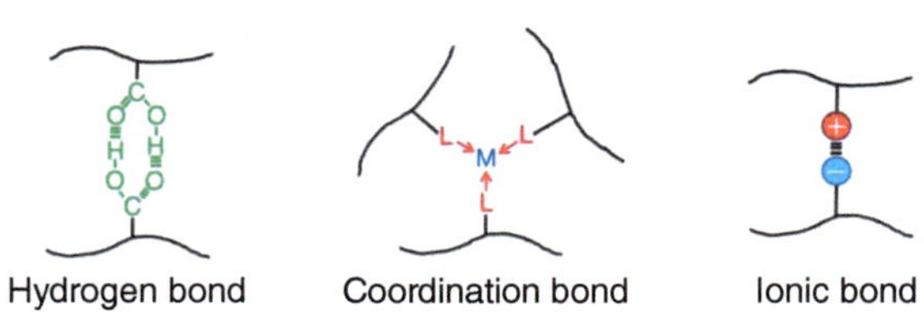

Figure 2.13 Diagram of reversible formation–dissociation bonds between polymer chains including hydrogen bonds, coordination bonds, and ionic bonds. Source: Takeoka et al. [34]/Taylor & Francis, CC BY 4.0.

can undergo repeated dissociation and form cross-linking structures because of elastic deformation. In other words, when reversible cross-linking structures are introduced, energy dissipates due to the deformation of the elastic material, which makes the elastic material highly tough. In addition, the stress–strain curve of the elastic material changes significantly exhibiting a strain toughening mechanism under a high-stress rate [34].

2.3 Testing and Characterization of Energy-Absorbing Foams

2.3.1 Morphology Characterization

In the previous section, we introduced how to analyze the energy absorption and cushioning properties of polymer foams through their microstructures and deformation behavior. In this section, we will focus on how to analyze and characterize the properties of energy-absorbing polymer foams.

Morphological parameters are critical factors influencing the mechanical properties of polymer foams. By investigating some basic parameters such as porosity, pore topology, and pore size, it helps understand the deformation and energy absorption mechanism.

Scanning electron microscopy (SEM) is a method of observation that lies between transmission electron microscopy and optical microscopy. It uses a narrow, focused beam of high-energy electrons to scan the sample, and collects, amplifies, and reimages various physical information through the interaction between the beam and the material. The pore size, pore density, and topological structure can be thus analyzed using the SEM images (Figure 2.14).

The cell size or cell diameter can be directly measured from the SEM image using image analysis software such as Image J and Image Pro Plus.

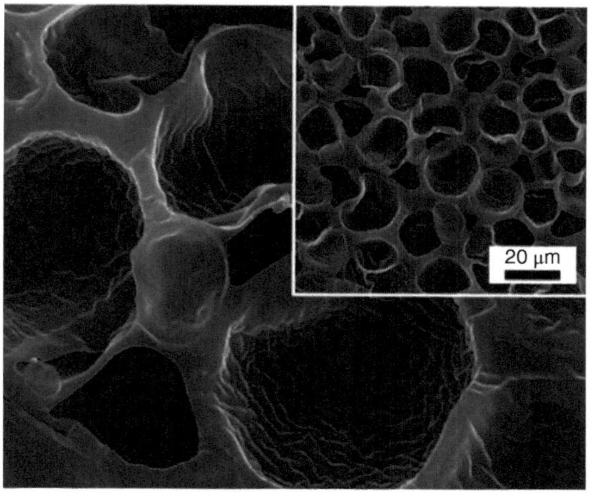

Figure 2.14 SEM characterization of cell size and cell density. Source: Hu et al. [35]/with permission of American Chemical Society.

The cell density N_f can be calculated using the following equation [35]:

$$N_f = \left(\frac{n}{A}\right)^{3/2} \tag{2.36}$$

where N_f (cm^{-3}) refers to the cell density of the sample, n is the number of cells in the micrograph, and A (cm^2) is the area of the micrograph.

The digital image correlation (DIC) method uses two digital images of the test piece before and after deformation and calculates the correlation to obtain deformation information in the region of interest. At high-strain rates, the combination of DIC technology and high-speed cameras can more accurately characterize the mesoscale deformation behavior of foam materials and obtain parameters such as elastic modulus and Poisson's ratio.

Due to the complex distribution of local strain components on the foam surface, it is difficult to obtain the correlation between global and local strain values. To facilitate a direct comparison of global and local strain values, the following equation is used to determine the effective local von Mises strain [36]:

$$\varepsilon_{eff}(x,y,t) = \left[\frac{2}{3}\left(\varepsilon_{xx}{}^2(x,y,t) + \varepsilon_{yy}{}^2(x,y,t) + 2\varepsilon_{xy}{}^2(x,y,t)\right)\right]^{1/2} \tag{2.37}$$

X-ray micro-computed tomography (μ-CT) scanning is used to obtain the real shape of the foam interior and reconstruct the true model of the foam without damage to the foam. This is suitable for exploring foam deformation and failure mechanisms under fixed shapes. Because the X-ray absorption of foam is low, low-resolution and high-resolution CT can be used to directly measure representative microstructures, including foam material deformation behavior (bending, buckling, etc.). The advantage of this modeling technique is that it can capture all the microstructural features of the foam, such as local density, pore size and shape distribution, and pore wall thickness. It can also be combined with simulation software for the simulation and prediction of foam material properties. For example, Scarpa and coworkers used μ-CT scanning to obtain a 3D solid finite element (FE) model to investigate the deformation mechanism inside a tensioned foam, and the results showed good agreement with experimental results obtained under tensile and compressive loading conditions (Figure 2.15) [37].

2.3.2 Quasistatic Compression Test

As introduced above, the stress–strain relationship of foam materials can be divided into three regions: linear, plateau, and densification, and the energy absorbed or dissipated by foam is roughly equal to the area under its stress–strain curve. Considering different time scales, the macroscopic compression of foam materials can be divided into quasi-static and dynamic compression based on the macroscopic loading rate (strain rate or loading velocity). For quasi-static compression testing, foam materials are compressed in macroscopic force balance, and universal material testing machines are usually used for quasi-static compression testing.

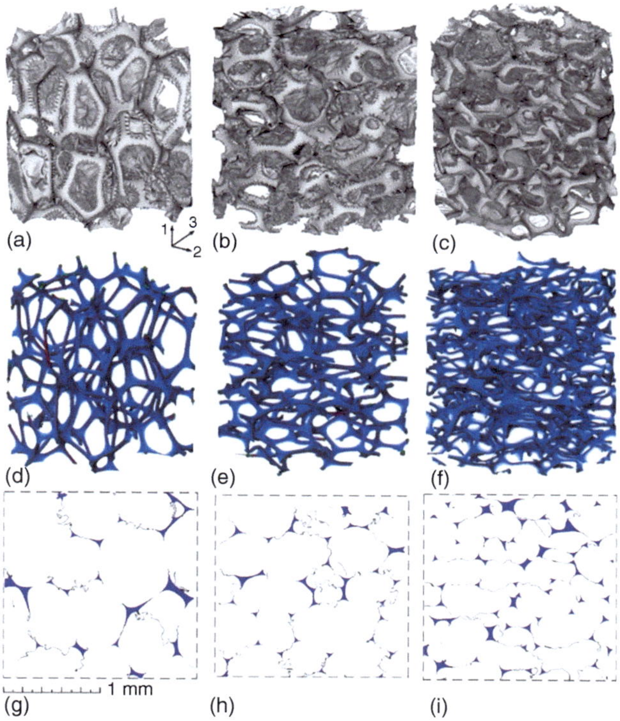

Figure 2.15 μ-CT scans and reconstructed models of original and stretched foams. Source: (a–c,g–i) Zhang et al. [37]/Elsevier/CC BY 4.0.

By integrating the stress–strain curve, the energy absorbed per unit volume can be obtained [38].

$$W = \int_0^{\varepsilon} \sigma(\varepsilon) d\varepsilon \tag{2.38}$$

where $\varepsilon = \varepsilon_d$ at the strain represented by ε_d, the unit volume energy absorption of foam reaches its maximum during the densification stage.

The energy absorption efficiency is the ratio of energy absorbed by the foam to the corresponding area, which is the ratio of the shaded area enclosed by the stress–strain curve to the corresponding stress (Figure 2.16) [39].

2.3.3 Dynamic Compression Test

Dynamic impact testing of foam materials is primarily used to assess their energy absorption and cushioning ability under high-speed impact. The dynamic compression characteristics of foam materials are crucial for their application in energy absorption, ballistic mitigation, and explosion/impact protection.

Drop Hammer Impact: The principle of the drop hammer impact test involves dropping a weight with a given mass and geometric shape from a given height

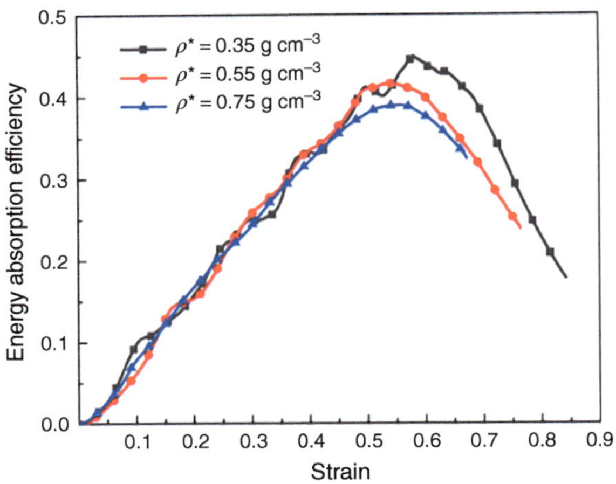

Figure 2.16 Energy absorption efficiency diagram of foams with different densities. Source: Guo et al. [39]/with permission of John Wiley & Sons.

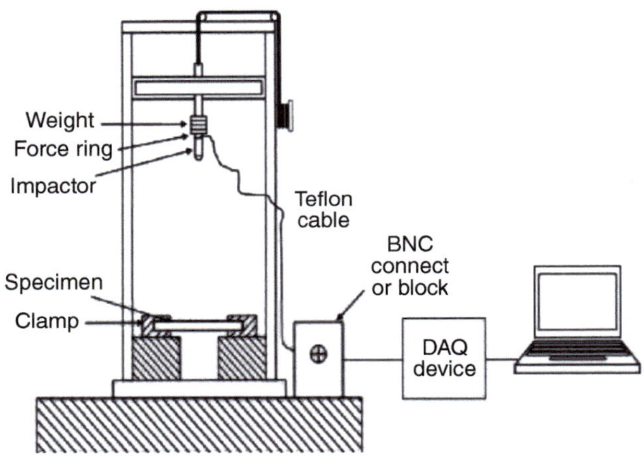

Figure 2.17 The layout of the drop-hammer impact test setup. Source: Yapici [40]/Scientific Research Publishing Inc, CC BY 4.0.

onto the specimen (Figure 2.17). By using a larger mass or a greater drop height, the impact energy can be increased. When the hammer falls from a height H, the output electrical signal from the accelerometer under the hammer is obtained as an a–t curve. The x–t curve can be obtained by double integrating the x–t curve.

The numerical principle behind it is the dynamic equation of the hammer (or small ball) $Ma(t) + Mg = F(t)$. Due to the impact stress of the foam plastic $\sigma(t) = F(t)/A$. Integrating $\frac{dv}{dt} = -a(t)$ twice yields the displacement $\Delta x(t)$ of the foam specimen under impact by the hammer. Thus, the strain $\varepsilon(t) = [\Delta x(t) / h] = \left[\int_0^t v_1 t - \int_0^t a(t)dt\right] dt/h$ of the foam specimen can be obtained. The stress–strain curve can be obtained by substituting $a(t)$ and eliminating the time parameter.

If we neglect mechanical energy loss during the process, the energy consumed during the impact unloading process (i.e. the energy consumed during the impact) is equal to the energy of the hammer rebounding

$$E = Mg(H - H_f) = Ah\left[\int_0^{\varepsilon_f} \sigma d\varepsilon + \int_{\varepsilon_f}^{\varepsilon max} (\sigma - \sigma_r) d\varepsilon\right] \tag{2.39}$$

Split-Hopkinson Pressure Bar (SHPB) is a type of experimental setup used for measuring the stress–strain relationship of materials under impact loading. The concept of the Hopkinson bar was originally proposed by Hopkinson in 1914 for measuring the pulse waveform under impact loading. In 1949, Kolsky improved the setup by dividing the pressure bar into two parts, with the specimen placed in between, which enabled the measurement of the stress–strain relationship of materials under impact loading. The theoretical basis of this technique is the one-dimensional stress wave theory, and the stress–strain relationship of the specimen is obtained by measuring the strain on the pressure bar. The main advantages of this technique are: (i) the measurement method avoids the difficulty of simultaneously measuring stress and strain at the same location of the specimen over time; (ii) the strain rate range involved in SHPB testing includes the strain rate at which the flow stress changes with the strain rate (100–10,000 s^{-1}); (iii) the incident waveform is easy to control by changing the impact velocity and the shape of the striker, the incident pulse waveform can be adjusted, which also adjusts the waveform acting on the specimen.

An ideal incident pulse must produce dynamic sample equilibrium and a nearly constant reflected pulse at a constant strain rate. Circular rubber pulse shapers are used to increase the duration of the loading wave to ensure multiple internal reflections occur inside the sample to reach stress equilibrium. Since the stress wave transmitted from the sample to the transmission bar is very small, semiconductor strain gauges are used to record the weak transmitted wave. The stress–strain curve is obtained from the strain history recorded by the strain gauges fixed on the incident and transmission bars using the following formula.

$$\dot{\varepsilon} = -\frac{2c_0}{l_s}\varepsilon_r \tag{2.40}$$

$$\varepsilon = \int_0^t \dot{\varepsilon} dt = \int_0^t \frac{2c_0}{l_s}\varepsilon_r dt \tag{2.41}$$

$$\sigma = \frac{A_0}{A_s}E\varepsilon_t \tag{2.42}$$

where l_s and A_s represent the original sample length and cross-sectional area, respectively; E, c_0, and A_0 represent Young's modulus, longitudinal wave velocity, and cross-sectional area of the bar, respectively; ε_r and ε_t represent the reflected signal and transmitted strain signal measured on the surface of the bar.

2.3.4 Rebound Performance and Hysteresis Testing

The rebound performance of foam materials is of great significance for buffering and sports applications. An ideal elastic material, when compressed, undergoes deformation under external force, and the work done by the external force is stored as potential energy. When the pressure is released, the potential energy is converted back into kinetic energy, and the material recovers its original shape and size. However, the stress–strain curve during compression and recovery does not overlap, indicating that the work consumed during compression is not equal to the work done by the material during recovery, resulting in a loss of energy known as hysteresis loss. The energy loss can be reflected by the following two testing methods.

In Rebound performance measurement, the drop-ball impact method is often used, which involves dropping a steel ball of specified diameter and mass onto a polymer foam from a specified height, and automatically calculating the percentage of the maximum height of the steel ball rebound to the height it was dropped from (i.e. rebound ratio) as an indicator of the polymer foam's rebound performance.

The hysteresis loss can fully reflect the ability of a material to absorb energy after being subjected to stress. The more severe the foam hysteresis phenomenon, the greater the hysteresis loss, which would result in superior blocking effects on mechanical energy. The testing principle of hysteresis loss involves using compression (with a pressure head area larger than the sample area) or indentation (with a pressure head area smaller than the sample area) to produce the deformation specified in the standard for the sample and immediately allowing the sample to recover. The ratio of the loss work to the loading work during this process is defined as the hysteresis loss. The main instruments required normally consist of a testing machine, a supporting plate, and a pressure head. The sample requirements for Method A (compression method) are a side length of 50 ± 2 mm and a thickness of 25 ± 2 mm, while the requirements for Method B (indentation method) are a side length of 380 ± 20 mm and a thickness of 50 ± 2 mm. Usually, the compression (indentation) direction during testing should be the same as the direction of force applied in the actual use of the product.

The basic experimental method for Method A (compression method) involves preloading the sample twice with a pressure head at a speed of 250 ± 25 mm min^{-1}, compressing it to 75–80% of its initial thickness, unloading the pressure at the same rate, and then allowing the sample to recover for 6 ± 1 min. The sample is then compressed at a speed of 50 ± 5 mm min^{-1} until it reaches 75 ± 5% of its initial thickness and immediately unloaded at the same speed until it is almost completely recovered.

For Method B (indentation method), the pressure head is preloaded at a speed of 100 ± 20 mm min^{-1} to compress the sample to 75 ± 2.5% of its initial thickness, and then the pressure is unloaded at the same rate, and the sample is allowed to recover for 4 ± 1 min. The sample is then indented at a speed of 100 ± 20 mm min^{-1} until it reaches 75 ± 2.5% of its initial thickness and immediately unloaded at the same speed until it is almost completely recovered while recording the load–displacement curve. The time interval between the end of loading and the start of unloading should not exceed 2 seconds. The hysteresis loss can be calculated according to the definition and the formula

$$H_{loss} = \frac{E_0 - E_r}{E_0} \times 100\% \tag{2.43}$$

where H_{loss} (%) is the hysteresis loss, E_0 and E_r are the loading and unloading work (J).

2.4 Preparation Methods of Energy-Absorbing Polymer Foam

Polymer foams, also known as porous polymer material, are an important type of polymer material that contains a large number of small foam cells within the polymer matrix. Polymer foams are prepared from polymers using mechanical, physical, and chemical foaming techniques. The mechanical foaming approach usually results in foams with large bubbles. For microcellular foams, chemical or physical blowing agents are used to generate cells within the polymer matrix. As introduced in Chapter 1, typical foaming processes contain batch foaming, extrusion foaming, and injection foaming. Physical foaming using supercritical fluids, such as carbon dioxide (CO_2) or nitrogen (N_2), as the blowing agents are highly desired for the large-scale production of energy-absorbing foams due to their environmentally friendly nature and high efficiency. The foaming process can usually be divided into pore formation, pore growth, and pore stabilization stages.

Cell nucleation: As the amount of gas gradually increases, the solution becomes oversaturated, and the gas escapes from the solution. The escaping gas forms the nucleus for pore formation.

Cell growth: When two cells of different sizes are close to each other, gas will diffuse from the smaller cell to a larger cell, and the two cells will merge. Due to nucleation, the number of pores increases, and the diameter of the pores expands. Therefore, pores can grow.

Cell stabilization: Due to the formation and growth of a large number of pores, the surface area and volume of the foam system continue to increase, and the pore walls become thinner. Therefore, the foam system becomes unstable. Pore stabilization is usually achieved by cooling or adding surfactants.

2.4.1 Open-Cell Foams

Superior buffering materials or systems are those materials or systems that can absorb more energy at the same load or the same energy at lower loads. Foam technology is ubiquitous across industries, and while it is mature and widely used, there are still areas for improvement. Ramirez and Gupta [41] compared PU foam with widely used EVA and TPU foams at similar densities. PU foam exhibited a 5% increase in buffering efficiency (Figure 2.18b,c,e, and f) and a 30% increase in energy absorption (Figure 2.18a and d)under quasi-static compression compared to TPU and EVA foams. In impact testing, PU foam demonstrated a 28.6% and 36.9% reduction in peak transmitted impact force at 5 and 7 J, respectively, compared to EVA and TPU foams of the same thickness (20 mm). The PU foam was produced using modified methylene diphenyl diisocyanate and a low molecular weight diamine by mixing with a high-speed electric stirrer to introduce small bubbles to create foam followed by curing at 80 °C under vacuum.

By controlling the cellular structure, the energy absorption properties of polymer foams can be tuned. Three-dimensional (3D) printing technology has been used to

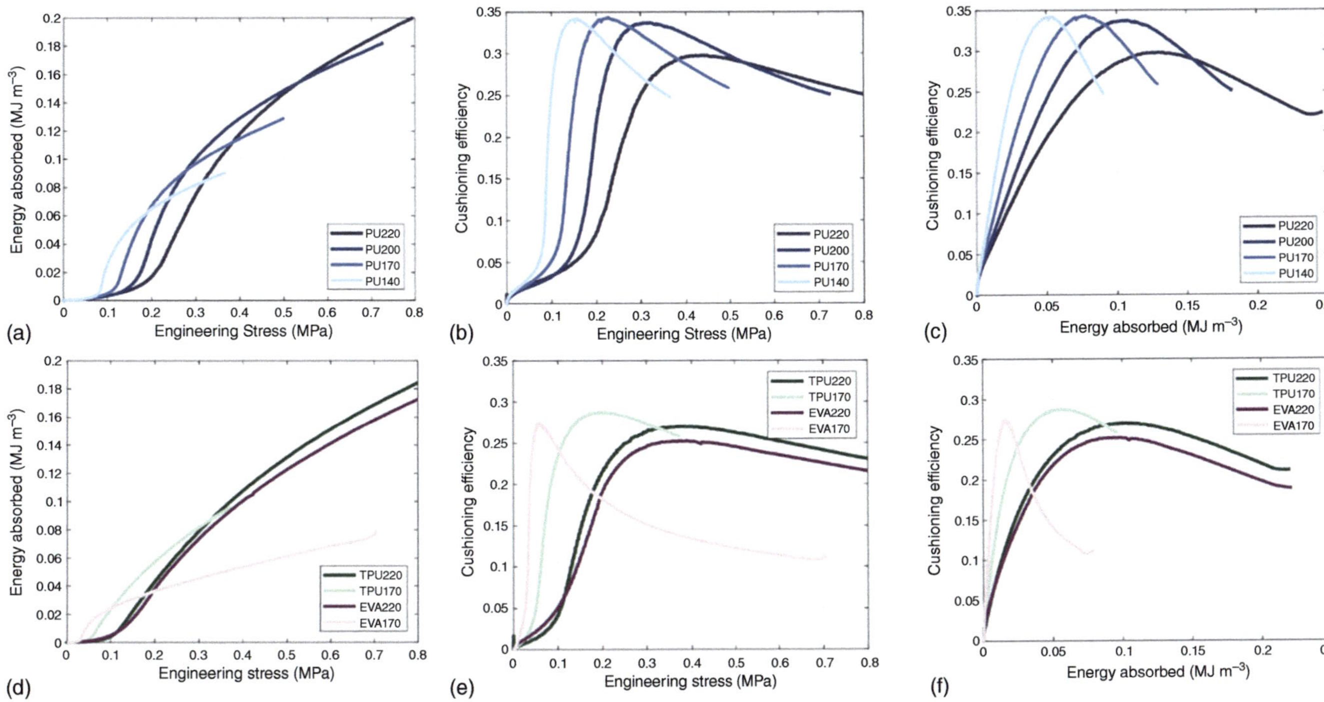

Figure 2.18 Quasi-static ($0.04\,s^{-1}$) cushioning properties of PU foams: (a) energy absorption as a function of stress; (b) cushioning efficiency as a function of stress; (c) cushioning efficiency as a function of energy absorbed. Quasi-static ($0.04\,s^{-1}$) cushioning properties of TPU and EVA foams: (d) energy absorption as a function of stress; (e) cushioning efficiency as a function of stress; (f) cushioning efficiency as a function of energy absorbed. Source: Ramirez and Gupta [41]/with permission of Springer Nature.

fabricate polymer foams with defined cellular structures. By utilizing 3D printing technology, lattice foams can be made with a uniform and ordered microstructure. Habib [42] used Multi Jet Fusion (MJF) technology to study the quasi-static energy absorption of six different polymer lattice structures with different unit cell topologies and deformation behaviors but the same relative density. The results show that by controlling and manipulating the micro topology of the lattice, the overall energy absorption of the lattice can be tuned to meet specific requirements.

2.4.2 Closed-Cell Foams

Physical foaming of thermoplastic polymers often leads to closed-cell foams due to their relatively high melt strength and melt elasticity compared to the foaming of thermoset precursors. Xu and coworkers [43] successfully prepared a lightweight polyethylene terephthalate (PET) bead foam part with excellent welding interface and good thermal insulation performance through the novel $scCO_2$ secondary foaming and molding strategy, and systematically investigated the corresponding influencing factors and mechanisms (Figure 2.19a). The compressive and tensile results (Figure 2.19b–d) showed that the foam undergoes a linear elastic zone initially. Between 7% and 60% compressive strain, the foam's cell wall compresses gradually, resulting in a slow increase in compressive stress to 1.27 MPa at 20% strain. Throughout this compressive stage, the foam continues to absorb external energy, suggesting potential applications in energy and shock absorption. Beyond 60% strain, the foam becomes compacted, leading to a rapid increase in compressive strength, showcasing its excellent mechanical performance under continuous external force, attributed to strong interfacial adhesion among neighboring beads.

The random pore structure of foam can lead to pore wall bending-dominated deformation, resulting in a rapid reduction in strength and stiffness as the porosity

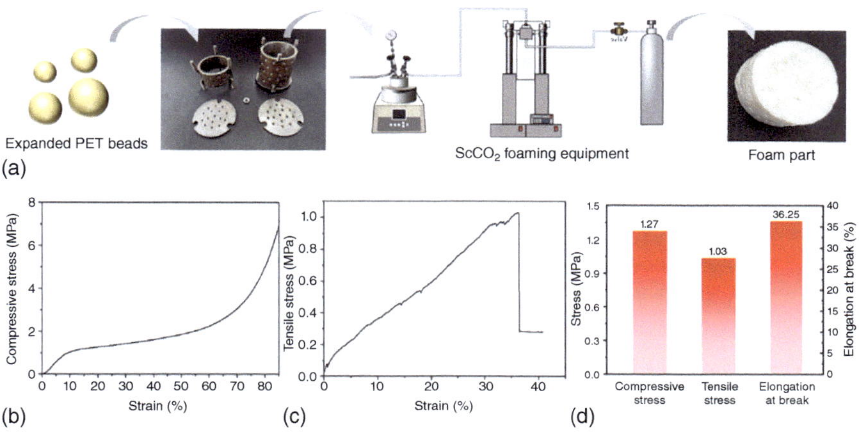

Figure 2.19 (a) Schematic illustration of the fabricating procedure for the PET foam part. Compressive (b) and tensile (c) stress–strain curve. (d) tensile stress and elongation at break. Source: Yang et al. [43]/with permission of John Wiley & Sons.

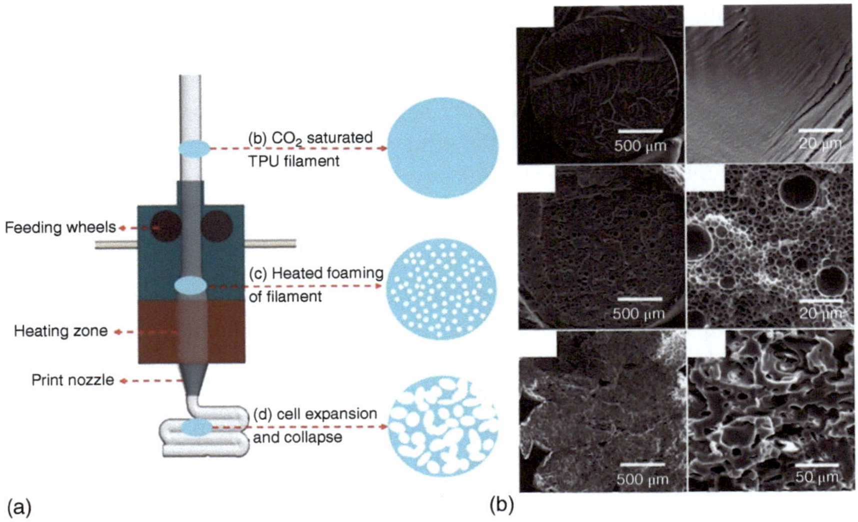

Figure 2.20 (a) Schematic illustration of the foaming of TPU filament during FDM printing; (b) SEM images of the extruded TPU filament with porous structure at different magnifications. Source: Zhang et al. [44]/with permission of John Wiley & Sons.

increases. In contrast, ordered lattice-type structures have ideal tensile-dominated properties. Shi and co-workers [44] proposed a facile fabrication strategy of PU tissue-engineering scaffold with a hierarchical structure, designed by a combination of fused deposition modeling (FDM) 3D printing and $scCO_2$ foaming (Figure 2.20a). The microcellular foaming and expansion process was indicated to compensate for mechanical loss because of the gaps of stacking layers by FDM printing, and the reported tensile strength of 28 MPa matched well with that of natural cartilage tissue constructs (Figure 2.20b). It was found that the microcellular cells introduced into the samples did not reduce the tensile properties compared with the unfoamed samples, so the microporous foaming process is beneficial for improving the mechanical properties of the scaffolds.

2.4.3 Composite Foams

Introducing a secondary phase or a filler component to the polymer matrix is an effective approach to altering the mechanical properties of polymer foams. For instance, soft PU foams are flexible but have poorer performance in absorbing impact energy, while hard PU foams have the risk of plastic collapse in the earlier stages of deformation although they have relatively high energy absorption capacity [45]. Linul et al. [46] proposed a method for obtaining enhanced semirigid PU foam using aluminum ultrafine fibers in a polymer matrix (Figure 2.21a,b). They found that adding AM to the foam can improve the elasticity and compressive strength of the enhanced semirigid PU foam. The compressive strength and energy absorption were improved by 61.81% and 71.29%, respectively, when 1.5% aluminum ultrafine fibers were added (Figure 2.21c).

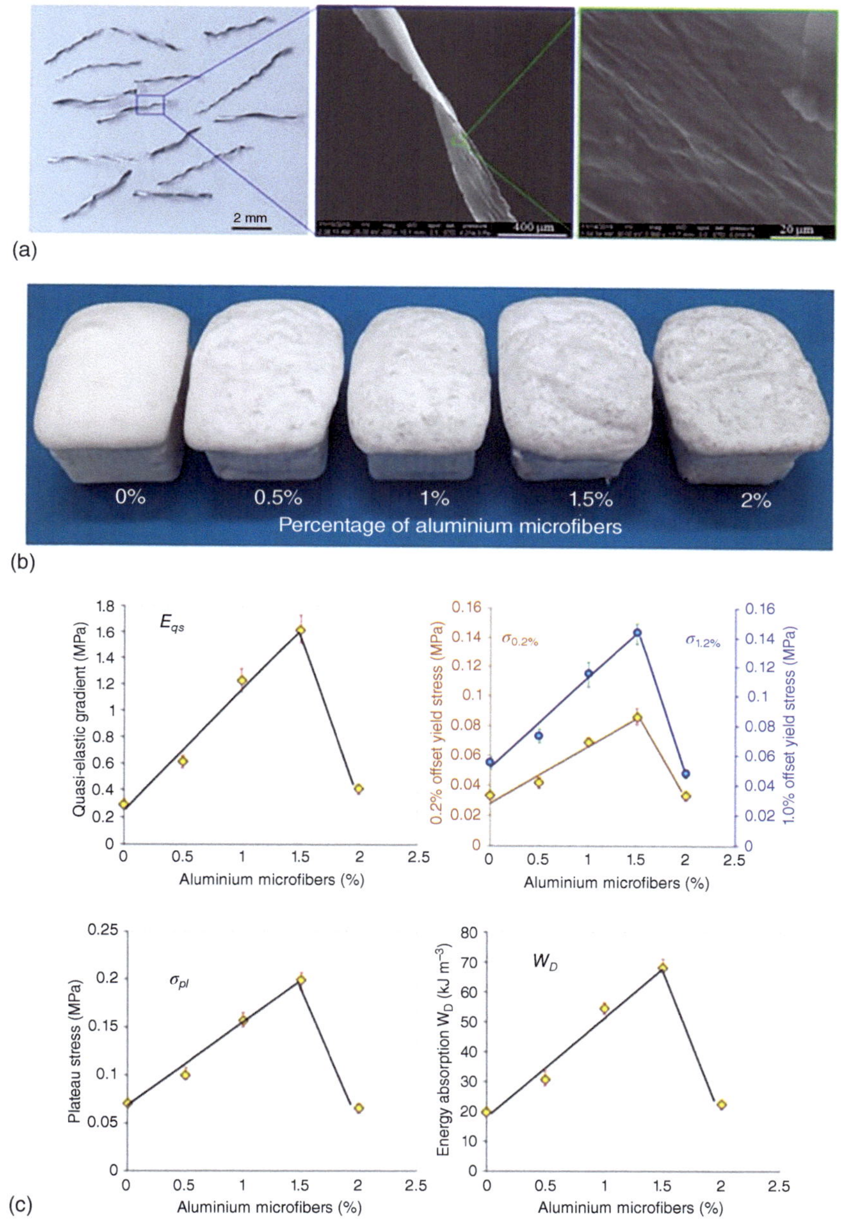

Figure 2.21 (a) Optical and SEM images of AMs; (b) Semirigid PU foam blocks obtained depending on the percentage of AMs. (c) Mechanical properties of investigated semirigid PU foams quasi-elastic gradient, 0.2% and 1.0% offset yield stress, plateau stress, energy absorption at densification strain. Source: Linul et al. [46]/MDPI/CC BY 4.0.

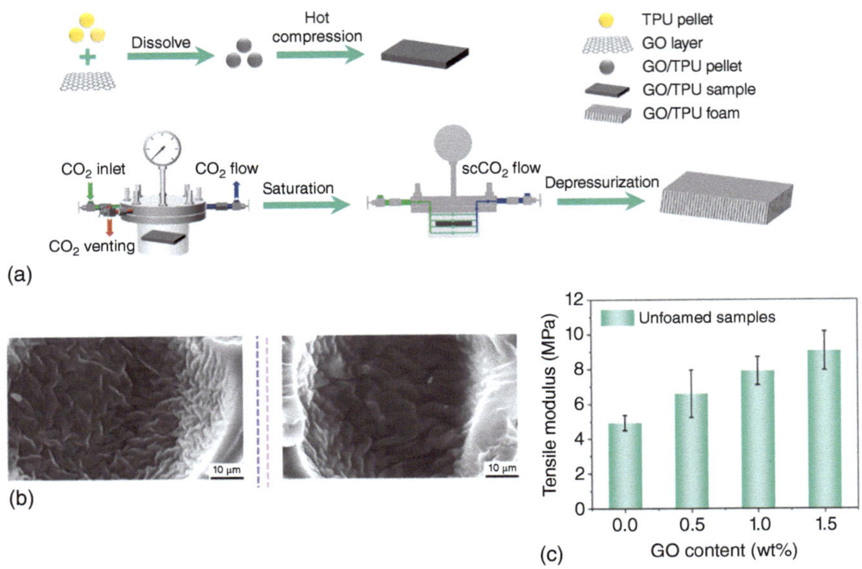

Figure 2.22 (a) Schematic illustration for the process of (a) fabrication of GO/TPU nanocomposite plates and preparing wrinkled GO/TPU foams via dynamic $scCO_2$ foaming; (b) SEM morphology of cell before and after adding GO; (c) Tensile modulus results of unfoamed TPU and GO/TPU composites. Source: Zhu et al. [47]/with permission of John Wiley & Sons.

Mi and co-workers [47] proposed a novel dynamic $scCO_2$ foaming method to prepare TPU foams containing graphene oxide (GO) nanofillers (Figure 2.22a). The addition of GO to the TPU matrix resulted in hydrogen bonds and physical cross-linking points. The dynamic $scCO_2$ foaming introduces a $scCO_2$ flow to the TPU molecules, which generates intramolecular stress and results in the subsequent release of stress intramolecular stress and the formation of micro-sized wrinkles on the cell surface (Figure 2.22b). The compressive properties of TPU/GO foams with wrinkles were higher than those without wrinkles. In addition, with the increase of GO content, the mechanical properties were significantly enhanced (Figure 2.22c), and it had excellent compressive properties and energy absorption characteristics.

2.4.4 Special-Structured Foams

When the substrate material is constant, the cellular structure of polymer foams is the sole factor affecting their mechanical performance. Auxetic foams, which have an NPR, are a type of foam material that expands in volume under tensile strain, and contracts in volume under compressive strain [28, 30]. Under compressive load, auxetic foams undergo lateral contraction, resulting in an increase in material density, and thus improving compressive strength and energy absorption [30].

Tang and co-workers [48] proposed a new method for directly preparing auxetic foam from polymer resin using a water-based phase change and CO_2 foaming agent. The method utilizes the synergy of the permeability difference between water and CO_2 during foaming to produce an auxetic foam (Figure 2.23a).

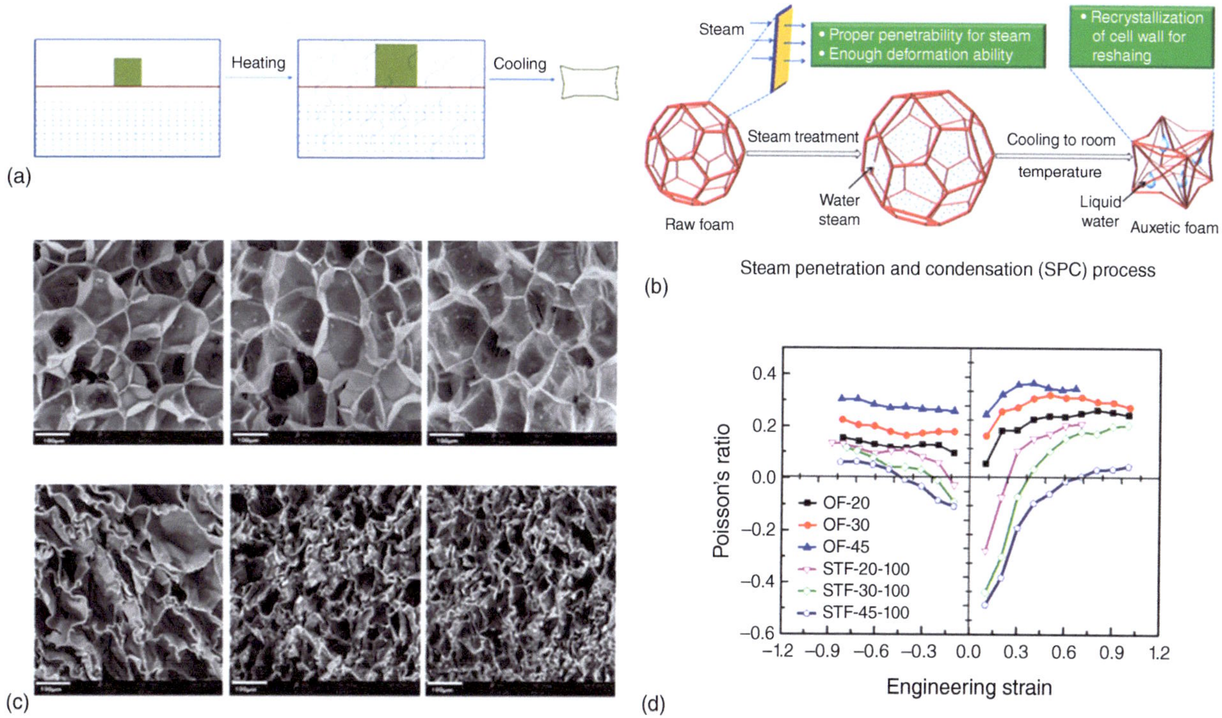

Figure 2.23 (a) Process for the preparation of NPR foam; (b) Schematic drawing for the possible formation mechanism of the reentrant structure from closed-cell foam; (c) SEM images of PE foam specimen before and after steam treatment; (d) relationship between Poisson's ratio and engineering strain of PE foam specimens at different treatment temperatures. Source: Fan et al. [48]/with permission of American Chemical Society.

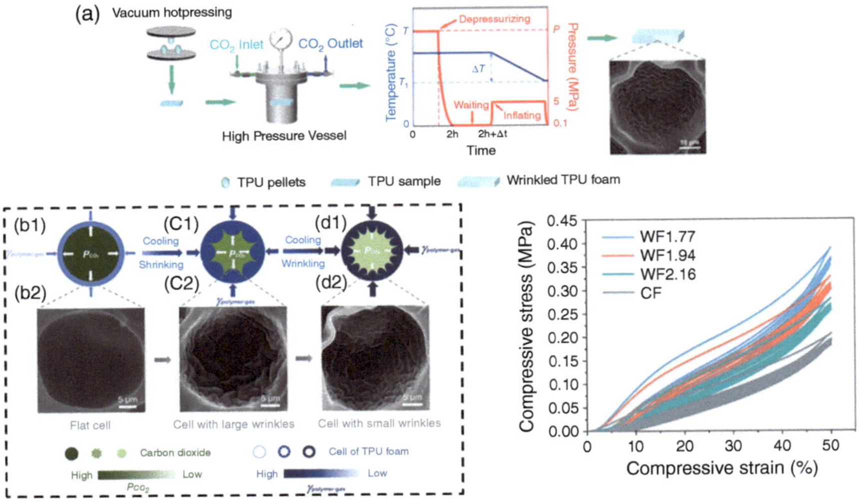

Figure 2.24 (a) Schematic illustration of wrinkly TPU foams fabrication strategy; (b–d) Schematic illustration of the formation mechanism of wrinkly structure in TPU foams; (e) compressive stress–strain curves with a constant compressive strain of 50%. Source: Hu et al. [35]/with permission of American Chemical Society.

As shown in Figure 2.23b, the matrix is expanded by the phase change of CO_2 and water during foaming, and the contraction of the cells was triggered by the liquidation of the water vapor, which caused the formation of the concave structured cellular structure (Figure 2.23c). The developed foam demonstrated NPR property (Figure 2.23d), good tensile cyclic stability and energy absorption properties. The method overcomes the bottleneck of directly preparing auxetic foam from high molecular weight resin via a simple approach.

Mi and co-workers [35] developed a gas cooling-assisted $scCO_2$ foaming method, which uses cold CO_2 flushing to quickly reduce the gas temperature inside the cells of the foam and lead to the formation of a special wrinkled cellular structure (Figure 2.24a). As shown in Figure 2.24b–d, the rapid outlet of high-temperature gas and inlet of low-temperature gas would cause a fast temperature change of foams. The fast cooling of the foam resulted in the release of intramolecular stress of TPU in the form of micro-sized maze-like wrinkles. The cyclic compression properties indicated that the performance of wrinkled foam was higher than that of traditional foam. The foam also had an energy absorption rate improvement of 147.0% (Figure 2.24e), which suggested that the dense wrinkle structure improves energy absorption while maintaining low-energy loss.

2.5 Applications of Energy-Absorbing Polymer Foams

Attributing to their excellent energy absorption capacity, polymer foams have a wide range of applications in practice.

2.5.1 Energy-Absorbing Foams in Sports

Sports and safety industries are major application fields that utilize energy-absorbing foams. In practice, foams with different properties are often combined to produce a laminated composite foam mat to realize the required performance. For example, exercise mats used to inhibit landing are mainly used in gymnastics, high jump, pole vault, and combat sports. In combat sports, mats need to be stiff enough to provide the best surface for complex movements. In addition, high impact damping capacity is required so that the athlete's load does not reach the risk limit of damaging health, and the mat must not sink when athletes are walking or jumping on it to prevent ankle injury.

Polymer foam products used for winter sports also have protective functions. They appear in the helmets and various protective clothing and helmets worn by athletes. The construction of energy-absorbing clothes is similar to that of protectors used in motorsports, which are usually sandwich structures made of a harder thermoplastic polymer outer layer and a softer polymer foam inner layer. The helmet must distribute the load over a large area of the skull and dissipate the impact energy applied to the human head. With the help of a helmet, the strike contact time could be prolonged, and the peak force could be reduced. Similarly, there are knee pads for ball games, which can better protect the joints and prevent fractures when they are hit. The boxing head cover needs to be double-sided protection, both the head of the one being hit and the hand of the one who is hitting. Therefore, a soft cover is required for cushioning and shock energy absorption.

In the field of sports shoes, polymer foams are essential parts for the shoe sole. In recent years, more and more $scCO_2$ of scN_2 foamed thermoplastic polymers have been used to replace traditional thermoset polymer foams due to the excellent elasticity and adjustable mechanical properties that are realized by the closed cellular structures with micro-size cells.

2.5.2 Energy-Absorbing Foams for House and Home

Household appliances use cushioned packaging materials to prevent the product from being damaged. In the choice of commonly used buffer packaging materials, there are mainly three kinds, EPS foam board, PE foam EPE, and PU foam. EPS foam board has good shock buffering performance, high compressive strength, poor water absorption performance, strong oil resistance, simple modeling, and low price. However, it is easy to produce plastic deformation after compression, poor weather resistance, brittle texture, and easy-to-shatter dust. PE foam has the advantages of low density, good flexibility, and a high recovery rate. Their impact resistance is better than polystyrene plastic. It is widely used in advanced furniture, household appliances, instruments, craft gifts, etc. PU foam has good resilience, good buffering performance, and a wide range of use temperature range, which is mainly used in mildew, anti-corrosion, insect prevention, sound insulation, heat insulation, and other aspects, widely used in furniture. Slow rebound PU foam is widely used for pillows and mattresses with a memory effect. Slow rebound PU foam means that after the foam is deformed by external action, it is slowly restored to its original shape. Slow rebound PU foam has very low resilience, so its repulsive force against external pressure is very small, which can greatly reduce the pressure at the contact point and increase human comfort.

2.5.3 Energy-Absorbing Foams for Transportation and Vehicles

Polymer foams are widely used in automobile, ship, aerospace, and other transportation fields because of their excellent cushioning energy absorption and light texture. The automotive foam market is generally categorized by type (PU foam, polyolefin (PO) foam, others), application (seats, door panels, bumper systems, others), and end use (automotive, light commercial vehicles, heavy commercial vehicles). According to a comprehensive Research report by Market Research Future (MRFR), the compound annual growth rate of automotive foam materials in 2021–2028 is 10.23%. The market is valued at $38.61 billion in 2021 and $80.27 billion in 2028.

Polymer foams with high performance are also needed under extreme conditions. Polymer foams is used in the 4.2-meter-diameter full-wave transparent rocket fairing of the carrier rocket designed by China Aerospace Science and Technology Corporation Limited. The main material for the front cone in this fully transparent rocket fairing scheme is a magical "foam" called polymethylacrylimide (PMI). Fairing is used to protect satellites and other payloads to prevent satellites from being affected by harmful environment such as air power, pneumatic heating, and sound vibration, and is an important part of the launch vehicle. PMI foam possesses superior tensile, compression, shear, bending, and other strength and modulus indexes to other foam materials under the same density condition. The thermal insulation sandwich layer of PMI foamed core material used in the fairing of the Mars lander has the inherent advantages of lightweight, high rigidity, high thermal conductivity, and low expansion rate. It not only combines the substantial weight reduction of the fairing and the stable operation of the engine but also realizes the thermal insulation and noise reduction during the severe landing impact and improves the reliability of the system. In the Marine field, polymer foams are also indispensable. A foam fender is a floating fender that uses light and elastic foam as a buffer medium to absorb impact energy. It is composed of a closed-cell EVA core and an external PU shell. The outer layer is strong enough to resist friction, and the enclosed battery inside prevents any water from penetrating the fender body. As an alternative to inflatable rubber fenders, foam-filled fenders are extremely durable.

2.5.4 Energy-Absorbing Foams for Security and Military

With terrorist attacks on civilian and military infrastructure on the rise worldwide, the protection of personnel and structures from the devastating effects of explosive waves has become critical. The US Department of State reported that there were more than 13,400 terrorist attacks worldwide in 2014, resulting in more than 32,700 deaths and 34,700 injuries. Yuan and co-workers [49] used a self-built gas explosion experimental platform to explore the protective effect of PU foam on gas

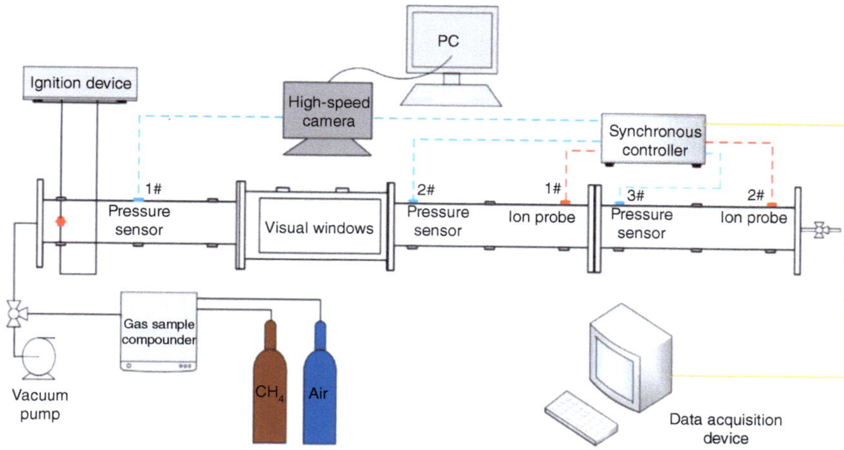

Figure 2.25 Diagram of an explosion experimental platform. Source: Li et al. [49]/MDPI/CC-BY-4.0/Public domain.

explosion (Figure 2.25). They found that the cell structure of PU foam can effectively quench the flame propagation and attenuate the explosion pressure, and the maximum explosion pressure and maximum rise rate are reduced by 86.2% and 84.7%, respectively. As a kind of personal protective equipment, body armor effectively protects the human body from gunshot damage by absorbing the impact energy of bullets. The impact energy of bullets is difficult to be fully absorbed by body armor, so that part of the impact energy is transferred to the human body, resulting in blunt injury. To improve the protective performance of body armor, the most feasible method is adding cushioning material on the back of the bulletproof plate to further weaken the impact strength of body armor. Therefore, energy-absorbing materials such as vinyl–vinyl acetate copolymer (EVA) and PU have been used as buffer plates to further reduce BABT.

References

1 Jun, L., Qin, Z., Mi, H.Y. et al. (2021). Fabrication of thermoplastic polyurethane foams with wrinkled pores and superior energy absorption properties by CO_2 foaming and fast chilling. *Macromolecular Materials and Engineering* 307: 2100600.

2 Holder, K.M., Cain, A.A., Plummer, M.G. et al. (2016). *Macromolecular Materials and Engineering* 301: 641.

3 Kohlhoff, D. and Ohshima, M. (2011). Open cell microcellular foams of polylactic acid (PLA)-based blends with semi-interpenetrating polymer networks. *Macromolecular Materials and Engineering* 296: 770–777.

4 Faruk, O., Błędzki, A.K., and Matuana, L.M. (2007). Microcellular foamed wood-plastic composites by different processes: a review. *Macromolecular Materials and Engineering* 292: 113–127.

5 Zhao, H., Elnasri, I., and Abdennadher, S. (2005). An experimental study on the behaviour under impact loading of metallic cellular materials. *International Journal of Mechanical Sciences* 47: 757–774.
6 Koohbor, B., Ravindran, S., and Kidane, A. (2021). In situ deformation characterization of density-graded foams in quasi-static and impact loading conditions. *International Journal of Impact Engineering* 150: 103820.
7 Avalle, M., Belingardi, G., and Montanini, R. (2001). Characterization of polymeric structural foams under compressive impact loading by means of energy-absorption diagram. *International Journal of Impact Engineering* 25: 455–472.
8 Chou, C.C., Zhao, Y., Chai, L. et al. (1995). Development of foam models as applications to vehicle interior. *SAE Transactions* 39: 335–348.
9 Gibson, L.J. and Ashby, M.F. (1982). The mechanics of three-dimensional cellular materials. *Proceedings of the Royal Society of London A Mathematical and Physical Sciences* 382: 43–59.
10 Gibson, L.J. and Ashby, M.F. (1989). Cellular Solids: Structure & Properties, Oxford: Pergamon Press, ISBN: 0-08-036607-4, 1988, 357 + ix pages, $35.00. *Advances in Polymer Technology* 9: 165–166.
11 Mills, N.J. (2007). *Polymer Foams Handbook: Engineering and Biomechanics Applications and Design Guide*. Elsevier.
12 Ley, A., Altschuh, P., Thom, V. et al. (2018). Characterization of a macro porous polymer membrane at micron-scale by confocal-laser-scanning microscopy and 3D image analysis. *Journal of Membrane Science* 564: 543–551.
13 Zhu, Y., Luo, G., Zhang, R. et al. (2020). Numerical simulation of static mechanical properties of PMMA microcellular foams. *Composites Science and Technology* 192: 108110.
14 Yasunaga, K., Neff, R.A., Neff, R.A. et al. (1996). Study of cell opening in flexible polyurethane foam. *Journal of Cellular Plastics* 32: 427–448.
15 Tomin, M. and Kmetty, Á. (2022). Polymer foams as advanced energy absorbing materials for sports applications—a review. *Journal of Applied Polymer Science* 139: 51714.
16 Mills, N.J., Fitzgerald, C., Gilchrist, A., and Verdejo, R. (2003). Polymer foams for personal protection: cushions, shoes and helmets. *Composites Science and Technology* 63: 2389–2400.
17 Rusch, K. (2003). Load-compression behavior of brittle foams. *Journal of Applied Polymer Science* 14: 1263–1276.
18 Li, Q.M. and Mines, R.A.W. (2002). Strain measures for rigid crushable foam in uniaxial compression. *Strain* 38: 132–140.
19 Sun, Y. and Li, Q.M. (2018). Dynamic compressive behaviour of cellular materials: a review of phenomenon, mechanism and modelling. *International Journal of Impact Engineering* 112: 74–115.
20 Rusch, K.C. (1970). Energy-absorbing characteristics of foamed polymers. *Journal of Applied Polymer Science* 14: 1433–1447.
21 Avalle, M., Belingardi, G., and Ibba, A. (2007). Mechanical models of cellular solids: parameters identification from experimental tests. *International Journal of Impact Engineering* 34: 3–27.

22 Jeong, K.Y., Cheon, S.S., and Munshi, M.B. (2012). A constitutive model for polyurethane foam with strain rate sensitivity. *Journal of Mechanical Science and Technology* 26: 2033–2038.

23 Jeong, K.Y. (2016). Constitutive modeling of polymeric foams having a four-parameter modulus function with strain rate sensitivity. *Journal of Mechanical Science and Technology* 30: 683–688.

24 Hössinger-Kalteis, A., Reiter, M., Jerabek, M., and Major, Z. (2020). Overview and comparison of modelling methods for foams. *Journal of Cellular Plastics* 57: 951–1001.

25 Simone, A.E. and Gibson, L.J. (1998). The effects of cell face curvature and corrugations on the stiffness and strength of metallic foams. *Acta Materialia* 46: 3929–3935.

26 Almanza, O., Masso-Moreu, Y., Mills, N.J., and Rodríguez-Pérez, M.A. (2004). Thermal expansion coefficient and bulk modulus of polyethylene closed-cell foams. *Journal of Polymer Science Part B-Polymer Physics* 42: 3741–3749.

27 Mills, N.J., Stämpfli, R., Marone, F., and Brühwiler, P.A. (2009). Finite element micromechanics model of impact compression of closed-cell polymer foams. *International Journal of Solids and Structures* 46: 677–697.

28 Lakes, R.S. (1987). Foam structures with a negative Poisson's ratio. *Science* 235: 1038–1040.

29 Evans, K.E., Nkansah, M.A., Hutchinson, I.J., and Rogers, S.C. (1991). Molecular network design. *Nature* 353: 124.

30 Evans, K.E. and Alderson, A. (2000). Auxetic materials: functional materials and structures from lateral thinking! *Advanced Materials* 12: 617.

31 Ma, Y., Liu, Z.F., Zhang, T.H. et al. (2023). Crushing behavior and energy absorption performance of a novel 3D folded structure with negative Poisson's ratio. *Advanced Engineering Materials* 25: 2201769.

32 Qin, Z., Yang, Y., Tian, Q. et al. (2023). Strain-hardening, impact protective and self-healing supramolecular polyurethane nanocomposites enabled by quadruple H-bonding, disulfide bonds and nanoparticles. *Chemical Engineering Journal* 467: 143434.

33 Tong, Z., Zhou, J., Wang, R.-Y., and Xu, J.-T. (2017). Interplay of microphase separation, crystallization and liquid crystalline ordering in crystalline/liquid crystalline block copolymers. *Polymer* 130: 1–9.

34 Takeoka, Y., Liu, S., and Asai, F. (2020). Improvement of mechanical properties of elastic materials by chemical methods. *Science and Technology of Advanced Materials* 21: 817–832.

35 Hu, J., Gu, R., Mi, H.-Y. et al. (2022). Self-reinforced thermoplastic polyurethane wrinkled foams with high energy absorption realized by gas cooling assisted supercritical CO_2 foaming. *Industrial & Engineering Chemistry Research* 61.

36 Koohbor, B., Ravindran, S., and Kidane, A. (2017). Effects of cell-wall instability and local failure on the response of closed-cell polymeric foams subjected to dynamic loading. *Mechanics of Materials* 116: 67–76.

37 Zhang, Q., Lu, W., Scarpa, F. et al. (2021). Topological characteristics and mechanical properties of uniaxially thermoformed auxetic foam. *Materials & Design* 211: 110139.

38 Linul, E., Şerban, D.A., Marsavina, L., and Sadowski, T. (2017). Assessment of collapse diagrams of rigid polyurethane foams under dynamic loading conditions. *Archives of Civil and Mechanical Engineering* 17: 457–466.

39 Guo, H., Song, F.C., Sun, M.Q., and Chen, Y. (2023). Enhancement effect of polyurea coating on the impact resistance of aluminum foam. *Journal of Applied Polymer Science* 140.

40 Yapici, A. (2009). Effect of low velocity impact damage on buckling properties. *Engineering* 1: 161–166.

41 Ramirez, B.J. and Gupta, V. (2019). Energy absorption and low velocity impact response of open-cell polyurea foams. *Journal of Dynamic Behavior of Materials* 5: 132–142.

42 Habib, F.N., Iovenitti, P., Masood, S.H., and Nikzad, M. (2018). Fabrication of polymeric lattice structures for optimum energy absorption using multi jet fusion technology. *Materials & Design* 155: 86–98.

43 Yang, S.L., Ma, W.J., Ren, X.F. et al. (2023). Lightweight polyethylene terephthalate bead foams with good interfacial adhesion and thermal insulation performance fabricated by supercritical carbon dioxide secondary foaming strategy. *Journal of Applied Polymer Science*.

44 Zhang, S., Shi, X.T., Miao, Z.Y. et al. (2022). 3D-printed polyurethane tissue-engineering scaffold with hierarchical microcellular foam structure and antibacterial properties. *Advanced Engineering Materials* 24: 2101134.

45 Wang, Y., Li, J., Xie, Y. et al. (2022). Fabrication of wrinkled thermoplastic polyurethane foams by dynamic supercritical carbon dioxide foaming. *The Journal of Supercritical Fluids* 180: 105429.

46 Linul, E., Vălean, C., and Linul, P.-A. (2018). Compressive behavior of aluminum microfibers reinforced semi-rigid polyurethane foams. *Polymers* 10: 1298.

47 Zhu, X., Li, X., Mi, H.-Y. et al. (2022). Graphene oxide/thermoplastic polyurethane wrinkled foams with enhanced compression performance fabricated by dynamic supercritical CO_2 foaming. *Journal of Applied Polymer Science* 139: e52485.

48 Fan, D., Shi, Z., Li, N. et al. (2020). Novel method for preparing a high-performance auxetic foam directly from polymer resin by a one-pot CO_2 foaming process. *ACS Applied Materials & Interfaces* 12: 48040–48048.

49 Li, C.H., Zhang, G.Y., and Yuan, B.H. (2023). Exceptional performance of flame-retardant polyurethane foam: the suppression effect on explosion pressure and flame propagation of methane-air premixed gas. *Materials* 16: 7602.

3

Thermal Insulation Polymer Foams

3.1 Overview of Thermal Insulation Foams

The massive consumption of fossil energy has brought severe global problems such as environmental pollution, global warming, energy shortages, climate change, and loss of biodiversity, which makes it urgent to save energy and reduce emissions. Thermal insulation materials have thermal conductivity less than or equal to $0.12\,W\,(m\,K)^{-1}$ and are used in a wide range of applications, including building materials (wall and floor insulations, pipe insulation), transportation of goods (frozen food and specific equipment), etc. The most common insulation materials are extruded or expanded polystyrene (PS), mineral wool, lightweight concrete, phenolic foam, etc. Compared to inorganic insulation materials such as mineral wool, polymer foam is in considerable demand in actual use because of its ease of processing, low cost, and good insulation properties.

The foam has a large volume of gas whose thermal conductivity is significantly lower than that of solid materials, which renders the foam superior insulation performance. The parameter characterizing the thermal insulation ability is mainly thermal conductivity, which refers to the ability of an object to conduct heat. It is defined as the heat transferred per unit temperature gradient per unit time via a unit thermal conductivity surface.

The thermal insulation performance of a polymer foam is affected by the composition and structure of the solid and gas phases. As illustrated in Figure 3.1 [2], the solid phase is referred to as phase I, the gas phase is referred to as phase II, and phase III indicates the interface solid–gas phase. Solid phase I can be single-component or multicomponent polymers. By designing a unique cellular structure to reduce air convection to reduce energy exchange, a foam with good thermal insulation can be produced. The foam can be rigid or flexible, depending on the modulus of the matrix material and whether the cells are open or closed [2].

In a polymer foam with open cells, the gas is interchangeable and quickly escapes or exchanges energy with air, which can lead to a serious decrease in thermal insulation performance. Nonetheless, the cell density and cell size are critical parameters affecting the thermal conductivity of the foam. According to the cell size and density, the foam can be divided into the following categories as listed in Table 3.1 [3].

Functional Polymer Foams: Green Fabrication Methods, Performance and Applications,
First Edition. Haoyang Mi.

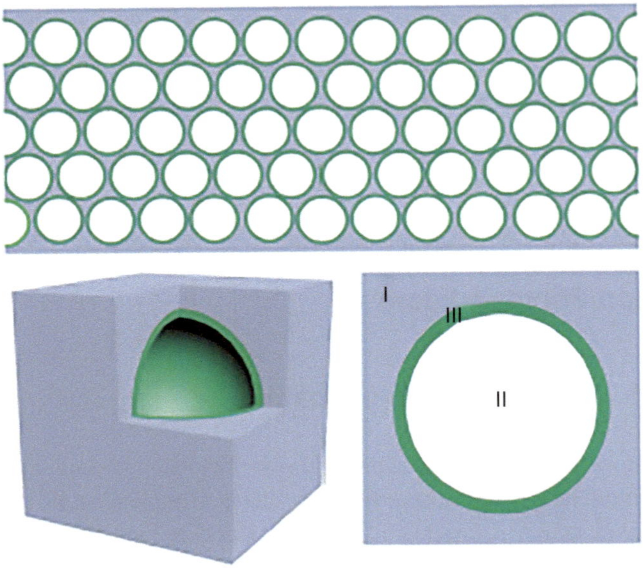

Figure 3.1 Illustration of a vesicle composed of three phases. Phase I represents the continuous component of the heat conduction process, and phase II is the gas phase. Phase III connects phases I and II, which is generally a solid wall. Source: Hu et al. [1]/with permission of John Wiley & Sons.

Table 3.1 Classification of polymer foams depends on the cell size and cell density.

	Macrocellular foams protective heat plates placed on both	**Microcellular foams**	**Ultra-microcellular foams**	**Nanocellular foams**
Average cell diameter (μm)	>100	1–100	0.1–1	0.001 < 0.1
Cell density	$<10^6$	10^6–10^9	10^9–10^{12}	$>10^{12}$

Source: Adapted from Pradeep [3].

In general, the foams with smaller cell sizes and high cell density possess lower thermal conductivity.

3.2 Fundamentals of Thermal Insulation

Classical heat transfer theory considers solid-phase conduction heat transfer, bubble-hole thermal convection (less impact), and gas-phase radiation heat transfer (more impact on low-density polymer foam) as the three basic ways of heat transfer in porous materials, as shown in Figure 3.2 [4]. Solid-phase heat transfer is the heat transfer process from the high-temperature to the low-temperature part. Thermal convection is the process of fluid transfer through the flowing medium,

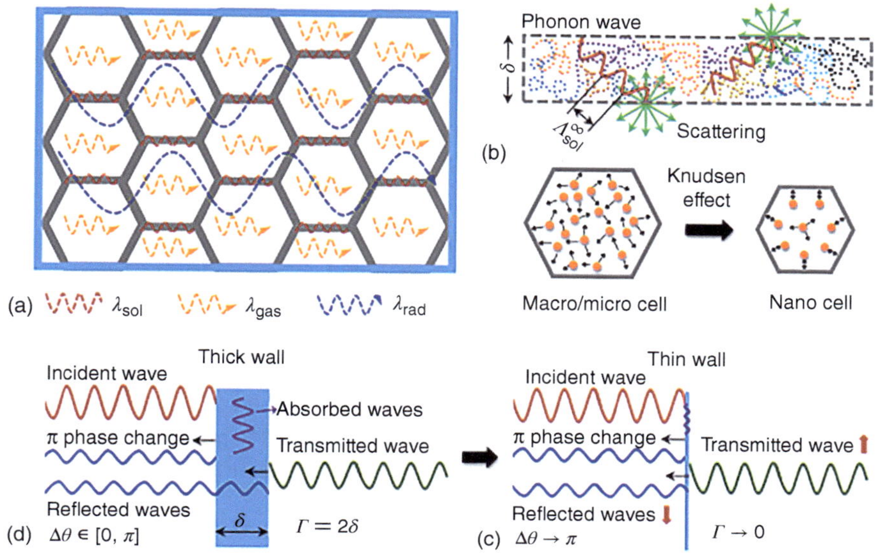

Figure 3.2 (a) Schematic diagram of the effect of various factors on thermal conductivity; (b) phonon vibrations at the cell boundary; (c) diagram of the motion of gas molecules in the cells; (d) thin wall interference leading to stronger transmission wave effectiveness. Source: Wang et al. [4]/with permission of Royal Society of Chemistry.

usually accompanied by the transfer of heat by mixing the medium. In natural convection, thermal convection is caused by the presence of density differences between hot and cold fluids. In artificial convection processes, mechanical agitation or externally applied pressure is generally required. Thermal radiation refers to the phenomenon of an object having a temperature that radiates electromagnetic waves. All objects with a temperature above absolute zero will produce thermal radiation; the higher the temperature, the greater the total energy radiated. General thermal radiation mainly relies on the longer wavelength of visible light and infrared light propagation. In the vacuum, thermal radiation is the only way to transfer heat.

For foams, Kuhn et al. [5] studied an approximately linear temperature profile along the thickness direction of the foam, where the interaction between heat conduction and heat radiation is negligible. Therefore, the effective thermal conductivity of polymer foam can be obtained by superimposing different terms in a linear way, and its total thermal conductivity λ_{tot} can be expressed as:

$$\lambda_{tot} = \lambda_{gas} + \lambda_{sol} + \lambda_{rad} + \lambda_{con} \tag{3.1}$$

The λ_{gas}, λ_{sol}, λ_{rad}, and λ_{con} represent the contributions of the gas phase, solid phase, thermal radiation, and thermal convection to the thermal conductivity, respectively. The heat conduction in solids mainly originates from the lattice vibration of solid molecules at their equilibrium positions, and the heat conduction in gases primarily comes from the collision of gas molecules; the heat conduction in solids and gases depends heavily on the cellular structure of foams, which also plays an important role in the overall heat conduction effect.

3.2.1 The Convection of Heat Transfer

Thermal convection of heat transfer is only possible when the Grashof number (Gr) is above 1000, and *Gr* describes the ratio of the buoyant force driving convection to the viscous force opposing it with an expression [1].

$$Gr = \frac{g \cdot a \cdot \Delta T \cdot d^3 \cdot \rho_g}{\eta^3} \tag{3.2}$$

where g is the acceleration constant of gravity (9.81 m s^{-2}), α is the expansion coefficient of the gas volume, ΔT is the temperature difference of a single cell, d is the diameter of the cell (assumed to be a sphere), and ρ_g and η denote the density and viscosity of the gas, respectively. In the case of atmospheric pressure (1 atm), Gr $\geq$ 1000 occurs only when the cell diameter is greater than 10 mm, so gas convection basically does not occur for polymer foams with microsized cells. The mechanism can also be illustrated by the Knudsen effect, a phenomenon in porous materials where the thermal conductivity of the gas decreases as its pore diameter approaches the mean free range of the gas.

3.2.2 Thermal Conduction of Gases and Solids

Simulation of heat transfer in foam can be represented using a cubic foam model, where the smallest repeating unit of the cubic foam is the individual foam unit. Given the thermal conductivity is infinite in the direction perpendicular to the heat transfer, the individual cubic units can be divided into parallel series, and the corresponding equivalent heat transfer diagram is given in Figure 3.3 [4]. The expression for the upper limit of the effective thermal conductivity $\lambda_{con,w}^{up}$ is

$$\lambda_{con,w}^{up} = \left[R_{con,w}^{low}(\delta_c + \delta_w)\right]^{-1} \tag{3.3}$$

where δ_c and δ_w denote the cell size and cell wall thickness, respectively. $R_{con,w}^{low}$ is the lower limit of effective thermal conductivity, which can be expressed by the equation:

$$R_{con,w}^{low} = \left(\frac{1}{R_{S1}} - \frac{1}{R_g}\right)^{-1} + R_{S2} \tag{3.4}$$

where R_{S1} and R_{S2} are the thermal conductivity of the solid element of the square loop and the thermal resistance of the top and bottom elements, respectively, and R_g is the thermal resistance of the gas in the cell unit, and the equation for each thermal resistance is expressed as:

$$R_g = \left(k_{gas}^{eff} \cdot \delta_c\right)^{-1} \tag{3.5}$$

$$R_{S1} = \frac{\delta_c}{k_{sol}^{eff}\left(2\delta_c\delta_w + \delta_w^2\right)} \tag{3.6}$$

$$R_{S2} = \frac{\delta_{wall}}{k_{sol}^{eff}\delta_w^2} \tag{3.7}$$

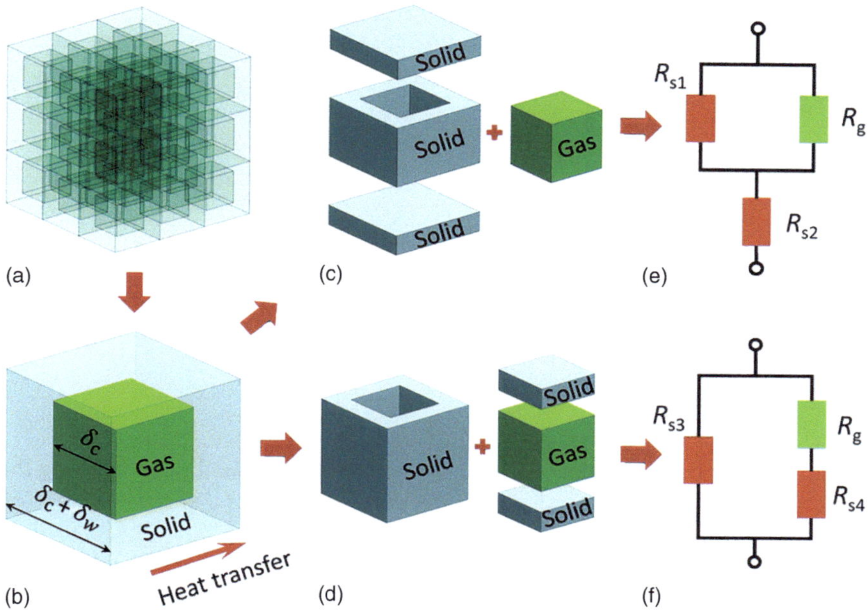

Figure 3.3 Thermal conductivity diagrams of closed cell foam: (a) cubic model; (b) schematic diagram of cells extracted from the model; (c) planar series model in the plane perpendicular to the heat transfer direction; (d) series-parallel model with zero thermal conductivity perpendicular to the heat transfer direction; (e,f) schematic diagram of corresponding thermal conduction model. Source: Wang et al. [4]/with permission of Royal Society of Chemistry.

Combining Eqs. (3.3–3.7) and taking $k_{gas}^{eff} \leq k_{sol}^{eff}$ into account, the equation for the upper thermal conductivity is obtained as follows:

$$\begin{cases} \lambda_{con,w}^{up} = \alpha_{up} k_{sol}^{eff} + \beta_{up} k_{gas}^{eff} \\ \alpha_{up} = 1 - \frac{\delta_c^3}{(\delta_c+\delta_w)^3 - \delta_c^2 \delta_w} \\ \beta_{up} = \frac{\delta_c^3 + \delta_c^2 \delta_w}{(\delta_c+\delta_w)^3 - \delta_c^2 \delta_w} \end{cases} \tag{3.8}$$

Assuming a thermal conductivity of 0 perpendicular to the heat transfer direction, which is the lower limit, the cells are analyzed using series-parallel connections, which are the corresponding equivalent heat map. The lower limit of the effective thermal conductivity $\lambda_{con,w}^{low}$ can be expressed as:

$$\lambda_{con,w}^{low} = \left[R_{con,w}^{up}(\delta_c + \delta_w)\right]^{-1} \tag{3.9}$$

where $R_{con,w}^{up}$ is the upper limit of the effective thermal resistance, which can be expressed as:

$$R_{con,w}^{up} = \left(\frac{1}{R_{S3}} + \frac{1}{R_{S4} + R_g}\right)^{-1} \tag{3.10}$$

R_{S3} and R_{S4} represent the solid thermal resistance in Figure 3.3f, respectively, which can be expressed by the following equation:

$$R_{S3} = \frac{\delta_c + \delta_w}{k_{sol}^{eff}\left(2\delta_c\delta_w + \delta_w^2\right)} \tag{3.11}$$

$$R_{S2} = \frac{\delta_w}{k_{sol}^{eff}\delta_w^2} \tag{3.12}$$

Combining Eqs. (3.5, 3.9–3.12) and considering $k_{gas}^{eff} \leqslant k_{sol}^{eff}$, the following model is obtained:

$$\begin{cases} \lambda_{con,w}^{low} = \alpha_{low}k_{sol}^{eff} + \beta_{low}k_{gas}^{eff} \\ \alpha_{low} = 1 - \frac{\delta_c^2}{(\delta_c+\delta_w)^2} \\ \beta_{low} = \frac{\delta_c^3+\delta_w(\delta_c+\delta_w)^2}{\delta_c(\delta_c+\delta_w)^2} \end{cases} \tag{3.13}$$

The aforementioned model considers the cell wall as a membrane, ignoring the role of cell struts in heat transfer, and, therefore, constructs a special cubic foam composed entirely of struts.

Assuming infinite and zero heat conduction in the plane perpendicular to the heat transfer direction, the cubic unit can be divided into two ways, as shown in Figure 3.4b and c, which refers to the open cell structure of porous materials. The upper and lower thermal conductivities are deduced from the relevant heat transfer

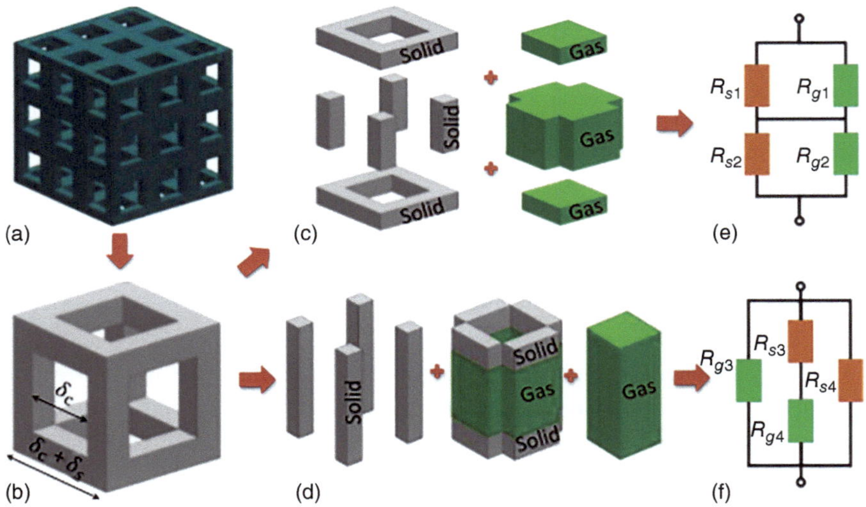

Figure 3.4 Schematic diagram of the heat transfer model consisting entirely of bubble struts: (a) reconstructed cell model consisting entirely of struts, (b) illustration of the cell unit, (c) parallel-series model with thermal conductivity at infinity in the plane perpendicular to the heat transfer direction, (d) series-parallel model based on zero thermal conductivity in the plane perpendicular to the heat transfer direction, (e) heat transfer diagram for the parallel-series model, and (f) heat transfer diagram for the series-parallel model. Source: Wang et al. [4]/with permission of Royal Society of Chemistry.

loop diagrams as follows:

$$\lambda_{con,s}^{up} = \frac{(2\delta_c + \delta_s)\delta_s^3\left(k_{sol}^{eff}\right)^2 + 2\delta_c\delta_s\left(\delta_c^2 + 3\delta_c\delta_s + \delta_s^2\right)k_{sol}^{eff}k_{gas}^{eff} + \delta_c^3(\delta_c + 2\delta_s)\left(k_{gas}^{eff}\right)^2}{\left(2\delta_c^3\delta_s + 3\delta_c^2\delta_s^2 + 2\delta_c\delta_s^3 + \delta_s^4\right)k_{sol}^{eff} + \left(\delta_c^4 + 2\delta_c^3\delta_s + 3\delta_c^2\delta_s^2 + 2\delta_c\delta_s^3\right)k_{gas}^{eff}} \tag{3.14}$$

$$\lambda_{con,s}^{low} = \frac{\delta_c\delta_s^2\left(k_{sol}^{eff}\right)^2 + \left(\delta_c^3 + 2\delta_c^2\delta_s + 2\delta_c\delta_s^2 + \delta_s^3\right)k_{sol}^{eff}k_{gas}^{eff} + \left(\delta_c^2\delta_s\right)\left(k_{gas}^{eff}\right)^2}{\delta_c(\delta_c + \delta_s)^2k_{sol}^{eff} + \delta_s(\delta_c + \delta_s)k_{gas}^{eff}} \tag{3.15}$$

where, δ_s is the thickness of the cell pillar, and the upper and lower thermal conductivities of the cell wall and the cell pillar can be obtained as:

$$\lambda_{con,s}^{up} = (1 - f_s)\lambda_{con,w}^{up} + f_s\lambda_{con,s}^{up} \tag{3.16}$$

$$\lambda_{con}^{low} = (1 - f_s)\lambda_{con,w}^{low} + f_s\lambda_{con,s}^{low} \tag{3.17}$$

f_s is the proportion of the volume of the cell pillar in the cell to the whole solid. It can be seen that the difference in thermal conductivity between the upper and lower limits of this model consisting entirely of struts is not significant.

3.2.3 Thermal Radiation

The third mode of heat transfer in foam is thermal radiation, in which heat is transferred via electromagnetic waves. Considering that electromagnetic waves can be scattered in a large number of interfaces present in the foam and eventually absorbed by the surrounding cell walls, radiation can contribute up to 40% of the total thermal conductivity [6–9]. On a microscopic level, incident electromagnetic waves are partially absorbed at the interface and partially absorbed as they pass through the walls of each cell, with reflection occurring at the solid–gas interface.

To reduce the thermal conductivity, the radiation can be attenuated by reflection at the cell surface and by absorption through the cell walls as it passes through the polymer foam. In the transmission of thermal radiation, it can be described using a simplified model of a cubic foam with uniform wall thickness (Figure 3.5) [4]. The cubic foam model consists of three parallel groups of plates, where the green plate is perpendicular to the heat transfer direction and can block the radiation by reflection and absorption. The yellow and gray plates are parallel to the heat transfer direction and can only absorb radiation but not reflect it. The focus here will be on the role of radiation on individual cells. All these plates can absorb radiation waves, for the optical thickness of the cells, in which the radiation waves are transmitted diffusely due to Mie scattering. The total amount of radiation waves passing through the three plates is the same.

For the absorption behavior of a single plate (Figure 3.6a), the incident wave is reflected at the gas–solid interface when it enters from the gas to the cell wall, and the refracted wave is reflected again at the gas–solid interface when it enters from the cell wall into another gas phase. The cell wall is thinner than the radiation wavelength

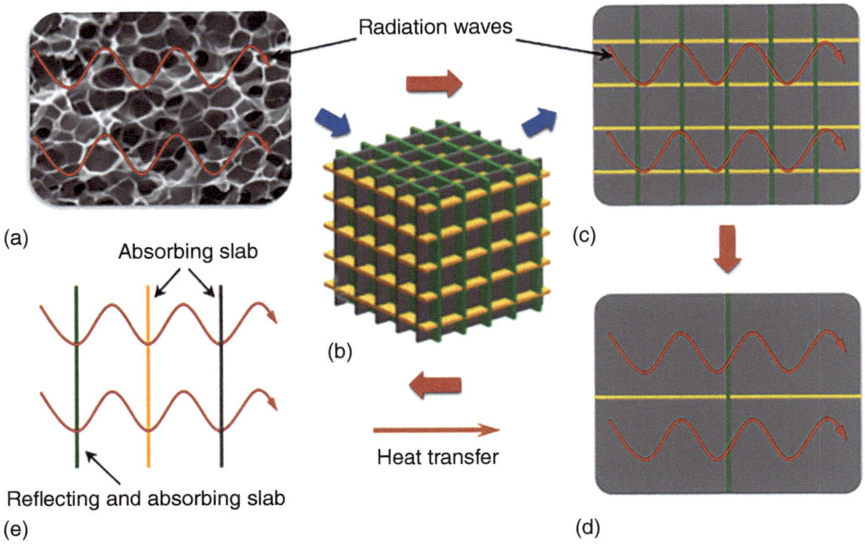

Figure 3.5 Schematic diagram of the simulated thermal radiation process through the polymer foam: (a) realistic pore structure of the polymer foam, (b) reconstructed three-dimensional cubic foam, (c) reconstructed two-dimensional cubic foam, (d) the extracted cell consists of three vertically intersecting planes, (e) the basic analytical model consists of three parallel plates: one reflecting and absorbing plate and two absorbing plates. Source: Wang et al. [4]/with permission of Royal Society of Chemistry.

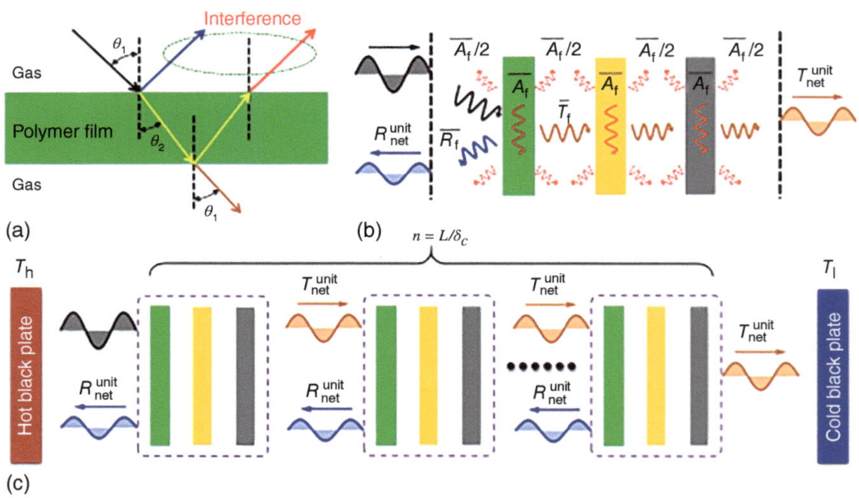

Figure 3.6 Schematic diagram of the propagation characteristics of thermal radiation waves in multilayer materials, (a) schematic diagram of cell wall radiation, (b) thermal radiation through a cell consisting of a reflective absorber plate and two pure absorber plates, and (c) A set of three plate cells sandwiched between a hot black plate and a cold black plate. Source: Wang et al. [4]/with permission of Royal Society of Chemistry.

for nano-sized foams, so the interference effect needs to be considered. In addition, the intensity of the radiation wave is attenuated by absorption when the radiation wave is transmitted through the cell wall.

The reflectivity of the cell wall $\mathfrak{R}_f$, considering interference and absorption, can be expressed by the Fresnel's equation.

$$\begin{cases} \mathfrak{R}_f = \frac{\rho_{gs}^2 e^{4\pi k_s \delta_W \cos\theta_2/\lambda} + \rho_{gs}^2 e^{-4\pi k_s \delta_W \cos\theta_2/\lambda} + 2\rho_{gs}\rho_{sg}\cos\omega_1}{e^{4\pi k_s \delta_W \cos\theta_2/\lambda} + \rho_{gs}^2 e^{-4\pi k_s \delta_W \cos\theta_2/\lambda} + 2\rho_{gs}\rho_{sg}\cos\omega_2} \\ \omega_1 = \phi_{sg} - \phi_{gs} + 4\pi n_s \delta_w \cos\theta_2/\lambda \\ \omega_2 = \phi_{sg} + \phi_{gs} + 4\pi n_s \delta_w \cos\theta_2/\lambda \end{cases} \tag{3.18}$$

where ρ_{gs} and ρ_{sg} are the reflection coefficient at the gas–solid interface and the reflection coefficient at the solid–gas interface, respectively. n_s and k_s are the actual polymer refractive index and the assumed refractive index, and ϕ_{gs} and ϕ_{sg} are the phase change of the reflected wave at the gas–solid interface and the reflected wave at the solid–gas interface, respectively, denoting the wavelength and the refraction angle at the gas–solid interface, which can be expressed as:

$$\theta_2 = \sin^{-1}(n_g \sin\theta_1/n_s) \tag{3.19}$$

n_g is the refractive index of the gas, and θ_1 is the angle of incidence radiation wave. To determine ρ_{gs}, ρ_{sg}, ϕ_{sg}, and ϕ_{gs}, it is necessary to consider the incident radiation waves perpendicular or parallel to the plane. For radiation perpendicular to the plane of incidence, these variables can be expressed as:

$$\rho_{sg} = -\rho_{gs} = \sqrt{\frac{(n_g\cos\theta_1 - n_s\cos\theta_2)^2 + k_s^2\cos^2\theta_2}{(n_g\cos\theta_1 + n_s\cos\theta_2)^2 + k_s^2\cos^2\theta_2}} \tag{3.20}$$

$$\phi_{sg} = \phi_{gs} = \tan^{-1}\frac{2n_g k_s \cos\theta_1\cos\theta_2}{(n_s^2 + k_s^2)\cos^2\theta_2 - n_g^2\cos^2\theta_1} \tag{3.21}$$

For radiation parallel to the plane of incidence, the aforementioned equation can be rewritten as follows:

$$\rho_{sg} = -\rho_{gs} = \sqrt{\frac{\left[(n_s^2 - k_s^2)\cos\theta_1 - n_g n_s\cos\theta_2\right]^2 + (2n_g k_s\cos\theta_1 - n_g k_s\cos\theta_2)^2}{\left[(n_s^2 - k_s^2)\cos\theta_1 + n_g n_s\cos\theta_2\right]^2 + (2n_g k_s\cos\theta_1 + n_g k_s\cos\theta_2)^2}} \tag{3.22}$$

$$\phi_{sg} = \phi_{gs} = \tan^{-1}\frac{2n_g k_s\cos\theta_1\cos\theta_2}{(n_s^2 + k_s^2)\cos^2\theta_1 - n_g^2\cos^2\theta_2} \tag{3.23}$$

For the transmittance Γ_f, the interference effect needs to be taken into account and calculated using Fresnel's formula:

$$\Gamma_f = \frac{\tau_{gs}^2\tau_{sg}^2 e^{-4\pi k_s\delta_w\cos\theta_2/\lambda}}{1 + \rho_{gs}^2\rho_{sg}^2 e^{-8\pi k_s\delta_w\cos\theta_2/\lambda} + 2\rho_{gs}\rho_{sg}e^{-4\pi k_s\delta_w\cos\theta_2/\lambda}\cos\varpi_2} \tag{3.24}$$

where τ_{gs} and τ_{sg} are the gas–solid interface transmission coefficient and the solid–gas interface transmission coefficient, respectively. Incident radiation waves perpendicular or parallel to the incident surface need to be considered, respectively. For the perpendicular case, it can be expressed as:

$$\tau_{gs} = \frac{2n_g \cos\theta_1}{\sqrt{(n_g\cos\theta_1 + n_s\cos\theta_2)^2 + k_s^2\cos^2\theta_2}} \tag{3.25}$$

$$\tau_{sg} = \frac{2\sqrt{n_s^2 + k_s^2}\cos\theta_2}{\sqrt{(n_g\cos\theta_1 + n_s\cos\theta_2)^2 + k_s^2\cos^2\theta_2}} \tag{3.26}$$

For the parallel case, τ_{gs} and τ_{sg} can be expressed as:

$$\tau_{gs} = \frac{2\left(n_s^2 + k_s^2\right)\cos\theta_1}{\sqrt{\left[\left(n_s^2 - k_s^2\right)\cos\theta_1 + n_s n_g\cos\theta_2\right]^2 + (2n_g k_s\cos\theta_1 + n_g k_s\cos\theta_2)^2}} \tag{3.27}$$

$$\tau_{sg} = \frac{2n_g^2\sqrt{\left(n_s^2 + k_s^2\right)}\cos\theta_2}{\sqrt{\left[\left(n_s^2 - k_s^2\right)\cos\theta_1 + n_s n_g\cos\theta_2\right]^2 + (2n_g k_s\cos\theta_1 + n_g k_s\cos\theta_2)^2}} \tag{3.28}$$

Thermal radiation follows Planck's law and is uniformly distributed at an angle of incidence of −90°. The average reflectance and transmittance of wavelengths can be expressed as:

$$\overline{\mathfrak{R}_f} = \frac{1}{\int_{\lambda_{\min}}^{\lambda_{\max}} f(\lambda)d\lambda} \int_0^{\pi/2}\int_{\lambda_{\min}}^{\lambda_{\max}} 2f(\lambda)\mathfrak{R}_f \sin\theta_1\cos\theta_1 d\lambda d\theta_1 \tag{3.29}$$

$$\overline{\Gamma_f} = \frac{1}{\int_{\lambda_{\min}}^{\lambda_{\max}} f(\lambda)d\lambda} \int_0^{\pi/2}\int_{\lambda_{\min}}^{\lambda_{\max}} 2f(\lambda)\Gamma_f \sin\theta_1\cos\theta_1 d\lambda d\theta_1 \tag{3.30}$$

$f(\lambda)$ is the Planck spectral distribution with the expression of:

$$f(\lambda) = \frac{2c^2\pi h}{\lambda^5\exp(hc/\lambda k_B)} \tag{3.31}$$

where c is the speed of light, h is Planck's constant, and $\overline{A_f}$ is Boltzmann's constant. Therefore, the average absorbance of the cell wall can be expressed as:

$$\overline{A_f} = 1 - \overline{\mathfrak{R}_f} - \overline{\Gamma_f} \tag{3.32}$$

As shown in Figure 3.6b, the incident radiation wave will be absorbed and reflected by the absorber plate, corresponding to the reflectivity and absorbance $\overline{\mathfrak{R}_f}$ and $\overline{A_f}$, each plate absorbs half of the radiation energy reflected forward. Taking into account the repeated reflection, absorption, and re-reflection, the radiation energy through the three plates can be expressed as follows:

$$\Gamma_{net}^{unit} = \frac{(1 + \overline{\Gamma_f} - \overline{\mathfrak{R}_f})(\overline{\Gamma_f} + \overline{\mathfrak{R}_f})}{2 - (1 - \overline{\Gamma_f} + \overline{\mathfrak{R}_f})(1 - \overline{\Gamma_f} - \overline{\mathfrak{R}_f})} \tag{3.33}$$

Reflected radiant energy is expressed as:

$$\mathfrak{R}_{net}^{unit} = 1 - \Gamma_{net}^{unit} = \frac{(1 - \overline{\Gamma_f} + \overline{\mathfrak{R}_f})}{2} + \frac{(1 + \overline{\Gamma_f} - \overline{\mathfrak{R}_f})(\overline{\Gamma_f} - \overline{\mathfrak{R}_f})}{2 - (1 - \overline{\Gamma_f} + \overline{\mathfrak{R}_f})(1 - \overline{\Gamma_f} - \overline{\mathfrak{R}_f})} \tag{3.34}$$

Assuming that a polymer foam consisting of n three-plate cells is sandwiched between two black surfaces with temperatures of T_h and T_1 (Figure 3.6c), the net radiant heat flux transmitted through the foam can be expressed as:

$$Q_{rad} = \frac{\sigma_{SB}\left(T_H^4 - T_1^4\right)}{1 + n\left(1/\Gamma_{net}^{unit} - 1\right)} \approx \frac{4\sigma_{SB}\overline{T}^3 \Delta T}{1 + n\left(1/\Gamma_{net}^{unit} - 1\right)} \tag{3.35}$$

$\overline{T} = \frac{T_h+T_1}{2}$ is the average temperature, the temperature difference $\Delta T = T_h - T_1$, and the Stefan–Boltzmann constant is σ_{SB}, so the effective radiative thermal conductivity λ_{rad} can be obtained from Fourier's law as follows:

$$\lambda_{rad} = \frac{4\sigma_{SB}\overline{T}^3 L}{1 + \frac{L}{\delta_c+\sigma_w}\left(1/\Gamma_{net}^{unit} - 1\right)} \tag{3.36}$$

3.3 Performance and Characterization

3.3.1 Property Characterization Methods

3.3.1.1 Thermal Conductivity Measurements

There are several methods for thermal conductivity measurements, including steady-state and nonstationary methods, each of which is only applicable to materials with certain characteristics. Non-steady-state methods can directly determine the thermal diffusivity, while steady-state methods are more suitable for measuring the thermal conductivity of dry materials.

Guarded hot plate (GHP), also known as the Poensgen instrument, is the most effective and commonly used method for measuring the thermal conductivity of insulation materials. The device comprises a cooling panel, a heating panel, a protective heater system, and an insulation layer. The heating panel is electrically heated, while the cooling panel is a Pelter cooler or liquid-cooled heat sink. Structurally symmetrical placement, with protective heat plates placed on both sides is shown in Figure 3.7 [10].

Place a uniform sample of the same thickness between the heating and cooling panels, select an appropriate temperature difference, and measure the power output

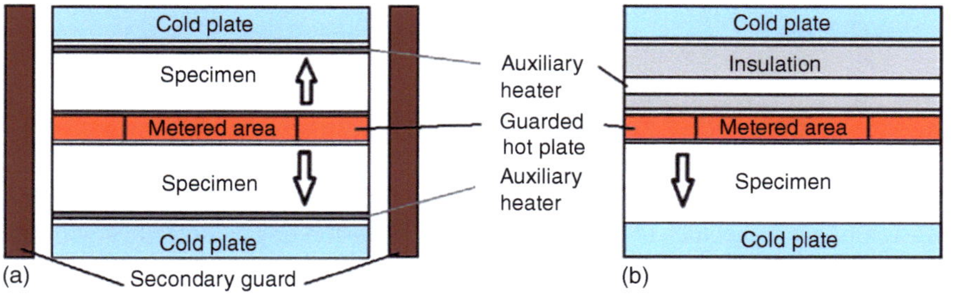

Figure 3.7 GHP device diagram for (a) two samples and (b) one sample. Source: Numan et al. [10]/IntechOpen, CC BY 3.0.

of the hot plate with an area of A when the entire system reaches thermal equilibrium. The thermal conductivity k_{eff} can be expressed by the following formula [10]:

$$k_{eff} = \frac{Q}{2} \frac{\Delta x}{A.\Delta T} \tag{3.37}$$

As the heat flow Q is the power of electric heater, the sample thickness is Δx. The sample temperature difference $\Delta \tau$ is the difference between the temperature of the hot plate and the temperature of the cold plate, the heat transfer area is A.

3.3.1.2 Porosity (p)Measurement for a Polymer Foam

Porosity, the volume fraction of air in the material, can be calculated by the following formula.

$$p = 1 - \frac{1}{\phi} \tag{3.38}$$

where, the expansion rate (ϕ) is equal to the density ratio of the solid to the foam, which can be tested using the ASTM D792-00 water drive method by dividing the density of solid material by the density of the foamed material.

$$\phi = \frac{\rho_s}{\rho_f} \tag{3.39}$$

where, ρ_s and ρ_s are the measured densities of solids and foams, respectively.

3.3.1.3 Thermal Deformation Measurement

The thermal deformation of materials can be characterized by the parameter CTE (coefficient of thermal expansion). The thermal expansion of solids is a phenomenon in which the kinetic energy of adjacent atoms in the solid and the average distance increase, and the thermal expansion occurs with the increase in temperature.

The commonly used testing method is the thermal expansion and contraction method. Typically, the wire material is fixed on the measuring device and placed in a constant temperature box to measure its length at different temperatures, and the linear thermal expansion coefficient is calculated. The linear coefficient of thermal expansion α can be expressed by a formula:

$$\alpha = \frac{\Delta L}{L \cdot \Delta T} \tag{3.40}$$

As the length is L, the $\frac{\Delta L}{\Delta T}$ is the coefficient of change in length for each degree Celsius change.

3.3.2 Factors Affecting Thermal Conductivity

3.3.2.1 Pore Size and Porosity

The contribution of the gas in the foam to the thermal conductivity comes from the collision of gas molecules in the cell, that is the collision of gas molecules with the solid wall in phase II and interfacial phase III. The collisions can be described by a kinetic theory model as follows.

$$k_{s,0} = \frac{1}{3} \cdot C_{g,0} \cdot v_{g,0} \cdot \Lambda_{g,0} \cdot \varepsilon \tag{3.41}$$

where, $C_{g,0}$, $v_{g,0}$, and $\Lambda_{g,0}$ denote the specific volume heat of the gas, the root mean square of the velocity of the gas molecules, and the mean free range of the gas molecules, respectively. According to kinetic theory, the mean free range of gas molecules is:

$$\Lambda_{g,0} = \frac{k_B \cdot T}{\sqrt{2} \cdot \pi \cdot P \cdot l^2} \tag{3.42}$$

where, l is the collision diameter of gas molecules, P, T, and k_B are gas pressure, temperature, and Boltzmann's constant, respectively. In an ideal gas, the thermal conductivity is independent of the gas pressure because the specific volume heat $C_{g,0}$ is proportional to the pressure, while the mean free range $\Lambda_{g,0}$ is inversely proportional to the pressure. In contrast, the motion of gas molecules in foam is limited by the cell size, and when the cell size is less than $\Lambda_{g,0}$, $\Lambda_{g,0}$ is reduced due to the collision of gas molecules with the solid matrix, since it is affected by the spatial confinement effect. Thus, it is necessary to use the apparent mean free range Λ_{eff} to express the corresponding thermal conductivity, the effective thermal conductivity of the gas in a confined space is expressed as:

$$k_{g,eff} = \frac{1}{3} \cdot C_{g,0} \cdot v_{g,0} \cdot \Lambda_{g,eff} \cdot \varepsilon \tag{3.43}$$

$\Lambda_{g,eff}$ can be determined by Mattiesson's rule that

$$\frac{1}{\Lambda_{g,eff}} = \frac{1}{\Lambda_{g,0}} + \frac{1}{d} \tag{3.44}$$

where, d is the cell size. Combined with Eqs. (3.41–3.44), $\Lambda_{g,eff}$ can be derived from the effective thermal conductivity.

$$k_{g,eff} = \frac{C_{g,0} \cdot v_{g,0} \cdot \varepsilon}{3} \left(\frac{\sqrt{2} \cdot \pi \cdot P \cdot l^2}{k_B \cdot T} + \frac{1}{d} \right)^{-1} or k_{g,eff} = \frac{k_{g,0}}{1 + \Lambda_{g,0}/d} \tag{3.45}$$

where $\Lambda_{g,0}/d$ is known as the Knudsen parameter (Kn) and is used to evaluate the effect of the change in the thermal conductivity of the gas when the cell size is smaller than $\Lambda_{g,0}$ is shown in Eq. (3.45), the effective thermal conductivity decreases as the K_n value increases. Therefore, increasing the mean free range of the corresponding volume of gas molecules or decreasing the cell size can effectively increase the K_n and effectively reduce the thermal conductivity of the gas in the cells.

Maintaining high porosity and small cell sizes can reduce the thermal conductivity of both the gas and solid phases at high temperatures, which makes foams a unique class of insulation materials with significantly enhanced thermal insulation properties compared to the corresponding solid materials [4, 11, 12].

For the thermal conductivity of gases, increasing the mean free range of gas molecules, reducing the cell size, and increasing the porosity are preferred means to improve the thermal insulation properties of polymer foams. When the cell density is high enough, the continuous heat transfer through the polymer matrix is disrupted, which enhances the collision of gas molecules and the vibration of solid molecules in the polymer matrix , thus, largely attenuating the heat transfer.

For example, the thermal conductivity of polyetherimide (PEI) foam could reach 0.015 W (m K)$^{-1}$, which is lower than that of air when the cell size is reduced to about 100 nm, since the small cells can limit the movement and collisions of gas molecules [11]. This phenomenon was first demonstrated by Notario et al. [13] in polymethyl methacrylate (PMMA)foam, which demonstrated the reduction of gas heat transfer when confined to a small space.

For solid heat conduction, the solid thermal conductivity depends mainly on the molecular vibrations in the material. Unlike the free motion of gas molecules, the molecules of a solid can only vibrate around their equilibrium position under thermal excitation, and the energy is quantized and discrete. Phonons are the smallest energy vibrations used to describe the simple harmonic vibrations of the lattice, and their transport properties in a solid determine the thermal conductivity of the material. The thermal conductivity of a solid can be described by a kinetic model as follows:

$$k_{s,0} = \frac{1}{3} \cdot C_{s,0} \cdot v_{ph,0} \cdot \Lambda_{s,0} \tag{3.46}$$

where $C_{s,0}$, $v_{ph,0}$, and $\Lambda_{s,0}$ denote the specific volume heat of the solid, the root mean square of the velocity of phonons in the solid, and the mean free range of phonons in the solid, respectively. The native polymer material usually has a short phonon mean free range and thus exhibits poor thermal conductivity. For nanoporous polymer foams, the boundary scattering on the cell walls and struts plays an important role in phonon transport and thermal conductivity.

The scattering of phonons affects the thermal conductivity when the average free range of phonons is comparable to the characteristic size of the solid, which is determined by the cell size d and the size parameter δ. Similar to the effect of spatial confinement of gas molecules on thermal conductivity, the size effect of solids can be described by the effective mean free range of phonons ($\Lambda_{s,eff}$)

$$\frac{1}{\Lambda_{s,eff}} = \frac{1}{\Lambda_{s,0}} + \frac{2}{d+\delta} \tag{3.47}$$

Combining the aforementioned equation, the solid thermal conductivity of porous foam is:

$$k_{s,eff} = \frac{1}{3} \cdot C_{s,0} \cdot v_{ph,0} \cdot \left(\frac{1}{\Lambda_{s,0}} + \frac{2}{d+\delta}\right)^{-1} \text{ or } k_{s,eff} = k_{s,0} \cdot \left(1 + \frac{2\Lambda_{s,0}}{d+\delta}\right)^{-1} \tag{3.48}$$

where, $k_{s,0}$ is the thermal conductivity of the solid. Since the solid occupies only a fraction of the foam, the thermal conductivity of the solid material to the foam can be expressed as:

$$k_s = (1-p) \cdot k_{s,eff} = (1-p) \cdot k_{s,0} \cdot \left(1 + \frac{2\Lambda_{s,0}}{d+\delta}\right)^{-1} \tag{3.49}$$

Therefore, when the porosity is fixed, reducing the cell size and increasing the mean free range can reduce the thermal conductivity of a solid. While making the cell size and mean free range constant, lower thermal conductivity of the solid can be achieved by increasing the porosity.

Sundarram and Li [11] reported that the thermal conductivity decreased from 0.25 to 0.008 W $(m K)^{-1}$ when the cell size of PMMA foam was reduced from 10 um to 10 nm at a porosity of 85%. They also found that the increase in the overall porosity of the foam facilitated the reduction in thermal conductivity but was influenced by the cell size. The thermal conductivity was lower with a larger cell size. Other researchers have reported that the trend of decreasing thermal conductivity with increasing porosity is not linear. Sonntag and coworkers [14] calculated that the thermal conductivity of PS foams was the lowest when the cell density was in the range of 0.1–0.2 g cm^{-3}. Wang et al. [15] and Tsuda et al. [16] investigated the effect of porosity on thermal conductivity by building a nanocellular foam model. They found that the thermal conductivity reduces with the increase in the porosity, the decrease in cell size, and the increase in the number of cell walls , due to the obstruction of phonon propagation. When the number of cell walls reaches a threshold, the cell walls become too thin to absorb infrared radiation, which makes the radiative thermal conductivity λ_{rad} dominant, and continuing to increase the number of cell walls leads to an increase in the overall thermal conductivity (Figure 3.8) [4].

3.3.2.2 Temperature

The measurement temperature is an important factor influencing the thermal conductivity of insulating materials. For example, Yu et al. [17] measured the thermal conductivity of PMMA foams at different temperatures, as shown in Figure 3.9 [17]. When the porosity is fixed at 0.8, the higher the temperature, the higher the total thermal conductivity, gas–solid phase conduction, and radiative thermal conductivity. Generally, the cell size has a greater effect on thermal radiation than thermal conduction, and the temperature has a greater effect on thermal radiation as well.

3.3.2.3 Material Refractive Index

The refractive indices of polymer materials are generally in the range of 1.4–2.0, which would also affect the thermal conductivity of porous polymers. Yu et al. [17] chose five models with refractive indices of 1.50, 1.75, 2.00, 2.25, and 2.50, respectively, according to the experimental measurement and law of Tsuda. All other conditions are maintained the same at a temperature of $T = 300$ K,

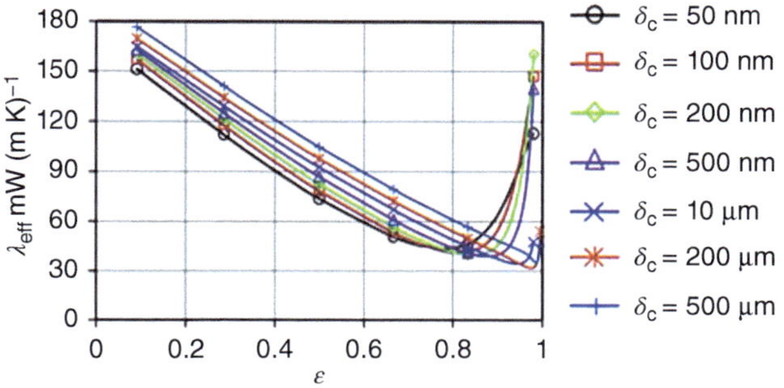

Figure 3.8 Trend of thermal conductivity with porosity for different cell sizes (δ_c is the cell sizes and ε is the void fraction). Source: Wang et al. [4]/with permission of Royal Society of Chemistry.

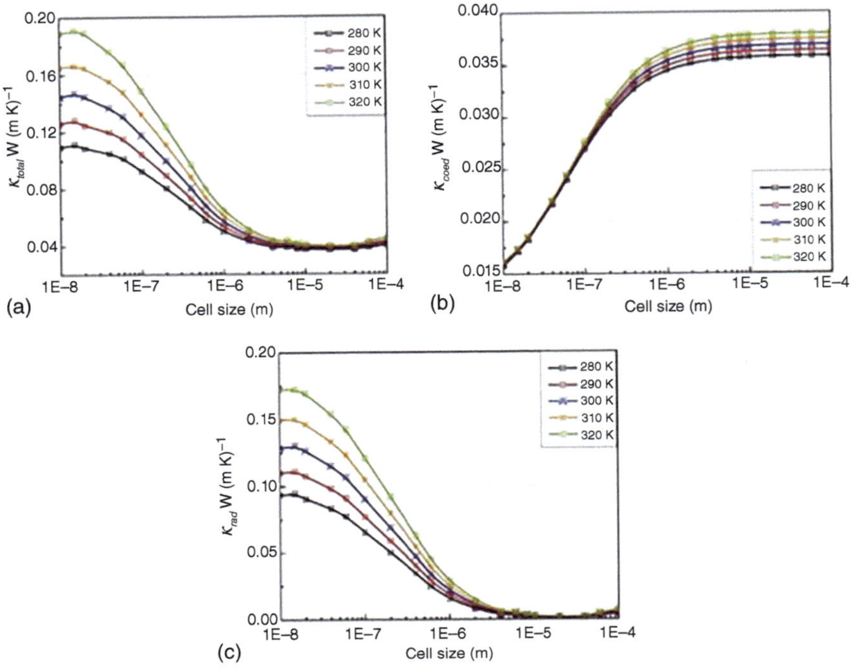

Figure 3.9 Variation of thermal conductivity of PMMA foam with pore size at different temperatures: (a) total thermal conductivity; (b) gas–solid phase thermal conductivity; and (c) radiative thermal conductivity. Source: Yu et al. [17]/with permission of Elsevier.

$\phi = 0.80$, and the pure solid chosen was PMMA with a thermal conductivity of 0.19 W (m K)$^{-1}$. It was found that (Figure 3.10) the total thermal conductivity and radiation thermal conductivity are increased with the increase of refractive index when the aperture diameter $\delta \leq 30$ nm. When the aperture diameter $\delta \geq 100$ nm, the thermal conductivity decreases with the increase of refractive index, according to the Fresnel reflection equation. When the cell size is larger, the energy flux is mainly in the form of propagating waves, which increases the thermal conductivity. In addition, the refractive index has little effect on the gas–solid thermal conductivity k_{cond}, and the curve is overlapped for different samples [17]. Therefore, for both gas-phase and solid-phase thermal conductivity, reducing the cell size is an effective way to obtain good insulation, while appropriately reducing porosity can reduce thermal conductivity.

3.4 Fabrication of Thermal Insulation Polymer Foams

3.4.1 Composite Polymer Matrices

According to the model described above, the gas thermal conductivity (λ_{gas}) is a function of the porosity, cell size, and thermal conductivity of the main gas. The solid thermal conductivity (λ_{solid}) is a function of the porosity, support, and thermal

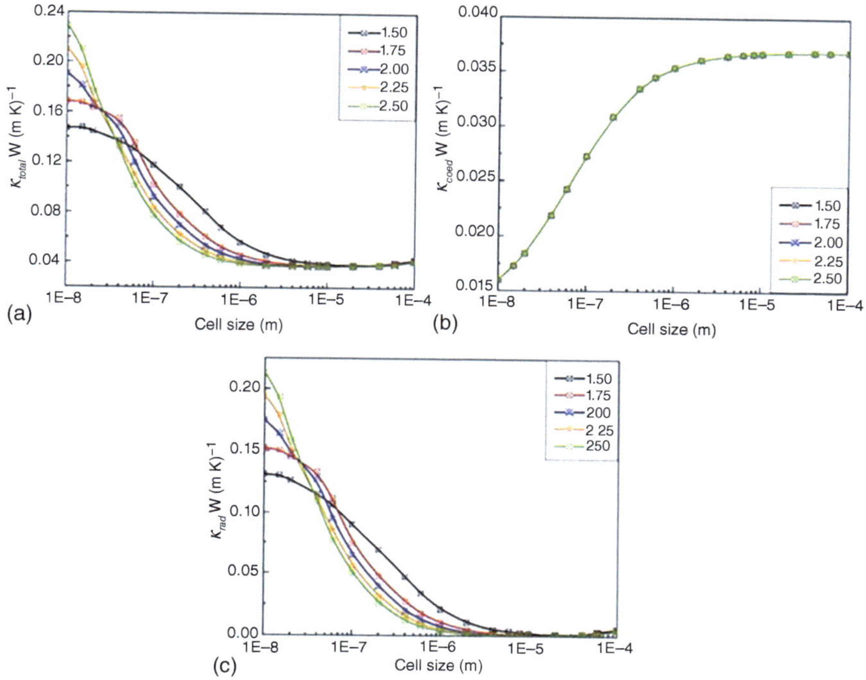

Figure 3.10 Variation of thermal conductivity of PMMA foam with pore size for different refractive indices: (a) total thermal conductivity; (b) gas–solid phase thermal conductivity; and (c) radiation thermal conductivity. Source: Yu et al. [17]/with permission of Elsevier.

conductivity of the main solid. The radiative thermal conductivity (λ_{rad}) is determined by the Rosseland extinction coefficient and is related to the cell wall thickness and the reflection index. There are two general ways to increase the porosity, namely, decreasing the volume fraction of the solid or increasing the number of cells while maintaining the cell size. Increasing the number of cells increases the amount of cell walls, and makes the heat conduction discontinuous, which in turn causes the reduction in thermal conductivity.

Foam morphology, gas properties, and polymer matrix are three main factors that need to be considered for the improvement of the thermal insulation properties of polymer foam. Moreover, filling the foam with a gas having a larger mean free range also reduces the thermal conductivity of the gas. However, some gases may be corrosive to the substrate polymer or be harmful to the environment and human health. Alcohols and other alkane gases as foaming agents have the disadvantage of high flammability although they can reduce the thermal conductivity of the foam. Therefore, CO_2 is being more and more used in the fabrication of thermal insulating polymer foams.

Since the thermal conductivity of solids is generally 10 times higher than that of gases, another effective way to reduce the thermal conductivity is to reduce the volume fraction of solid matrix material in the foam, that is, to increase the porosity, which usually leads to the decrease in the volume fraction of solids and λ_{sol}. For the choice of the polymer matrix material, in general, using a lower thermal conductivity polymer as the matrix material can reduce the solid thermal conductivity of the foam. For example, at room temperature, the thermal conductivity of pure polypropylene is 0.33 W $(m\,K)^{-1}$, the thermal conductivity of pure polyurethane is 0.18 W $(m\,K)^{-1}$, and the polyurethane foam has lower thermal conductivity than the polypropylene foam with the same cell size and porosity.

The matrix properties can be improved by changing the main structure of the matrix and adding fillers to form composites. Improving the structure of the matrix can improve the thermal conductivity and thermal expansion properties of the matrix. The addition of functional fillers can improve the thermal properties of the composite foam [18]. The key factor is the interfacial compatibility between matrix and filler, and improving the interfacial compatibility can reduce the surface energy of the filler, which can be improved by various processes such as heat treatment, physical, and chemical modification, etc.

The size, amount, and type of filler can significantly affect the material properties, Romesberg et al. [19] added 10% carbon black particles in PS foam and slightly reduced the thermal conductivity. In contrast, Gong et al. [20] achieved a reduction in the thermal conductivity λ of PS foams to 0.030 W $(m\,K)^{-1}$ by adding nanofillers with high IR absorbance, such as carbon nanotubes (CNTs). The reduction of thermal conductivity can be achieved even in very small amounts as long as the additives have a sufficiently high IR absorption. For example, Zhang et al. [21] added 0.1% w/w of graphite flakes to PS foam, which reduced the thermal conductivity of PS composite from 0.0357 to 0.033 W $(m\,K)^{-1}$. This is because the nanofillers in the polymer matrix get closer to each other and compress the nanofillers within the cell walls as the polymer matrix is foamed (Figure 3.11) [22]. In this case, nanoparticles dispersed in a polymer matrix can be effective in shielding IR radiation [21, 23]. Another reason for using nanofillers to reduce thermal conductivity is that solid particles can reduce the free energy activation barrier for bubble nucleation at the filler–polymer interface, which facilitates the formation of smaller pores, resulting in a higher pore density and better control of diameter homogeneity.

Inorganic/metal particles were first used by Hansen and Martin [24] as early as the 1960s to promote heterogenous cell nucleation. For example, Thomassin and coworkers [25] showed that MWCNTs can act as nucleating agents to increase the cell density in PMMA/MWCNTs foams by 1 to 2 orders of magnitude. Zhao et al. [26] demonstrated that the addition of 5% w/w polytetrafluoroethylene (PTFE) nanofibers to polypropylene (PP) foam can improve thermal insulation performance by increasing the expansion ratio of PP foam. PTFE nanofibers reduced the cell size by an order of magnitude and increased the expansion ratio by three times. With enhanced morphological changes, the thermal conductivity reduced from 0.060 to 0.0365 W $(m\,K)^{-1}$. Silica nanoparticles are typical fillers for polymer-based composites [27], which can result in strong compatibility with the polymer chains and form a better cellular structure that is beneficial for thermal insulation [28].

The influence of nanofillers on the morphology of polymer foams mainly depends on the size of the fillers, and their surface chemistry. Goren et al. [29] investigated

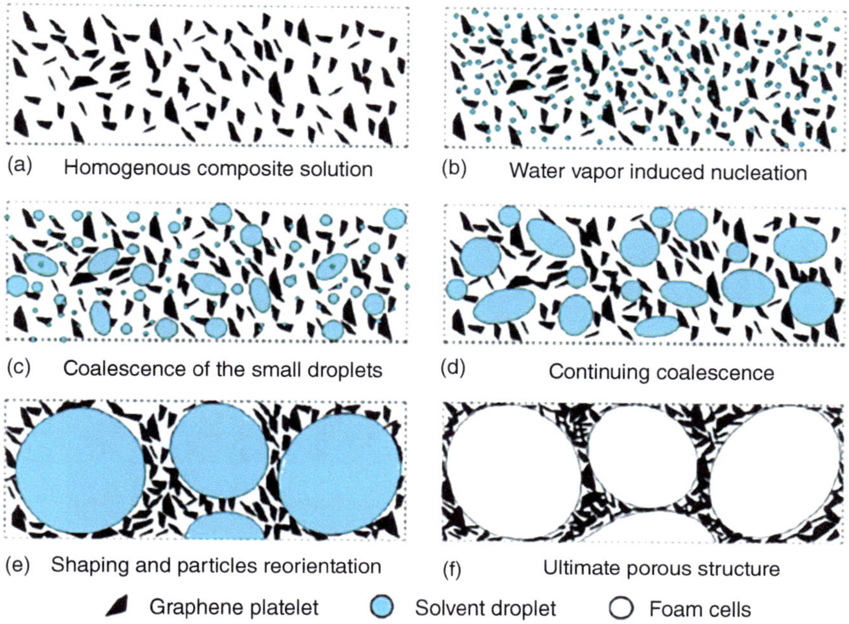

Figure 3.11 Sketch of bubble pore formation of foam containing solid nanofillers: (a) homogeneous composite solution, (b) water-vapor-induced nucleation, (c) droplet aggregation, (d) accelerated coalescence, (e) plasticization of graphene sheets and reorientation of particles, and (f) final formation of porous structure. Source: Ling et al. [22]/with permission of American Chemical Society.

PMMA foams and found that decreasing the size of nanofillers increased the cell density by 20–30% compared with unfilled PMMA. In another study, Sergeeva and coworkers [30] added fillers with different sizes (70 and 250 um) into PU and analyzed the effect of filler size on the cell size of the foam. It was found that the cell size was reduced by 35–50% compared to unfilled PU foams, but no obvious relationship between filler size and pore size reduction was found. Goren et al. [29] investigated the effect of the surface chemistry of fillers on PMMA foam morphology. They added fluoroalkylated functionalized silica nanoparticles to PMMA for supercritical CO_2 foaming and found that the reduction in surface energy due to surface functionalization reduces the cell size. Similar findings were reported by Vancso and coworkers [31] who used nucleated-shell silica-polydimethylsiloxane (PDMS) nanoparticles as nucleating agents and PMMA as a substrate to obtain nanofoams with a cell size of 400 nm. The small cell size was obtained because of the reduced nucleation energy of PDMS (Figure 3.12) [31].

3.4.2 Fabrication of Bimodal Foams

Bimodal cell structure polymer foams refer to foams with two different cell sizes and distributions. Due to the bimodal cell size distribution of porous polymer, the foams typically have good barrier properties, excellent absorption and

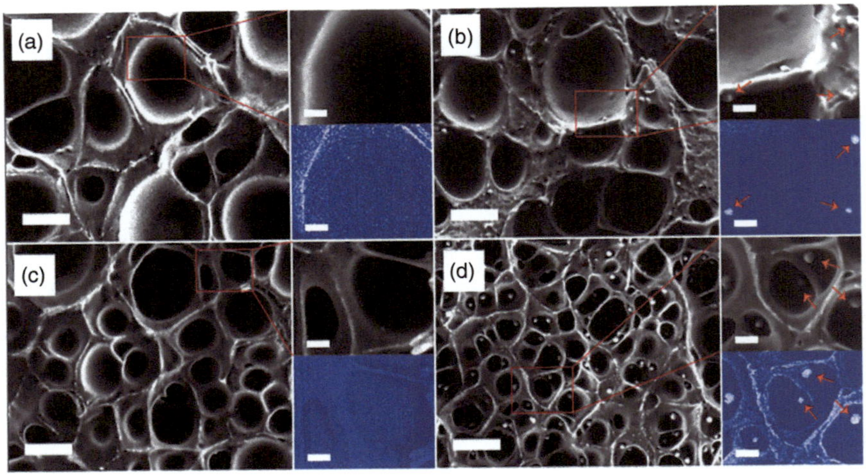

Figure 3.12 SEM images of PMMA foam with nanoparticles of: (a) 20 nm, (b) 80 nm diameter and core-shell nanoparticles of (c) 20 nm, and (d) SEM images of 80 nm SiO_2-PDMS core-shell nanoparticles of PMMA foam. Source: Liu et al. [31]/with permission of American Chemical Society.

penetration properties, and can meet special filtration, sound absorption, and lining applications.

To fabricate polymer foams with bimodal cell structures, various approaches have been developed. For example, Zhang et al. [32] used a dual foaming agent to fabricate a bimodal cell-structured PS foam. CO_2 and water were used as co-blowing agents, nano clay and activated carbon were added in PS as nucleation agents. It was found that the addition of nano-clay increased the nucleation rate of CO_2, while the addition of activated carbon effectively reduced the foaming time of water, which resulted in the production of many small cells by CO_2, and large cells by water vapor as illustrated in Figure 3.13 [33]. Compared with the PS foams with uniform cell structure, the bimodal foams prepared by this method exhibit better compression and thermal insulation properties.

Mi and coworkers [34] developed a multiple soaking temperature (MST) method to prepare PLA foams with a bimodal open-cell structure. As shown in Figure 3.14, after a high-temperature soaking stage, the PLA sample underwent a low-temperature soaking stage and an elevated-temperature soaking stage. Since the solubility of $scCO_2$ in the polymer matrix decreases with the increase in temperature, the saturated $scCO_2$ would precipitate in the elevating temperature stage, which caused the formation of $scCO_2$ aggregates in the PLA matrix. In the final foaming process, the rapid depressurization triggered the nucleation of micro-sized cells, and the aggregated $scCO_2$ regions formed cells with much larger cell sizes due to the excess CO_2 concentration and the diffusion of CO_2 into these large voids. The developed bimodal PLA foam showed reduced density and demonstrated the ability to be used in tissue engineering.

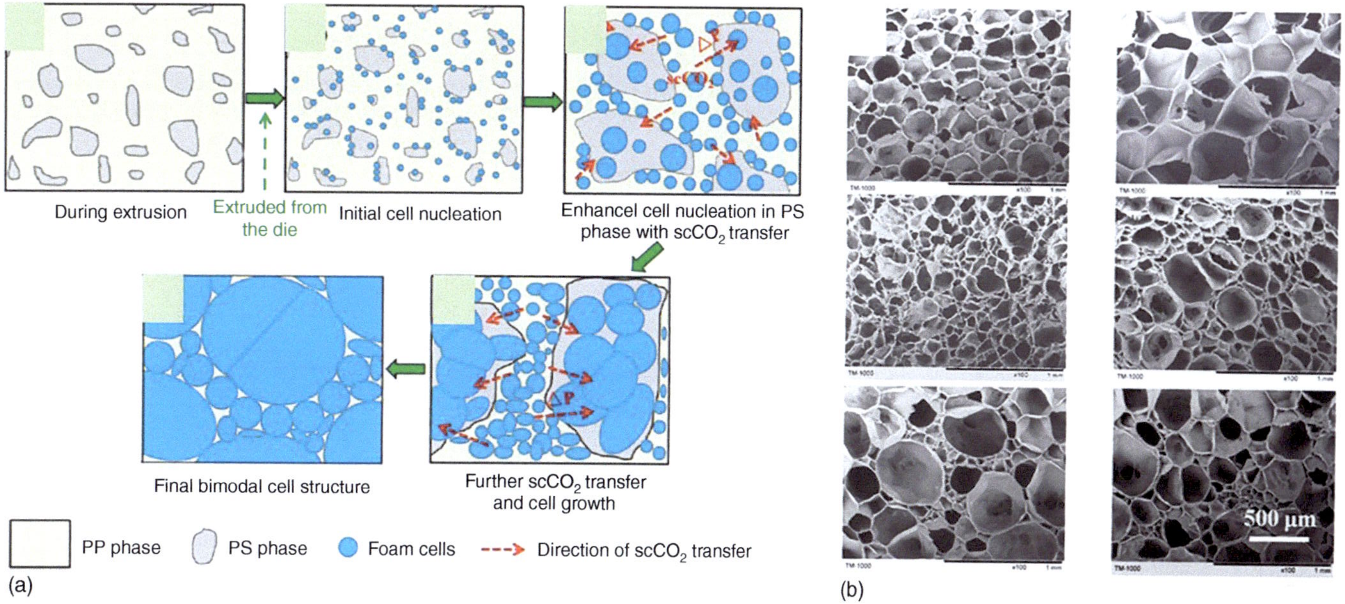

Figure 3.13 (a) Schematic illustration of the bimodal structure formation mechanism; (b) the SEM of Bimodal cell structure. Source: Wang et al. [33]/with permission of Elsevier.

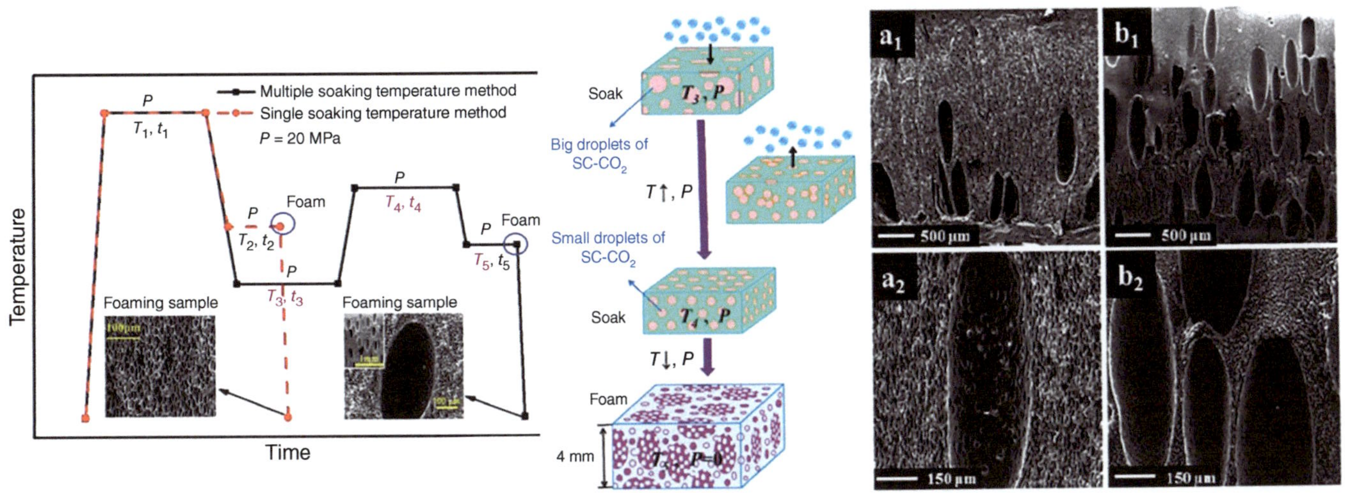

Figure 3.14 Schematic illustration of the mechanism for the multiple soaking temperature (MST) method to produce PLA foams with bimodal cell structure, and the SEM images of the prepared PLA foams. Source: Huang et al. [34]/with permssion of Elsevier.

In addition, fabricating polymer blends takes advantage of different polymer phases, which is also capable of producing foams with bimodal structure. Wang et al. [33] prepared a bimodal PP/PS blended foam by scCO_2 foaming. Due to the difference in CO_2 solubility and melt strength between PP and PS, the large cells were mainly formed in the PS phase, while small cells were mostly produced in the PP phase. The higher solubility in the PP phase leads to a higher scCO_2 concentration gradient at the PP/PS phase interface, which causes the diffusion of scCO_2 from the PP phase into the PS phase, leading to an increased nucleation capacity in the PS phase and the growth of cells. Therefore, the cells in the PS phase grow larger and larger, while the growth of small cells in the PP phase is restricted, which eventually leads to the formation of a bimodal cellular structure. Similarly, Qu et al. [35] prepared PS/thermoplastic polyurethane (TPU) bimodal foams using melt blending and solid intermittent foaming. TPU presented an island phase in the PS matrix, and they concluded that the generation of bimodal cell structure in the PS/TPU blends is depended on the viscosity, morphology of the polymer blends, and the solubility and diffusivity of the CO_2. They also found that the introduction of TPU resulted in a lower thermal conductivity. Luo et al. [36] prepared PMMA/SiO_2 bimodal foams using scCO_2 foaming, which had a more homogeneous cell structure, smaller cell size, and higher cell density compared to pure PMMA foams. With the addition of 0.5% mass fraction of SiO_2, the thermal conductivity was reduced by 62.7% compared to pure PMMA.

3.4.3 Fabrication of Closed-Cell Foams

Cai et al. [37] proposed a multiscale model to predict the effect of the internal structure of closed-cell insulation (phenolic resin) on dynamic heat transfer. The relationship between multiple structural parameters and the effective transmission coefficient was analyzed using Boltzmann's method to reconstruct the closed-cell insulation. As in Figure 3.15, the authors obtained scanning electron microscope (SEM) maps of the samples using the Voronoi diagram division method, depicting porous closed-cell pores with a porosity of 85%. The reconstructed network with the solid phase in the black area and the vesicles in the white area is created based on the Voronoi diagram splitting method and the four-parameter generation (QSGS) method.

The effective thermal conductivity can be obtained by Eq. (3.50):

$$\rho_m c_m \frac{\partial T}{\partial t} = \frac{\partial}{\partial X}\left(K_{eff}\frac{\partial T}{\partial X} + L_v\delta_v\frac{\partial(\varphi P_{v,sat})}{\partial X}\right) - \left(K_{eff}\frac{\partial T}{\partial X} + L_v\delta_v\frac{\partial(\varphi P_{v,sat})}{\partial X}\right)$$
$$= h_{surf}(T_\infty - T_{surf}) + L_v h_m(P_{v,\infty} - P_{v,surf}) \tag{3.50}$$

where, ρ_m is the material density (kg m^{-3}) and c_m is the specific heat capacity (J $(kg\ K)^{-1}$). K_{eff} is the effective thermal conductivity (W $(m\ K)^{-1}$), and h_{surf} is the convective heat transfer coefficient (W $(m^2\ K)^{-1}$). L_v is the latent heat of evaporation. $P_{v,sat}$ is the saturated vapor pressure.

Zhao et al. [38] prepared closed-cell polyvinylidene fluoride (PVDF) foams with a specific microporous structure, with an ultralow dielectric coefficient and a low thermal conductivity (0.027 W $(m\ K)^{-1}$), using a batch foaming method. The excellent

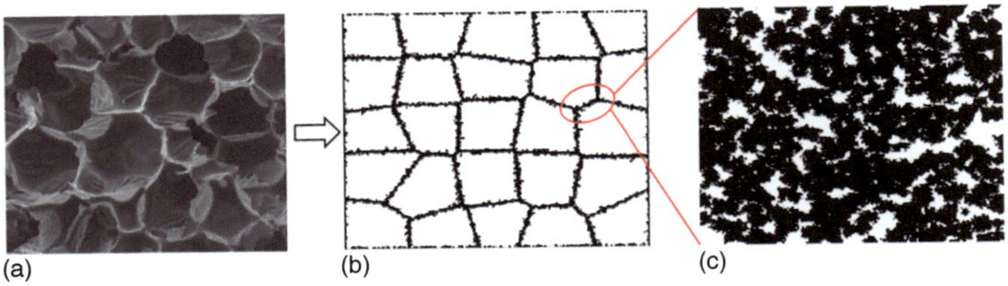

Figure 3.15 On-the-fly reconstruction model of closed cell insulation: (a) SEM image of phenolic foam, (b) cell reconstruction using Voronoi diagram splitting method, and (c) skeleton reconstruction using QSGS method . Source: Cai et al. [37]/with permission of Elsevier.

thermal insulation properties are attributed to the low density and porous structure of PVDF foams that expand the specific surface area for heat transfer, and the closed-cell foam makes it difficult for gas molecules to interact within the pores, thus, hindering the energy exchange between gas molecules.

3.4.4 Fabrication of Polymer Foams with Honeycomb Structures

Honeycomb structures are mostly used in metallic thermal protection systems to provide high-temperature thermal protection and to withstand aerodynamic forces. Compared with metallic honeycomb materials, polymer honeycomb structures have a lower thermal conductivity and are good insulating materials. Because of the honeycomb structure, the walls and ribs of the cells form a closed cell structure with a high rate of closure, and the cells are independent of each other, so they can effectively reflect and absorb all kinds of energy and have a good thermal insulation effect.

For semicrystalline PP foam, Yang et al. [39] produced recyclable PP foam with high mechanical strength and good thermal insulation properties by $scCO_2$ foaming using β-NAs as the nucleation agent. The prepared PP foam showed a honeycomb structure with a high expansion ratio, low thermal conductivity (26.4 mW $(m\,K)^{-1}$), and good tensile and compressive strength. By introducing β-NAs, the crystallization of PP became faster, the crystal size decreased, and the viscoelasticity increased. In addition, the honeycomb structure greatly increased the path length for heat transfer and reduced the heat transfer efficiency (Figure 3.16a–c).

3.4.5 Fabrication of Nanocellular Polymer Foams

Microcellular foaming products have small cell sizes, and a large number of cells can reduce material loss and achieve good insulation and sound absorption effects, as well as lower dielectric constants. Microcellular injection molding or injection foaming is a technique to produce polymer foams with complex or defined geometry. It is a scalable approach for the practical production of plastic products with weight reduction. However, injection foaming generally suffers from a low expansion ratio and poor surface quality. How to produce polymer foams with uniform small cells is also a challenge for injection foaming.

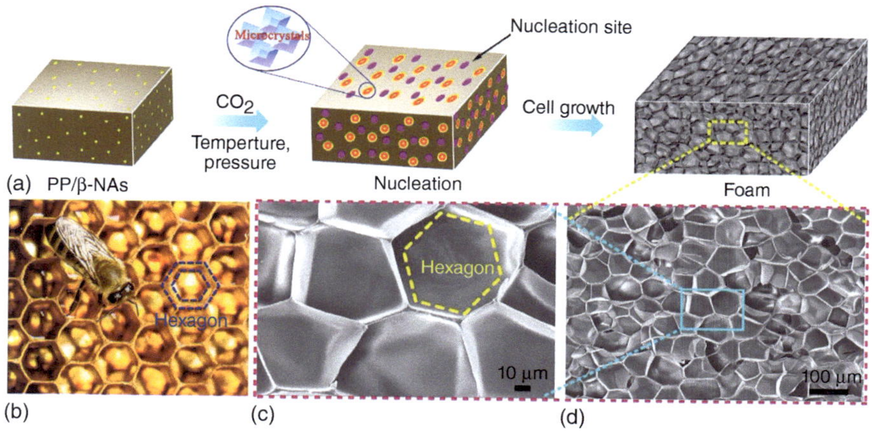

Figure 3.16 Schematic diagram of the fabrication process of the honeycomb PP foam and its corresponding microstructure. Source: Yang et al. [39]/with permission of Elsevier.

High-pressure injection foaming with mold opening offers an approach to overcome these challenges [40]. In the process, the polymer and gas system was injected into the mold cavity followed by a high-pressure packing stage, in which the gas was dissolved back into the polymer matrix. Then, the foaming of the material is triggered by the mold opening process, which suddenly releases the high-pressure state in the mold cavity, and the material is foamed to fill the enlarged mold cavity. In this process, the foaming and filling processes are completely independent, and the foaming temperature is different from the melt temperature, which is largely determined by the mold temperature. Using this approach, Wang et al. [41] fabricated lightweight, heat-insulating, and low dielectric coefficient high-impact polystyrene (HIPS) foam. By controlling the pressure holding time, cooling time, mold temperature, and mold opening distance, they adjusted the cell structure in a wide range, and prepared microporous HIPS foam with a weight reduction of up to 60%. As shown in Figure 3.17, the prepared foam showed a uniform cellular structure with a cell size of tens of micrometers. The foam exhibited a low thermal conductivity of 60 mW $(m\,K)^{-1}$. In addition, the thermal conductivity is mainly affected by the expansion ratio of the foam which can be effectively controlled by changing the mold temperature.

In another study, Wang et al. [40] developed a composite nanocelluar foam composed of PP and PTFE nanofibrial network. When subjected to external shear and tensile effects, PTFE microparticles were in situ deformed into a 3D network structure with fiber size in the range of 200–600 nm. As illustrated in Figure 3.18, the PTFE network affected the crystalline structure and induced the formation of much smaller crystalline regions and enhanced the viscosity of the melt. With the presence of the high aspect ratio fibers, the PP/PTFE composite formed a nanocellular morphology with cell size less than 500 nm due to the enhanced melt strength, and the heterogenous nucleation behavior of the PTFE network.

Apart from injection foaming, batch foaming is capable of producing polymer foams with fine cell structures without needing chemical solvents. In batch foaming,

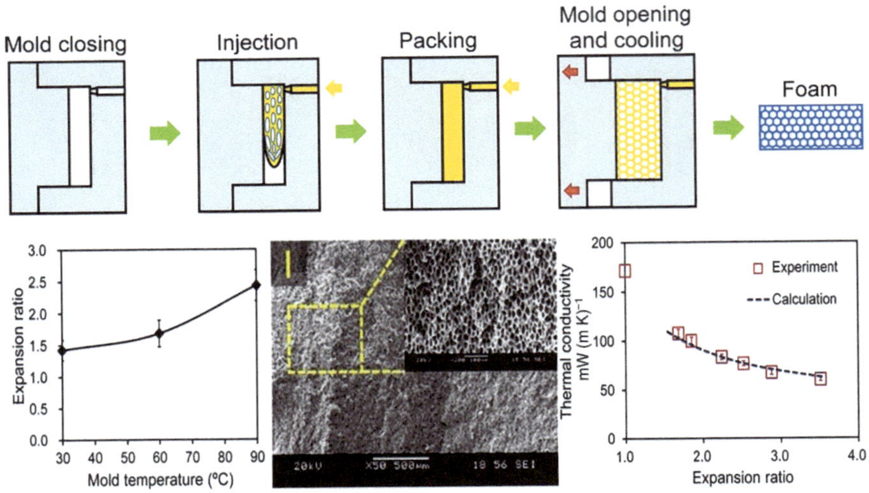

Figure 3.17 Schematic illustration of the high-pressure injection foaming process with mold opening, and results of expansion ratio dependence on mold temperature, the cellular structure of the fabricated foam, and the thermal conductivity of foams with different expansion ratios. Source: Wang et al. [41]/with permission of Royal Society of Chemistry.

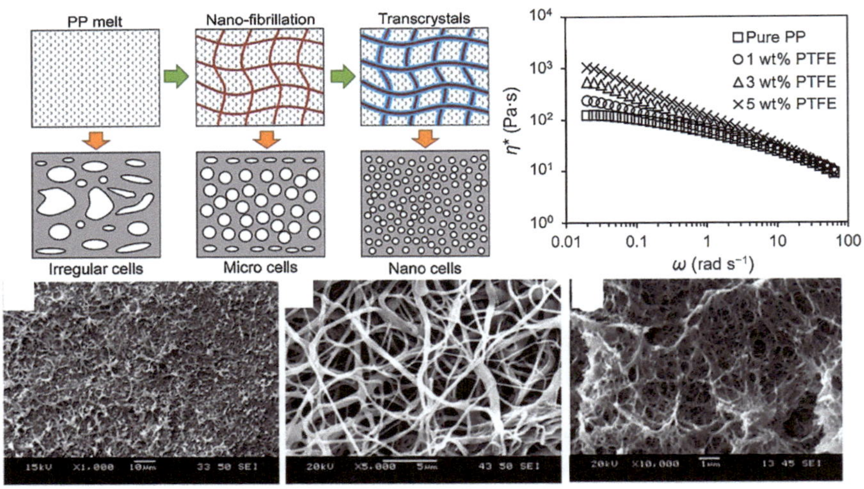

Figure 3.18 Illustration of the nanocellular structure formation mechanism, the viscosity results, the PTFE fibrous network formed in the PP matrix, and the nanocellular morphology of the foam. Source: Wang et al. [40]/with permission of Elsevier.

cell nucleation is triggered by the gas thermodynamic instability condition due to the fast change of solubility of the pressurized gas in the polymer matrix. There are basically two types of batch foaming methods, namely, one-step and two-step methods. In one-step foaming, the foaming is induced by rapid pressure drop at high temperatures after the polymer and foaming gas are fully saturated, which

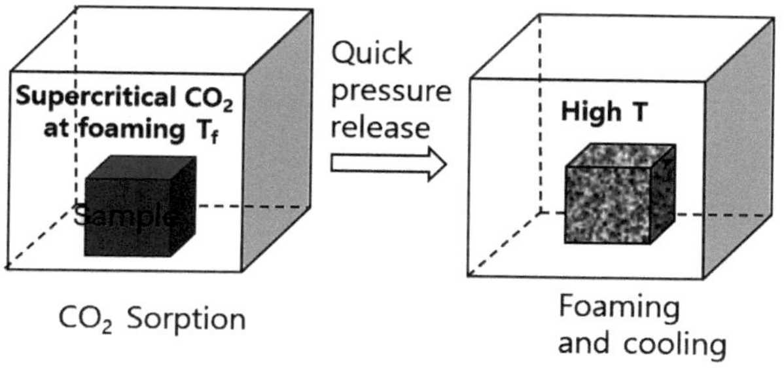

Figure 3.19 Scheme of the experimental setup used for the one-step batch foaming process. Source: Kim et al. [42]/with permission of American Chemical Society.

causes thermodynamic instability to induce cell nucleation and cell growth. The basic foaming process for one-step batch foaming is illustrated in Figure 3.19 [42].

The two-step foaming method is also known as the temperature-induced foaming method. In two-step foaming, the sample is saturated at a relatively low temperature and the foaming is triggered by transferring the sample to a hot environment such as a hot oil bath and hot oven. The cell nucleation and cell growth are triggered when the matrix polymer is sufficiently softened. To fabricate polymer foams with fine cellular structure, Wang et al. [43] prepared a PMMA foam with a porosity of 95.6% and an average cell size of 4.5 μm through a one-step $scCO_2$ foaming process. The PMMA foam has a relatively uniform cell distribution and cell size. A closed cell structure can be obtained at a lower foaming temperature (135 °C), and an open cell structure can be obtained at a higher foaming temperature (247.5 °C). The thermal conductivity of foam is mainly dominated by the solid phase. As the porosity of foam increases, the content of solid phase decreases linearly, which causes a linear decrease in the thermal conductivity. As the porosity further increases, the thermal conductivity will eventually increase with the porosity, because radiation heat transfer is dominant at that time. An ultralow thermal conductivity of 29.9 mW $(m\,K)^{-1}$ was achieved by optimizing the cellular structure.

In addition, Wang et al. [44] also fabricated adjustable nanoporous PMMA/TPU foams by one-step method and two-step method. The tensile and impact toughness of the nano-sized PMMA/TPU foam was significantly enhanced because TPU and nano-sized cell structure in the PMMA matrix promoted the motion ability of molecular chains. The nanoscale porous PMMA/TPU material with an average pore size of 205 nm and an expansion rate of 8 has an extremely low thermal conductivity of 24.8 mW $(m\,K)^{-1}$, due to the strong infrared absorption ability and low-radiation thermal conductivity of the PMMA matrix.

3.5 Other Thermal Insulation Polymer Foams

3.5.1 Microcellular Polyimide (PI) Foams

Polyimide (PI) is a high-performance polymer with excellent mechanical, thermal, and chemical properties. Due to these outstanding characteristics, PI has been widely applied in high-tech fields such as aerospace, electronics, advanced warships, and rail transit [45–47]. PI foam and aerogel have been developed recently

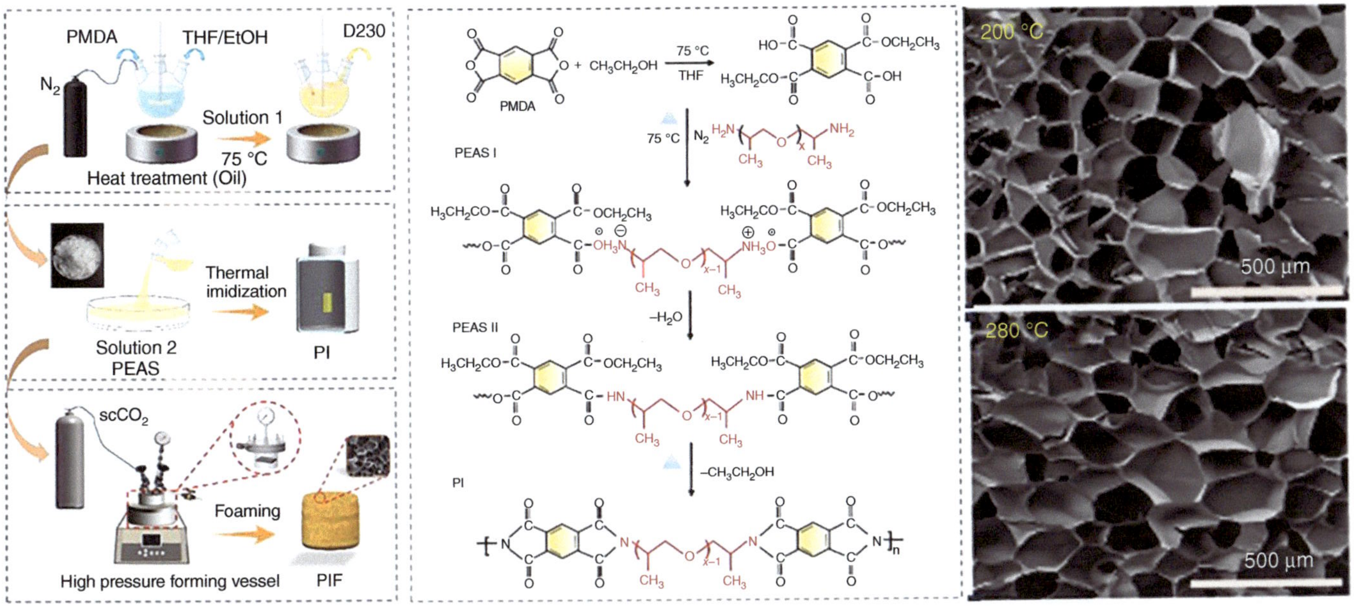

Figure 3.20 Schematics of PI foam synthesis, the corresponding reactions, and the microstructure of PI foams prepared via $scCO_2$ foaming. Source: Liu et al. [49]/with permission of Elsevier.

to adapt to harsh operating conditions. For example, new aircraft engine housings need foam that can withstand high temperatures above 300 °C to insulate and reduce noise.

Polyimide foam (PIF) shows great advantages in superior performance and industrial compatibility [48]. In addition to the imide ring in the PI chain skeleton, further foaming techniques have also generated a pore structure with gas as the dispersed phase, giving PIF many characteristics such as lightweight, good insulation, flame retardancy, corrosion resistance, radiation resistance, chemical stability, etc. Mi and coworkers [49] developed a PIF with uniform and controllable cellular structure by controlled $scCO_2$ foaming. In the study, the PI polymer was synthesized by poly(amic acid) (PAA) and polyester ammonium salt (PEAS) precursor solutions using pyromellitic dianhydride (PMDA) as dianhydride reagents and polyether amine Jeffamine D230 as aliphatic diamine reagent via solution polymerization. Then, the material is foamed using $scCO_2$ as the blowing agent. The PIF synthesized at a thermal imidization temperature of 200 °C exhibited a small cell diameter of 123.8 μm, and a high compressive strength of 0.4 MPa under 10% strain at high temperature and pressure (Figure 3.20).

3.5.2 Phenolic Foams Chilled Water Piping

Central air conditioning in the hotel is generally water-cooled, its chilled water piping will usually have the problem of condensate reflux, which is a critical requirement for pipe insulation. The insulation layer is an important part of the chilled water system in the central air conditioning. The pressure difference in air pressure along the pipe will constantly cause moisture to soak into the insulation layer, which makes the pipe more likely to corrode and produce mold.

Phenolic foams with a closed-cell structure that could prevent moisture infiltration have been used for chilled water piping. If part of the closed-cell structure is damaged, moisture will only penetrate into the damaged area, which allows the moisture penetration coefficient to be close to 0. The low moisture penetration coefficient is essential for the insulation layer to maintain low thermal conductivity. Typically, phenolic foams are made by mixing, foaming, and curing of phenolic resin containing surfactants, foaming agents, and assimilating agents. The phenolic foam insulation part will be glued together with the moisture-proof vapor barrier, then transported to the site and installed. In addition to chilled water piping, phenolic foams have also been used in the thermal insulation of hot-water systems to prevent heat loss.

Since traditional phenolic foam-based insulation layers use surfactant and various chemical agents in the production, other foams that are produced via environmentally friendly physical foaming methods, such as PP, PS, and PE, have been more and more used in thermal insulation systems.

3.5.3 Spray Polyurethane Foam

Rigid polyurethane foam is a kind of low-density microporous foam with closed-cell structure, which has low thermal conductivity and excellent performance, so it is considered a promising energy-saving material in the field of heat insulation and cold insulation.

Compared with other thermal insulation materials, PU foams have the characteristics of light-specific gravity, small thermal conductivity, low water absorption, oil resistance, alkali resistance, and acid resistance. PU foams are convenient for realizing rapid on-site spray foaming, which makes them ideal for various fields including construction, renovation, decoration sites, refrigerators, and electromobiles. The thermal insulation performance of PU foams mainly depends on the cellular structure, porosity, and matrix material of the foam. Foams with closed cells, fine cell size, and uniform cell distribution usually possess high thermal insulation properties. In addition, the spray PU foams are required to be flame retardant in many applications for the sake of fire risk management. Therefore, flame retardant agents are often added to the precursor reagents of the PU foam.

3.5.4 Thermal Insulation Aerogels

In the field of aerospace, the extreme environment of outer space poses many threats to the survival of astronauts, and personal protection is needed to protect their safety. To prevent space debris and solar radiation, Apollo-era astronauts used lunar suits made up of various materials, including Kevlar and Nomex. Kevlar has high mechanical ultimate tensile strength and modulus, providing bulletproof and fracture protection [50]. Nomex provides minimal bulletproof protection, but with high thermal resistance, it protects astronauts from extreme temperature differences in space [51, 52]. The thermal insulation of Nomex derives from its meta-aramid, kinked polymer backbone that inhibits polymer packing and results in slow phonon heat transport along unordered bonds throughout the fiber [52].

Nowadays, aerogels have been more and more used in our daily lives due to their excellent thermal insulating properties. Chen et al. [53] investigated the effect of addition of silica aerogel particles on the geopolymer foam. Four kinds of silica aerogel with different particle sizes were introduced to geopolymer foam aerogel render (GFAR) to improve their thermal insulation properties. It was found that large aerogel particles showed superior performance at the same loading content since large aerogel particles were more likely to be exposed to the cell surface of the GFAR. The optimum GFAR reached a high thermal conductivity of 0.133 W $(m\ K)^{-1}$ (Figure 3.21).

PI aerogels are often used for thermal insulation applications as well. Different from chemical or physical foaming approaches, the porous structure of PI aerogels is formed by phase separation from PI precursor solutions, and the solvents are normally removed by freeze drying or $scCO_2$ drying. For example, Zheng et al. [54] developed a PI aerogel via freeze-drying and thermal imidization. The developed PI aerogel possesses a highly cross-linked polymer network using p-phenylenediamine (PDA), biphenyltetracarboxylic dianhydride (BPDA), and the hexaaminotriptycene (HMT) as the cross-linker. As shown in Figure 3.22, the aerogels had a highly porous structure with a low density of 0.069 g cm^{-3}, a low volume shrinkage ratio of 10%, and an ultralow thermal conductivity of 25 mW $(m\ K)^{-1}$.

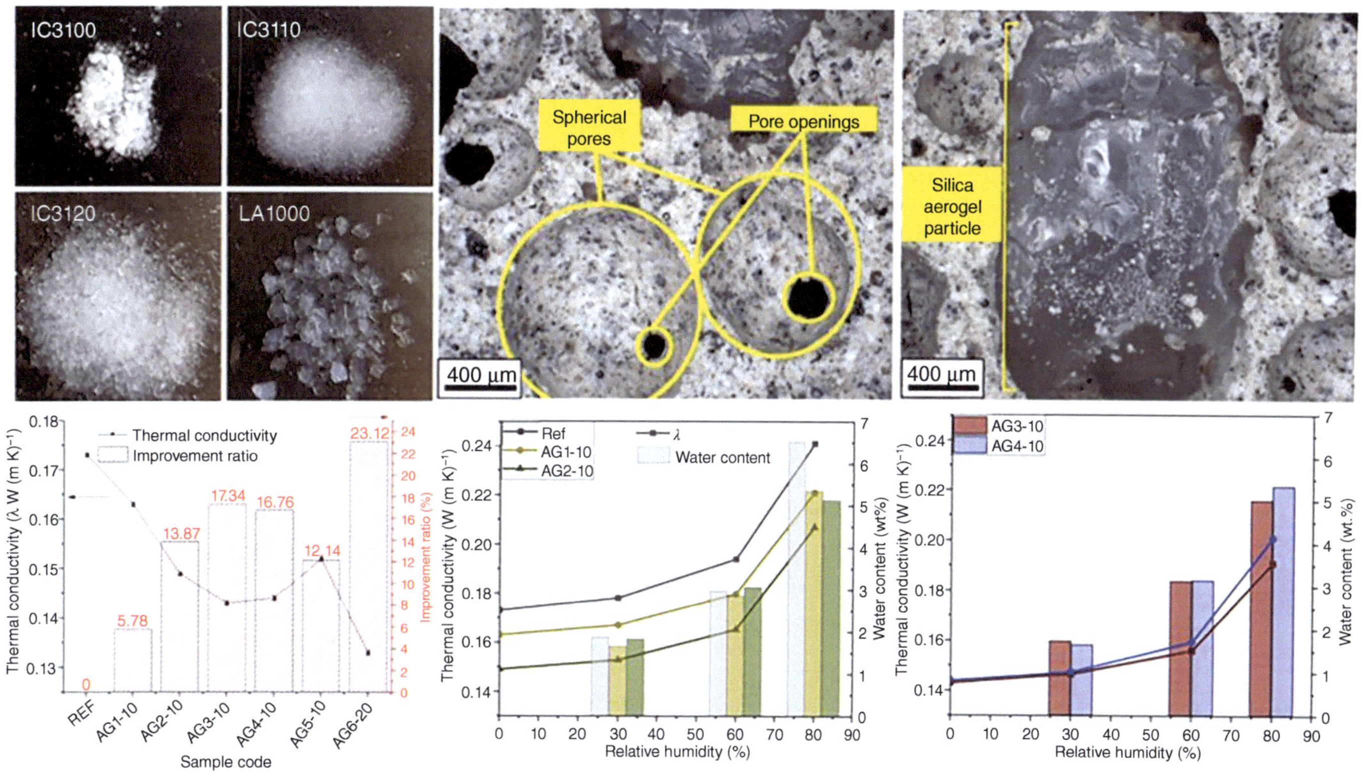

Figure 3.21 Four kinds of aerogel particles added to the GFAR; the morphology of the porous structure of GFAR and aerogel particles exposed to the cell surface; the thermal conductivity of different samples and the dependence on the relative humidity. Source: Chen et al. [53]/Elsevier/CC BY 4.0.

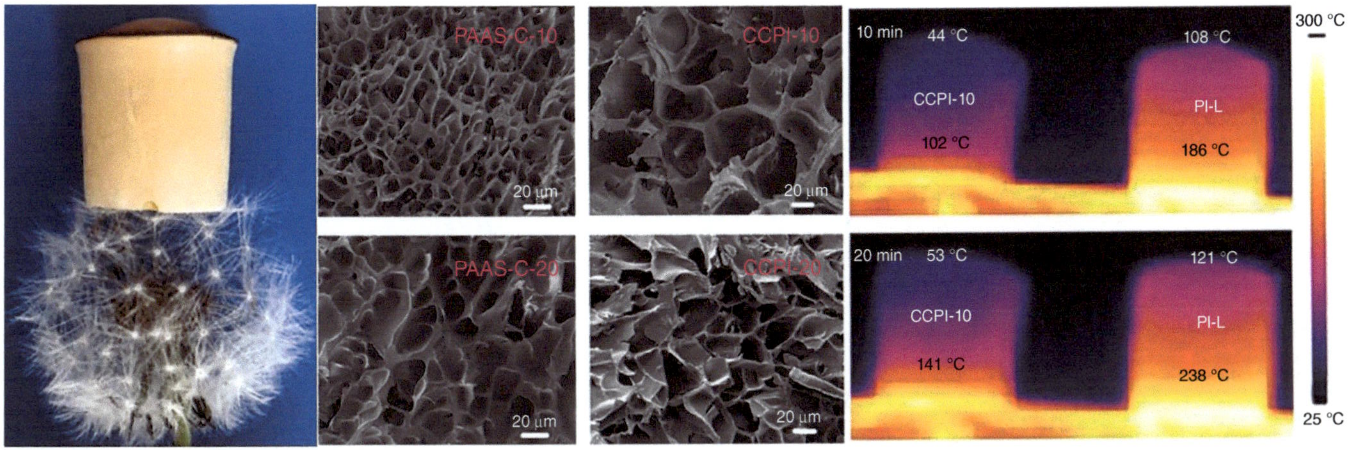

Figure 3.22 The photo and SEM images of PI aerogels synthesized by freeze-drying and thermal imidization and their thermal insulation properties. Source: Zheng et al. [54]/with permission of American Chemical Society.

References

1 Hu, F., Wu, S., and Sun, Y.J.A.M. (2019). Hollow-structured materials for thermal insulation. *Advanced Materials* 31: 1801001.

2 Azdast, T. and Hasanzadeh, R. (2021). Increasing cell density/decreasing cell size to produce microcellular and nanocellular thermoplastic foams: a review. *Journal of Cellular Plastics* 57: 769–797.

3 Pradeep, S.A. (2016). An investigation of bimodal cellular distributions via supercritical fluid assisted (SCF) foam injection molding. All Theses. Clemson University.

4 Wang, G., Wang, C., Zhao, J. et al. (2017). Modelling of thermal transport through a nanocellular polymer foam: toward the generation of a new superinsulating material. *Nanoscale* 9: 5996–6009.

5 Kuhn, J., Ebert, H.-P., Arduini-Schuster, M. et al. (1992). Thermal transport in polystyrene and polyurethane foam insulations. *International Journal of Heat and Mass Transfer* 35: 1795–1801.

6 Chavan, P.V., Rathod, P.V., Lee, J. et al. (2024). Active and passive modulation of solar light transmittance in a uniquely multifunctional dual-band single molecule for smart window applications. *Journal of Energy Chemistry* 88: 293–305.

7 Hoang, T., Truong, H., Han, J. et al. (2023). Room temperature roll-to-roll additive manufacturing of polydimethylsiloxane-based centrifugal microfluidic device for on-site isolation of ribonucleic acid from whole blood. *Materials Today Bio* 23: 100838.

8 Vafaei, B. and Ghahremaninezhad, A. (2023). Investigating the effect of hydrogel characteristics on the self-healing of cementitious materials. *Materials and Structures* 56: 177.

9 Zaharescu, T., Marius, B., Nicolescu, C.M., and Blanco, I. (2024). The contribution of silica nanoparticles to the stability of styrene-isoprene-styrene triblock copolymer (SIS). II. Energetic peculiarities. *Radiation Physics and Chemistry* 215: 111318.

10 Numan, Y. (2016). The review of some commonly used methods and techniques to measure the thermal conductivity of insulation materials. In: *Insulation Materials in Context of Sustainability* (ed. A. Amjad and A. Asaad). Rijeka: IntechOpen.

11 Sundarram, S.S. and Li, W. (2013). On thermal conductivity of micro-and nanocellular polymer foams. *Polymer Engineering and Science* 53: 1901–1909.

12 Xie, X., Yang, K., Li, D. et al. (2017). High and low thermal conductivity of amorphous macromolecules. *Physical Review B* 95: 035406.

13 Notario, B., Pinto, J., Solorzano, E. et al. (2015). Experimental validation of the Knudsen effect in nanocellular polymeric foams. *Polymer* 56: 57–67.

14 Forest, C., Chaumont, P., Cassagnau, P. et al. (2015). Polymer nano-foams for insulating applications prepared from CO_2 foaming. *Progress in Polymer Science* 41: 122–145.

15 Wang, G., Wang, L., Mark, L.H. et al. (2018). Ultralow-threshold and lightweight biodegradable porous PLA/MWCNT with segregated conductive networks for

high-performance thermal insulation and electromagnetic interference shielding applications. *ACS Applied Materials & Interfaces* 10: 1195–1203.

16 Tsuda, S., Yamaguchi, S., Kanamori, Y., and Yugami, H. (2018). Spectral and angular shaping of infrared radiation in a polymer resonator with molecular vibrational modes. *Optics Express* 26: 6899–6915.

17 Yu, H., Zhang, H., Buahom, P. et al. (2021). Prediction of thermal conductivity of micro/nano porous dielectric materials: theoretical model and impact factors. *Energy* 233: 121140.

18 Zhang, Z., Dai, X., Li, L. et al. (2021). Current status of research on the modification of thermal properties of epoxy resin-based syntactic foam insulation materials. *Polymers* 13: 3185.

19 Romesberg, F.E. and Debenedictis, M.A. (1989). Polystyrene foam containing carbon black. EP0372343A1, filed date 1989-11-27.

20 Gong, P., Wang, G., Tran, M. et al. (2017). Advanced bimodal polystyrene/multi-walled carbon nanotube nanocomposite foams for thermal insulation. *Carbon* 120: 43475.

21 Zhang, C., Zhu, B., and Lee, L.J. (2011). Extrusion foaming of polystyrene/carbon particles using carbon dioxide and water as co-blowing agents. *Polymer (London)* 52: 1847–1855.

22 Ling, J., Zhai, W., Feng, W. et al. (2013). Facile preparation of lightweight microcellular polyetherimide/graphene composite foams for electromagnetic interference shielding. *ACS Applied Materials & Interfaces* 5: 2677–2684.

23 Yang, Y., Gupta, M.C., Dudley, K.L., and Lawrence, R.W. (2005). Novel carbon nanotube–polystyrene foam composites for electromagnetic interference shielding. *Nano Letters* 5: 2131–2134.

24 Hansen, R.H. and Martin, W.M. (1965). Novel methods for the production of foamed polymers ii. nucleation of dissolved gas by finely-divided metals. *Journal of Polymer Science Part B: Polymer Letters* 3: 325–330.

25 Raquez, J.-M., Park, C.B., Jerome, C. et al. (2016). CO_2-blown microcellular non-isocyanate polyurethane (NIPU) foams: from bio- and CO_2-sourced monomers to potentially thermal insulating materials. *Green Chemistry* 18: 2206–2215.

26 Zhao, J., Zhao, Q., Wang, C. et al. (2017). High thermal insulation and compressive strength polypropylene foams fabricated by high-pressure foam injection molding and mold opening of nano-fibrillar composites. *Materials & Design* 131: 1–11.

27 Kaya, G.G. and Deveci, H. (2020). Synergistic effects of silica aerogels/xerogels on properties of polymer composites: a review. *Journal of Industrial and Engineering Chemistry* 89: 13–27.

28 Faghihian, H., Nourmoradi, H., and Shokouhi, M. (2014). Removal of copper (II) and nickel (II) from aqueous media using silica aerogel modified with amino propyl triethoxysilane as an adsorbent: equilibrium, kinetic, and isotherms study. *Desalination and Water Treatment* 52: 305–313.

29 Goren, K., Chen, L., Schadler, L.S., and Ozisik, R. (2010). Influence of nanoparticle surface chemistry and size on supercritical carbon dioxide processed nanocomposite foam morphology. *Journal of Supercritical Fluids* 51: 420–427.

30 Krishnamurthi, B., Bharadwaj-Somaskandan, S., Sergeeva, T., and Shutov, F. (2003). Effect of wood flour fillers on density and mechanical properties of polyurethane foams. *Cellular Polymers* 22: 371–381.

31 Liu, S., Eijkelenkamp, R., Duvigneau, J., and Vancso, G.J. (2017). Silica-assisted nucleation of polymer foam cells with nanoscopic dimensions: impact of particle size, line tension, and surface functionality. *ACS Applied Materials & Interfaces* 9: 37929–37940.

32 Zhang, C., Zhu, B., Li, D., and Lee, L.J. (2012). Extruded polystyrene foams with bimodal cell morphology. *Polymer* 53: 2435–2442.

33 Wang, K., Pang, Y., Wu, F. et al. (2016). Cell nucleation in dominating formation of bimodal cell structure in polypropylene/polystyrene blend foams prepared via continuous extrusion with supercritical CO_2. *Journal of Supercritical Fluids* 110: 65–74.

34 Huang, J.-N., Jing, X., Geng, L.-H. et al. (2015). A novel multiple soaking temperature (MST) method to prepare polylactic acid foams with bi-modal open-pore structure and their potential in tissue engineering applications. *The Journal of Supercritical Fluids* 103: 28–37.

35 Qu, Z.J., Mi, J.G., Jiao, Y. et al. (2019). Microcellular morphology evolution of polystyrene/thermoplastic polyurethane blends in the presence of supercritical CO_2. *Cellular Polymers* 38: 68–85.

36 Luo, G.Q., Gu, X.L., Zhang, J. et al. (2017). Microstructural, mechanical, and thermal-insulation properties of poly(methyl methacrylate)/silica aerogel bimodal cellular foams. *Journal of Applied Polymer Science* 134: 44434.

37 Cai, S., Xia, L., Xu, H. et al. (2022). Effect of internal structure on dynamically coupled heat and moisture transfer in closed-cell thermal insulation. *International Journal of Heat and Mass Transfer* 185: 122391.

38 Zhao, B., Zhao, C.X., Wang, C.D., and Park, C.B. (2018). Poly(vinylidene fluoride) foams: a promising low-k dielectric and heat-insulating material. *Journal of Materials Chemistry C* 6: 3065–3073.

39 Yang, C., Zhang, Q., Zhang, W. et al. (2021). High thermal insulation and compressive strength polypropylene microcellular foams with honeycomb structure. *Polymer Degradation and Stability* 183: 109406.

40 Wang, G., Zhao, G., Zhang, L. et al. (2018). Lightweight and tough nanocellular PP/PTFE nanocomposite foams with defect-free surfaces obtained using in situ nanofibrillation and nanocellular injection molding. *Chemical Engineering Journal* 350: 1–11.

41 Wang, G.L., Zhao, G.Q., Dong, G.W. et al. (2018). Lightweight, thermally insulating, and low dielectric microcellular high-impact polystyrene (HIPS) foams fabricated by high-pressure foam injection molding with mold opening. *Journal of Materials Chemistry C* 6: 12294–12305.

42 Kim, D.H., Lee, B.J., and Park, B.K. (2023). Influence of the foaming process on the burning behavior of the PET-PEN copolymer. *ACS Omega* 8: 19556–19566.

43 Wang, G., Zhao, J., Wang, G. et al. (2017). Low-density and structure-tunable microcellular PMMA foams with improved thermal-insulation and compressive mechanical properties. *European Polymer Journal* 95: 382–393.

44 Wang, G., Zhao, J., Mark, L.H. et al. (2017). Ultra-tough and super thermal-insulation nanocellular PMMA/TPU. *Chemical Engineering Journal* 325: 632–646.

45 Zhang, H.T., Fan, X.P., Chen, W.J. et al. (2022). A simple and green strategy for preparing flexible thermoplastic polyimide foams with exceptional mechanical, thermal-insulating properties, and temperature resistance for high-temperature lightweight composite sandwich structures. *Composites Part B: Engineering* 228: 109405.

46 Wang, L.L., Hu, A.J., Fan, L., and Yang, S.Y. (2013). Approach to produce rigid closed-cell polyimide foams. *High Performance Polymers* 25: 956–965.

47 Li, J., Zhang, G., Li, J. et al. (2017). Preparation and properties of polyimide/chopped carbon fiber composite foams. *Polymers for Advanced Technologies* 28: 28–34.

48 Yun, S., Sheng, X., Wang, S. et al. (2023). Preparation and properties of high-temperature-resistant, lightweight, flexible polyimide foams with different diamine structures. *Polymers* 15: 2609.

49 Liu, H., Wang, X., Mi, H.-Y. et al. (2024). Synthesis and fabrication of lightweight microcellular thermoplastic polyimide foams using supercritical CO_2 foaming. *Polymer* 290: 126548.

50 Tanner, D., Fitzgerald, J.A., and Phillips, B.R. (1989). The Keviar Story—an advanced materials case study. *Advanced Materials* 1: 151–156.

51 Gonzalez, G.M., Ward, J., Song, J. et al. (2020). Para-aramid fiber sheets for simultaneous mechanical and thermal protection in extreme environments. *Matter* 3: 742–758.

52 Chae, H.G. and Kumar, S. (2006). Rigid-rod polymeric fibers. *Journal of Applied Polymer Science* 100: 791–802.

53 Chen, Y., Klima, K.M., Brouwers, H.J.H., and Yu, Q.L. (2022). Effect of silica aerogel on thermal insulation and acoustic absorption of geopolymer foam composites: the role of aerogel particle size. *Composites Part B: Engineering* 242: 110048.

54 Zheng, S., Jiang, L., Chang, F. et al. (2022). Mechanically strong and thermally stable chemical cross-linked polyimide aerogels for thermal insulator. *ACS Applied Materials & Interfaces* 14: 50129–50141.

4

Acoustic Absorption Polymer Foams

4.1 Overview of Sound Absorption and Noise Reduction Foams

Noise is a series of sounds with different frequencies and intensities that combine irregularly and may cause anxiety or excessive intensity and harm health. Noise pollution has been regarded as one of the most serious environmental problems. Long-term exposure to noise pollution could cause psychological and physical diseases, often manifested as irritability, impatience, poor sleep, and other neurological diseases [1]. Means to control noise mainly include sound source structure vibration reduction and acoustic wave propagation suppression (e.g. sound insulation and absorption), that is, active control to identify the noise source and take actions on the transmission path, such as wrapping sound-absorbing materials [2].

Sound-absorbing materials reduce noise by transmitting energy and converting it into heat. Generally, sound-absorbing materials can be divided into porous sound-absorbing materials and resonant sound-absorbing materials. Porous sound-absorbing material is a kind of solid containing cavities and channels that construct a highly porous structure. The adiabatic processes would occur in porous materials when encountering high-frequency sound, resulting in heat loss due to friction as sound waves travel through irregular pores. At low frequencies, porous elastic materials absorb sound through energy loss caused by heat exchange, which is an isothermal process.

Porous sound-absorbing materials mainly include sound-absorbing fibrous materials, sound-absorbing granular materials, and sound-absorbing foams. Sound-absorbing foams can be divided into open-cell foams and closed-cell foams according to the opening and closing states of the cells. As a multifunctional material composed of solid phase and gas phase, polymer foams have been widely used in the industrial field in recent years because of their low density, low cost, and ease of fabrication, especially excellent acoustic absorption properties. The polymer foams produced using physical blowing agents are formed through the change of the physical form of the foaming agents, namely, through the expansion of compressed gas, liquid volatilization, or solid dissolution [3].

Functional Polymer Foams: Green Fabrication Methods, Performance and Applications,
First Edition. Haoyang Mi.

4.2 Fundamentals of Acoustic Absorption of Polymer Foams

4.2.1 Propagation and Absorption of Sound Waves

Noise reduction methods can be divided into passive noise reduction and active noise reduction. The difference between passive noise reduction and active noise reduction is determined by whether external energy needs to be applied in the process of noise reduction [4]. As illustrated in Figure 4.1, active noise control is effective in the reduction of sound waves at low frequencies, while passive noise control is effective in attenuating sound waves at high frequencies [5]. In practice, passive noise control is more cost-efficient and energy-efficient as it does not need any power source. Sound-absorbing polymer foams are generally passive noise reduction materials, which reduce noise by transmitting energy and converting it into heat. Thus, the investigation of the acoustic absorption material is mainly focused on passive noise absorption foams. Sound has the propagation characteristics of waves, as shown in Figure 4.1, when the sound is incident with a material, there are four types of waves including sound incidence, reflection, diffraction, and transmission waves [6].

When sound waves impact polymer foams, part of the energy is lost and converted into heat due to the damping by the air in the cells and the solid cell walls. The loss of sound waves in the propagation process is mainly in three ways, namely, viscous absorption, heat conduction absorption, and molecular relaxation absorption.

The viscous absorption mainly occurs in the adjacent medium, the particle motion state produces velocity gradient change, resulting in constant friction between particles, so that sound energy is converted into heat energy. Heat conduction and absorption refer to the existence of a temperature gradient in particles with uneven distribution in the medium, which causes heat exchange between adjacent particles, and part of the energy is converted into heat during the exchange process. Since the particle vibration and the sound wave propagation period in the medium are not synchronous, the phase difference between the two would cause the loss of sound energy. This process is referred to as molecular relaxation absorption.

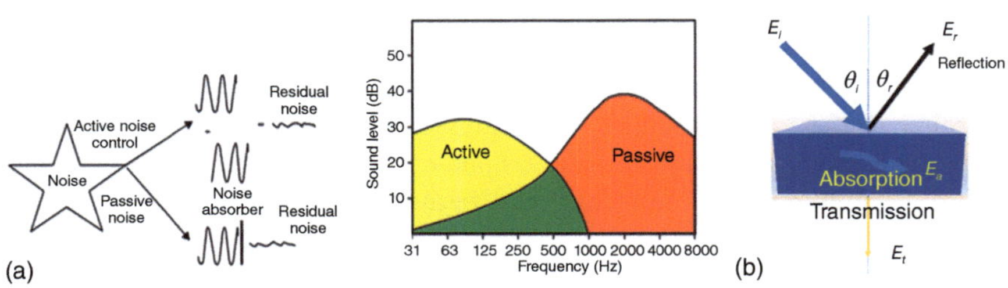

Figure 4.1 (a) Active and passive noise control mechanisms, frequency response region of the active and passive noise control methods, and (b) passive sound absorption process. Source: (a) Rajappan et al. [5]/with permission of Science Alert. (b) Adapted from Rajappan et al. [5].

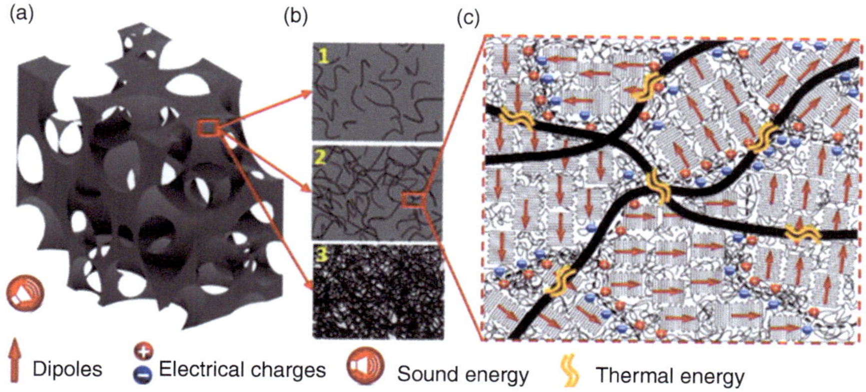

Figure 4.2 Schematic illustration of (a,b) the piezo-conductor hybrid foams with different concentrations of nanofillers, and (c) mechanisms of sound energy conversion to electricity with local piezoelectric effect and conversion of electrical energy to thermal energy from resistive loss. Source: Rahimabady et al. [7]/with permission of AIP Publishing LLC.

Under the alternating force of the sound wave, the material doped with piezoelectric particles can generate alternating voltage inside so that the energy of the sound wave can be converted into electric energy more efficiently. For polymer foams with damping and sound absorption properties, the change of strain lags behind the change of stress under the alternating sound wave, which would result in hysteresis and permanent deformation under the action of stress, thus, contributing to the energy consumption (Figure 4.2) [7].

4.2.2 Sound Absorption Principle and Models

The ratio between the absorbed sound energy and the sound energy incident on the surface of the material is called the sound absorption coefficient (α) (SAC), which is one of the most important and commonly used acoustic characteristics for sound absorption materials. Generally, α of a material is between 0 and 1. The larger the coefficient is, the better the sound absorption performance of the material will be. Rahimabady et al. demonstrated that the SAC of polyvinylidene fluoride (PVDF)/single wall carbon nanotube (SWCNT) composite foam could be enhanced by increasing the SWCNT content, and the α is higher at high sound frequencies, as shown in Figure 4.3a [7]. As illustrated in Figure 4.3b, a polymer foam is composed of micro-sized cells that are filled with air and solid matrixes acting as the cell walls and struts. The cells can take different geometries and can be either closed-cells or open-cells. The outer surface of polymer foams is the interface contact with the outside environment. When the sound wave incident on the polymer foam, the sound wave can enter the inner part of the foam through the microcells, causing the air in the cells to vibrate. Although the structure of polymer foams may be different, the sound absorption mechanism is similar.

The reduction of sound energy by the foam mainly owing to (i) air flows in the cell which produces friction with the cell walls so that the sound energy is converted

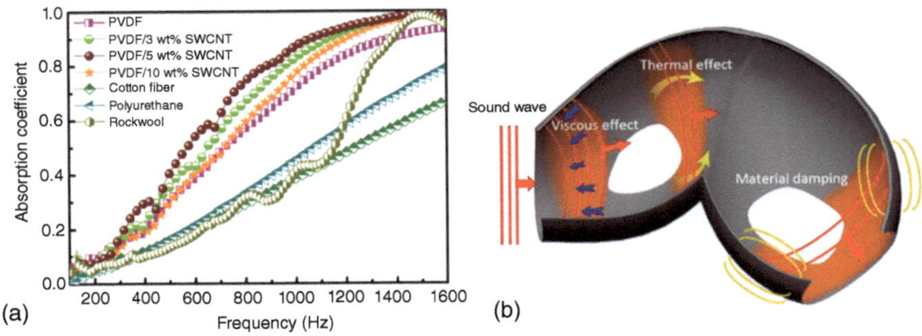

Figure 4.3 (a) Sound absorption coefficients of the PVDF/SWCNT piezo-conductor hybrid foams with different SWCNT concentrations, and (b) illustration for common sound energy dissipation mechanisms in a porous sound-absorbing material. Source: Rahimabady et al. [7]/with permission of AIP Publishing.

into heat and then dissipated; (ii) The air in the cells is intermittently compressed and released, resulting in energy reduction through the process of energy conversion; (iii) Sound energy can be converted into mechanical energy and heat through the resonance of the cell walls. Therefore, sound-absorbing polymer foams typically have two kinds of sound-reduction functions, namely, structural damping in the fluid domain (i.e. air) and material damping in the solid domain (i.e. polymer matrix).

For structural damping, due to the complexity of micro-scale porous structure, acoustic wave propagation in porous media is mainly dissipated in the form of viscous friction on the interconnected pores and heat exchange on the solid–fluid boundary. In terms of material damping, sound waves transmitted from air to solids are attenuated in vibration modes by molecular friction [8]. Structural damping and material damping are sometimes called “viscous inertia and thermal damping” and “viscoelastic frame damping,” respectively. Viscous inertia, also known as viscous dissipation, is expressed by the coefficient of viscosity. Damping usually refers to the ability of a material to lose vibration energy when subjected to vibration deformation. The greater the damping, the more energy input to the system can be lost in a short time, which means that the time for the system to re-rest from an excited vibration is shorter [9].

It is found that the cell size, cell density, porosity, thickness, and flow resistance of sound-absorbing polymer foams are important factors affecting their sound-absorbing performance. In order to regulate the sound absorption performance of polymer foams, it is necessary to establish an accurate sound absorption model to realize the rational design of porous structures of the polymer foams.

At present, the sound absorption models of porous materials are divided into theoretical sound absorption model, empirical sound absorption model, and semiempirical sound absorption model [10]. The theoretical sound absorption model explains the physical process of sound wave dissipation in porous materials. By studying the propagation process of sound waves in the cylindrical through hole and considering the air viscosity and heat conduction effect based on the cylindrical through hole,

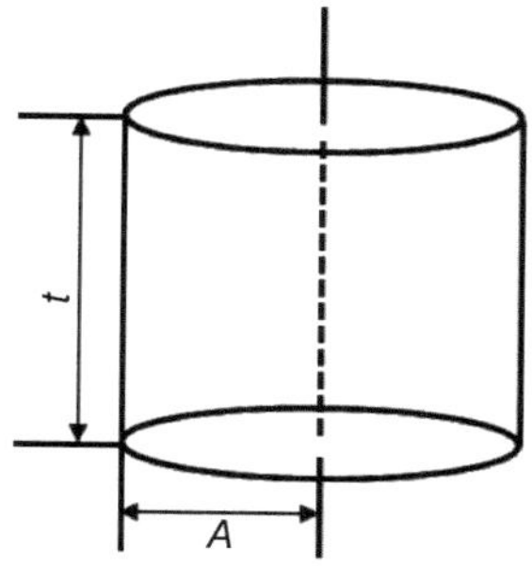

Figure 4.4 Schematic of the sound absorption circular tube model.

a sound absorption model of the circular tube is proposed combined with the porosity, average aperture, and material thickness.

In the sound absorption circular tube model, the radius of the circular tube is set as A and its length as t, both of which are far smaller than the wavelength of the sound wave. In the cylindrical coordinate system, the tube axis is the z-axis, and the distance from a point in the tube to the axis is the diameter coordinate r. The air in the tube is considered to be a large number of cylindrical layers as illustrated in Figure 4.4 [11].

According to Euler's equation of motion and continuity, the equation of motion of air in a tube can be obtained as follows:

$$j\omega\rho_0 u - \frac{\eta}{r}\frac{\partial}{\partial r}\left(r\frac{\partial u}{\partial x}\right) = \frac{\Delta p}{t} \tag{4.1}$$

where $\eta = 1.85 \times 10^{-5}$ κg s^{-1} m^{-1} is the coefficient of viscosity for air (at 15 °C), ρ_0 is the air density, ΔP is the sound pressure difference at both ends of the microtubule, u is the particle velocity of air along the axial direction, r is the radial diameter, t is the tube length, d is the tube diameter. The angular frequency of sound wave vibration is $\omega = 2\pi f$. The complex value density of the sound wave in the tube can be obtained as:

$$\rho = \rho_0\left[1 + \left(3^2 + \frac{k^2}{2}\right)^{-\frac{1}{2}} + \frac{8\eta}{jw\rho_0 A^2}\left(1 + \frac{k^2}{32}\right)^{\frac{1}{2}}\right] \tag{4.2}$$

In a round tube, the above equation can be simplified to

Density:

$$\rho = \rho_0 X \tag{4.3}$$

Specific acoustic impedance:

$$Z = \sqrt{K_T\rho} = \rho_0 c_0 X^{\frac{1}{2}}\left(\gamma - (\gamma - 1)\frac{1}{X}\right)^{-\frac{1}{2}} \tag{4.4}$$

Modulus of compression:

$$K_T = P_0\left(1 - \frac{\gamma - 1}{\gamma}\frac{1}{X}\right)^{-1} \tag{4.5}$$

Sound velocity:

$$c = \sqrt{K_T/\rho} = c_0(\gamma X - \gamma - 1)^{-\frac{1}{2}} \tag{4.6}$$

In formula,

$$k = d\sqrt{\frac{\omega\rho_0}{4\eta}} \tag{4.7}$$

The sound absorption circular tube model assumes the cells in the foam are tubular holes. However, there are a large number of open and closed cells with a range of cell diameters in a polymer foam, and the cells may have different geometries. Thus, a structural factor is introduced into the circular tube sound-absorbing model for correction. The structure factor is a parameter determined by the micro-morphology of the cells in the foam, which makes the modified model more accurate [12].

The typical empirical model is the Delany-Bazley model built by a large number of tests and experience summary of wool samples with different densities. The basic principle of the model is the power law relation, and the Z_C and γ are related to the static flow resistance of porous materials that are expressed by the following equations [13].

$$Z_c(\omega) = \rho_0 c_0 \left[1 + 0.0571(\rho_0 f/\sigma)^{-0.754} - i0.087(\rho_0 f/\sigma)^{-0.732}\right] \tag{4.8}$$

$$\gamma = \frac{\omega}{c_0}\left[1 + 0.098(\rho_0 f/\sigma)^{-0.7} - i0.189(\rho_0 f/\sigma)^{-0.595}\right] \tag{4.9}$$

$$z_s = z_c \coth(-ik_c l_s) \tag{4.10}$$

In the formula, ρ_0 is air density; ω is the circular frequency; c_0 is the speed of sound in a porous environment; f is the frequency; σ is the static flow resistance. Empirical models usually consider porous materials from a macroscopic perspective, thus ignoring the wave propagation characteristics in the pores. However, the introduction of the Z_c and γ values can provide an accurate budget of orders of magnitude, so the empirical model is still widely used.

The semiempirical sound absorption model combines the advantages of the theoretical model and the empirical model. The effective density ($\rho(\omega)$) and bulk modulus ($K(\omega)$) are calculated by combining the theoretical sound absorption model proposed by Zwikker and Kosten and the experimental tests [9, 14]. The Johnson–Champoux–Allard (JCA) model for rigid porous materials is the classical semiempirical sound absorption model. It gives the formula for calculating the $\rho(\omega)$ and $K(\omega)$ of the porous material by defining macroscopic acoustic parameters related to material microstructure geometry, fluid motion, and heat conduction [14, 15].

$$\rho_e(\omega) = \frac{\alpha_\infty \rho_0}{\phi}\left(1 + \frac{\sigma\phi}{i\omega\rho_0\alpha_\infty}\sqrt{1 + \frac{4i\alpha_\infty^2 \eta\rho_0\omega}{\sigma^2\Lambda^2\phi^2}}\right) \tag{4.11}$$

$$K_e(\omega) = \frac{\gamma p_0}{\phi}\left(\gamma - (\gamma - 1)\left(1 + \frac{8\eta}{i\Lambda'^2 N_{pr}\omega\rho_0}\sqrt{1 + \frac{i\rho_0\omega N_{pr}\Lambda'^2}{16\eta}}\right)^{-1}\right)^{-1} \tag{4.12}$$

where η is the viscosity of air, ρ_0 is the air density, γ is the ratio of the specific heat capacity, ω is the angular frequency, $\dot{N}_{pr}$ is the Prandtl number, and P_0 is the atmospheric pressure.

The characteristic impedance Z_c, the wavenumber k, and the corresponding surface impedance Z_s are determined as:

$$Z_c = \sqrt{K_e(\omega) \cdot \rho_e(\omega)} \tag{4.13}$$

$$k = \omega\sqrt{\rho_e/K_e} \tag{4.14}$$

$$Z_s = -jZ_c \cot(k_c d) \tag{4.15}$$

where d is the board thickness, and Z_s is the surface impedance. Finally, the SAC could be calculated by Eq. (4.15):

$$\alpha = 1 - \left| \frac{Z_s - \rho_0 c_0}{Z_s + \rho_0 c_0} \right|^2 \tag{4.16}$$

where c_0 represents the impedance of the air.

The JCA model can describe the acoustic viscosity and thermal reflection characteristics of porous materials with arbitrarily shaped pores. The model is widely used in acoustic research and engineering owing to its accuracy in the prediction of acoustic properties.

Based on the JCA equivalent fluid model, Park et al. investigated the sound absorption properties of a novel bimodal foam structure prepared from polylactic acid (PLA) and polyethylene glycol (PEG). The inversion method based on the JCA model and impedance tube normal incidence absorption coefficient is used to estimate the curvature and characteristic length. The results (Figure 4.5) show that the inversion results are in good agreement with those of direct measurement of normal incidence absorption coefficient [16].

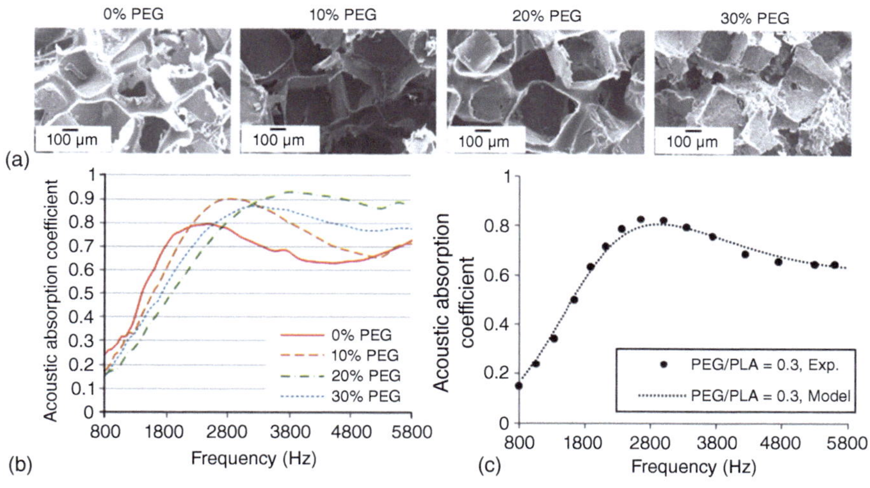

Figure 4.5 (a) SEM micrographs of PLA foams with different percentages of PEG, (b) sound absorption coefficient versus frequency for different foams, and (c) the acoustic modeling results of PLA/PEG foam.

4.3 Characterization and Influencing Factors for Sound

4.3.1 Characterization of Sound Absorption Properties

4.3.1.1 Sound Absorption Coefficient (α) Measurement

The SAC (α) is the most important characteristic of the sound-absorbing material. When sound waves strike a surface, the total sound energy can be expressed as follows:

$$E_i = E_r + E_a + E_t \tag{4.17}$$

where E_i is the total incident sound energy, E_r is the reflected sound energy, E_a is the absorbed sound energy, and E_t is the transmitted sound energy.

The α is the ratio of E_a to E_i on the material surface calculated using the following formula:

$$\alpha = \frac{E_a}{E_i} \tag{4.18}$$

The sound pressure is measured by two microphones at two different locations close to the sample. α is then calculated according to ASTM E1050-12 or ISO10534 by calling a transfer function based on the pressure difference between the two microphone positions [17]. As illustrated in Figure 4.6, in the SAC measurement, materials are usually tested at 125, 250, 500, 1000, 2000, and 4000 Hz audio frequencies, and the arithmetic mean value is calculated as the average SAC ($\overline{\alpha}$). In general, the materials with a $\overline{\alpha}$ above 0.2 is recognized as a sound-absorbing material. The material with $\overline{\alpha}$ above 0.5 is called efficient absorption material, which is identified as an ideal absorption material.

Since the SACs in different frequency bands could be significantly different, the separate measurement of α in a specific absorption frequency band is often used for accurate reflection of the sound absorption properties in particular circumstances. The impedance tube method based on the relationship between the incident sound

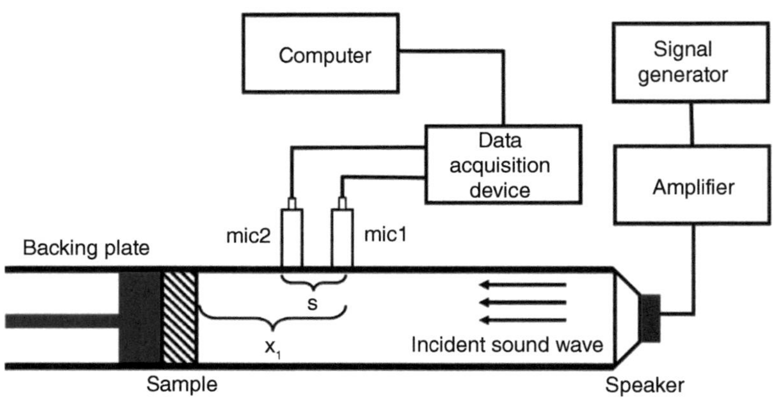

Figure 4.6 Schematic diagram of sound absorption measurement system. Source: Chanlert and Ruamcharoen [17]/IOP Publishing/CC BY 3.0.

and the reflected sound on the surface of the sample is extensively used to measure SACs of acoustic materials as a function of frequency. Sometimes, the average SAC is called noise reduction coefficient (NRC), which is the average sound absorption of the absorption coefficients at all frequencies.

4.3.1.2 Porosity Measurement

Porosity is the ratio of the volume or weight occupied by the pores of a porous material. It is usually calculated by the following formula:

$$V_f = 1 - \frac{\rho_f}{\rho_s} \tag{4.19}$$

where V_f refers to the porosity of the foam, ρ_f is the bulk density of the foam, and ρ_s is the solid density of the polymer matrix. The V_f usually takes the average of at least three measurements.

4.3.1.3 Cell Diameter, Cell Density, and Open-Cell Content

Cell diameter and density are the basic parameters used to statistically describe cellular morphology. The cell diameter is usually measured from the SEM image and the average value usually requires at least 50 cells from different regions. The cell density can be calculated from the number of cells in a unit area using the following equation:

$$\mathit{Cell\ density} = \left(\frac{n}{A}\right)^{3/2} \tag{4.20}$$

where n is the number of cells in the micrograph image, A is the area of the micrograph, which is the ratio of the density of the solid material and foamed sample.

The content of open cells in a foam is often measured from the SEM image and calculated by the percentage of open cells to the total amount of cells in the image. In addition, the open-cell content of foams can be measured using a gas pycnometer following the ASTM: D6226-10.

4.3.1.4 Tortuosity α_∞ Measurement

Tortuosity (α_∞) is an important parameter for determining sound absorption properties of the materials. It is the measure of the elongation of the passageway through the pores, compared to the thickness of the sample (Figure 4.7).

It can be defined as the ratio of the square of the actual length (L) of pores to the body length (l) of the material as Eq. (4.21):

$$\alpha_\infty = \frac{L}{l} \tag{4.21}$$

Allard [14] uses ultrasonic velocity measurement to deduce the analytical relationship between the phase velocity c_b of sound waves in the porous media and the speed of sound in air c_0, which is simply written as:

$${\alpha_\infty}^2 = \left(\frac{c_0}{c_b}\right)^2 (1 - 2\varphi) \tag{4.22}$$

where φ is the loss angle and can be determined from the damping of the pulse.

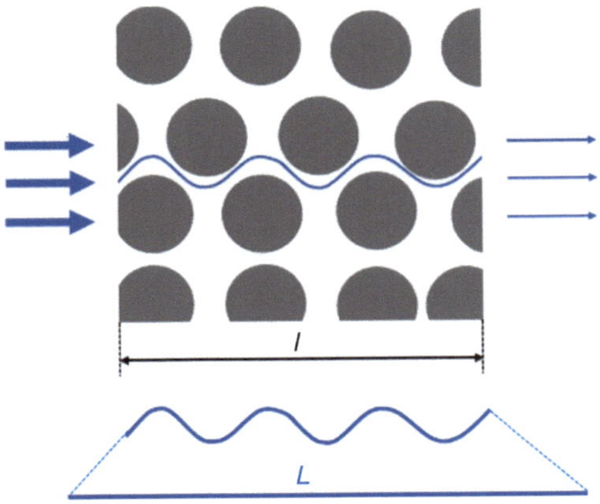

Figure 4.7 Schematic drawing of the effect of tortuosity on sound transmission.

4.3.1.5 Airflow Resistance Measurement

The airflow resistivity is used to evaluate the ability of air can pass through a testing porous material, which is influential to the sound absorption property of the material. It is usually measured with the test system of flow resistance instrument adopting the water tank method in direct current (DC) method as shown in Figure 4.8 [18]. The differential pressure levels above and beneath the specimen surface are measured by controlling the transmit time of one-way flow through the cylindrical tube. The airflow resistance (R) is expressed in Pa s m^{-3} and is calculated by the following equation:

$$R = \Delta p / qv \tag{4.23}$$

where Δp is the pressure difference on both sides of the specimen, qv is the volume velocity of airflow through the specimen.

4.3.2 Factors Affecting Sound Absorption Performance

4.3.2.1 Effect of Cellular Morphology

The microstructure of a polymer foam is the main factor affecting the sound absorption performance. The parameters characterizing the cellular morphology include porosity, cell diameter, cell density, and open-cell content. Porosity is the ratio of the volume or weight occupied by the pores of a porous material. When sound waves propagate inside the foam material, porosity is always considered a major factor.

Usually, the higher the porosity of foam, the larger the specific surface area, the greater the friction effect between air and pore wall, the better the sound attenuation, and the better the sound absorption performance will be. Increasing the porosity of polymer foam, the 3D porous structure would be more complex, and the sound wave would have a complex reflection during the propagation through the foam, which is beneficial for dissipating the sound energy. However, when investigating the effect of porosity, the cell diameter and cell morphology need to be considered at the same time, since a foam with a large cell size and high porosity would have a low flow resistance, which means that the sound energy transmitted through the foam (E_t) would be high, and the absorbed sound energy (E_a) would be low.

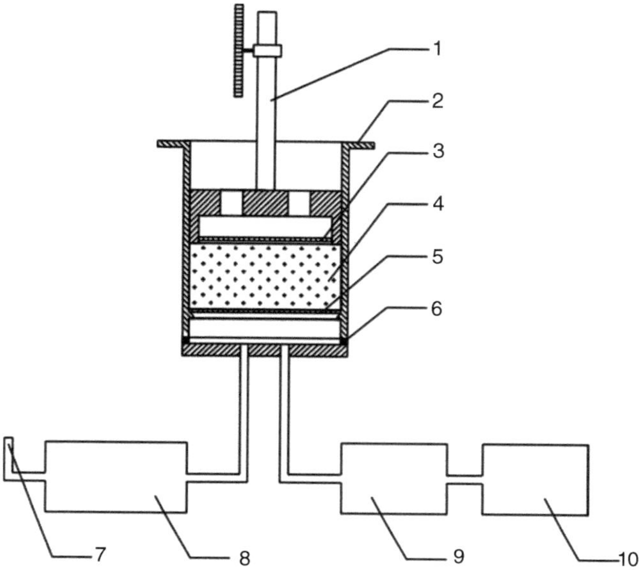

Figure 4.8 Airflow resistance measurement equipment illustration: (1) piston with thickness indicator, (2) measure specimen cylinder barrel, (3) grid or perforation plate, (4) specimen, (5) grid or perforation plate for supporting the specimen, (6) seal ring, (7) opening to the atmosphere, (8) differential pressure measuring equipment, (9) flow meter, and (10) air supply. Source: Adapted from ISO [18].

Generally, the sound absorbency of foam materials will increase when the porosity increases up to 7%. However, when the porosity value exceeds 70%, the value of absorption usually remains constant. Other structural factors such as cell size, cell-open rate, and cell density will play major roles. Enough pores on the surface of the foam will allow sound waves to penetrate the porous material for maximum energy dissipation [19]. Mosanenzadeh et al. [20] fabricated PP and PLA foams with different porosity and open-cell content and they found that the foams with a high open-cell content of around 93% showed the highest acoustic absorption coefficient. They also found that the absorption is poor at frequencies around 1000 Hz and lower when the thickness of the foam is limited to around 15 mm. They also verified the availability of the JCA model in the simulation of acoustic absorption performance.

Doutres et al. investigated the effect of closed cell content on the acoustic behavior of PU foams. The JCA porous model was used to estimate the sound absorption behavior of the PU foams. It was found that the presence of cell membranes increased the airflow resistivity, tortuosity, and the ratio between the thermal and viscous characteristic lengths. However, the thermal and viscous characteristic lengths was decrease with the increase of cell membranes. This feature made the foam with high closed cell content shown improved sound absorption efficiency at low frequencies, while their performances were reduced at high-frequency bands (Figure 4.9) [21].

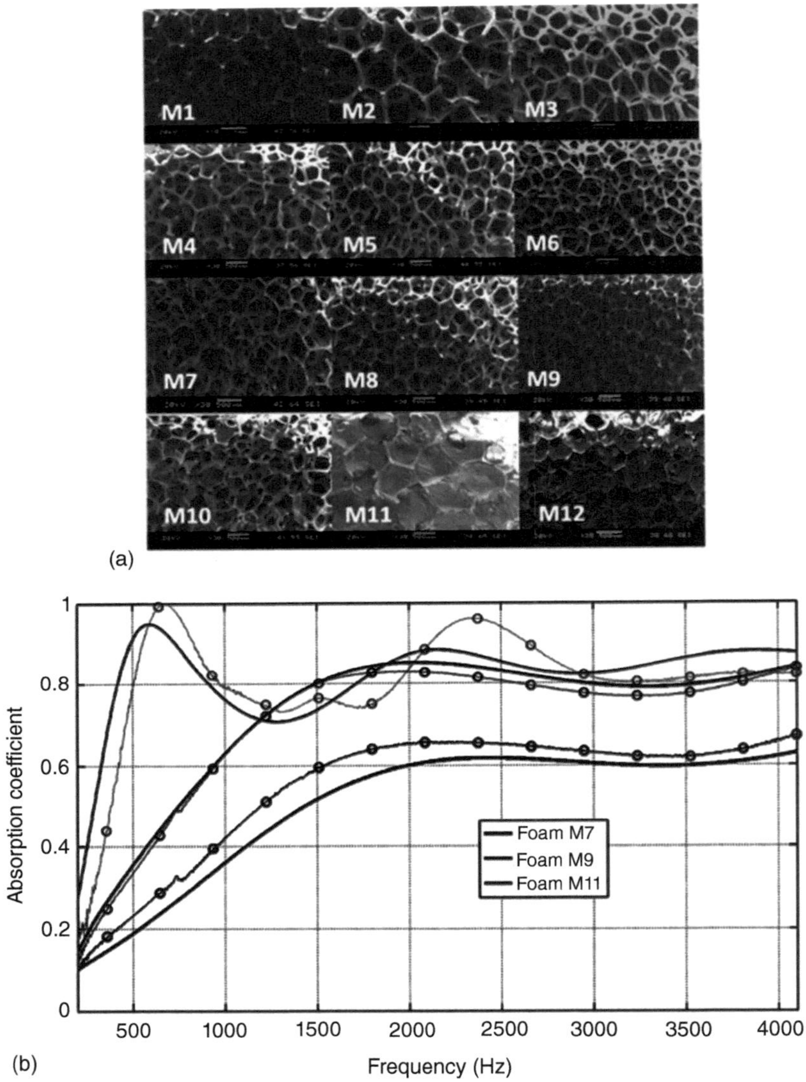

Figure 4.9 (a) SEM images of PU foams with different open-cell content, and (b) the absorption coefficient of PU foams with different open-cell content. Source: Doutres et al. [21]/with permission of AIP Publishing LLC.

4.3.2.2 Macro Shape and Geometry of Polymer Foam

The geometry of polymer foam has a significant influence on the sound absorption performance of polymer foam. Beranek investigated the sound absorption properties of five types of structures as silencer interior decoration structures, namely, (i) lamellar structures, (ii) lamellar structures, (iii) blanket layer structures, (iv) packed and semi-rigid pyramid structures, (v) wedge structures, and (vi) exponential conical pyramid and wedge structures [22]. They found that the most effective decoration was the wedge-shaped construction, which achieved the minimum total absorber depth required and absorbed at least 99% of the normal incident acoustic energy [23]. Since then, wedge structure has become the classic and most widely used structure form.

Sound-absorbing wedge includes concave and convex structures. The length and geometric arrangement of the wedge can be changed according to the needs. Sound waves are transmitted into the wedge structure and absorbed efficiently over a wide frequency range according to the gradient principle. At the top of the wedge plane, the main body of the sound absorption is very small, and the rest is air. The acoustic impedance increases gradually from near-air impedance to porous material impedance, and the relative change is not obvious [24]. In the case of a convex structure, the SAC may be different depending on the position of the microphone as illustrated in Figure 4.10 [25]. Therefore, it is desired to reduce the size of the convex wedges in practice to minimize this effect.

Sun and coworkers [26] investigated the effect of sharp holes of polyimide (PI) foams on the sound absorption properties. As shown in Figure 4.11, the concave wedge-shaped cavity structure increases the proportion of sound waves absorbed into the material compared to the foam material without the wedge-shaped cavity structure. The impedance of air acoustic medium and acoustic material matches well due to the gradual transition of sound waves in the split-structured PI foam. They also found that the SAC reached around 0.9 from 900 to 6300 Hz for the slip-structured PI foam with high pore distance, which almost eliminated the resonance sound absorption characteristics and ensured a high-efficiency sound absorption behavior of the foam.

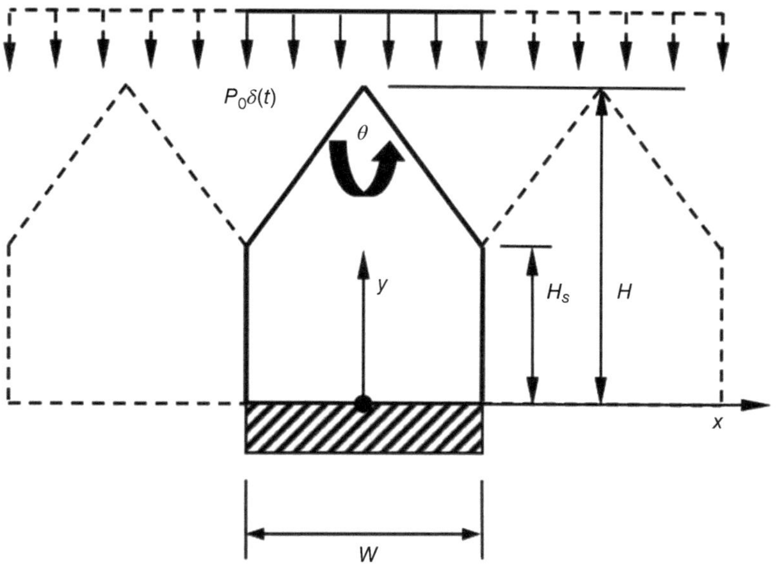

Figure 4.10 Illustration for acoustic wedge subjected to an impulsive pressure loading. Source: Tsay and Yeh [25]/with permission of Sage Publications.

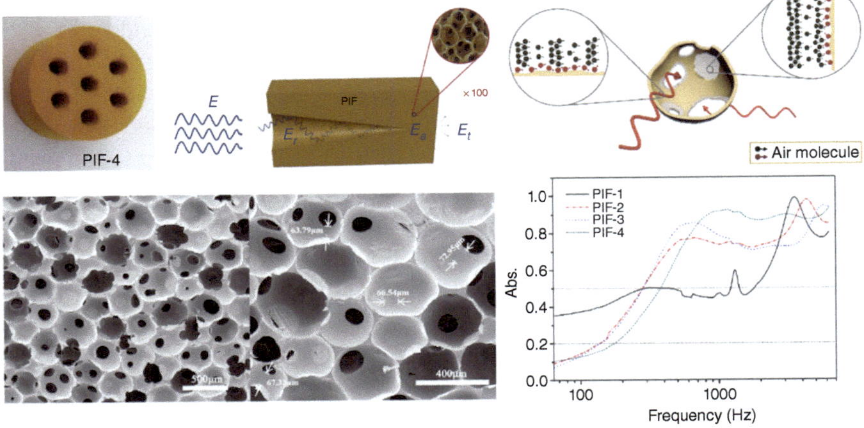

Figure 4.11 The sound absorption mechanism of slip-structured wedge cavity made of highly porous PI foam, the microstructure of the PI foam, and the acoustic absorption coefficient of PIF with increasing pore distance. Source: Ren et al. [26]/with permission of Elsevier.

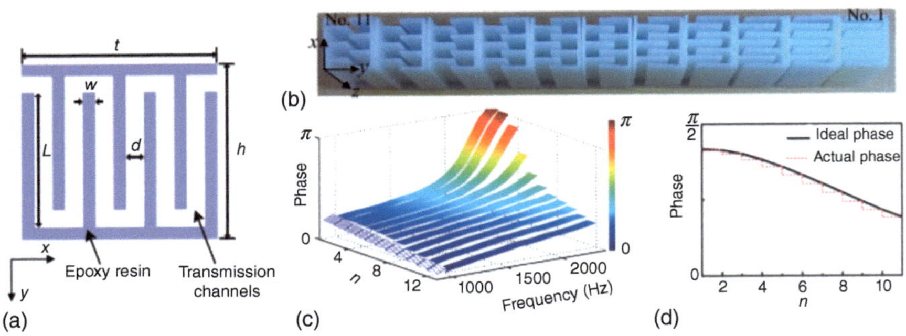

Figure 4.12 (a) Schematic cross-sectional illustration of a unit cell of the metasurface, (b) photograph of a realistic metasurface consisting of 11 units, (c) transmitted phase changes through the 11 unit cells as a function of frequency, and (d) designed gradient phase profile of the metasurface. Source: Yuan et al. [27]/with permission of IOP Publishing.

Cheng and coworkers [27] developed a coiling-up acoustic metasurface to manipulate sound radiation patterns. As shown in Figure 4.12a and b, the metasurface, which is fabricated by epoxy resin frames via 3D printing, has a zigzag transmission channel of sound waves isolated from adjacent cells. The full-wave simulation results (Figure 4.12c) indicated that the transmission phase shifts from 36.7° at 1000 Hz to 180° at 2400 Hz for the central unit cell, while the phase only shifts from 0° to 42.8° over the same frequency range for the edge unit cell. The metasurface could generate secondary sound sources with ideal gradient phase shift and showed the ability of controlling sound radiation such as by cylindrical-to-plane-wave conversion, plane wave focusing, and effective tunable acoustic negative refraction (Figure 4.12d).

4.3.2.3 Resistance of Airflow

Flow resistance represents the steady-state flow resistance through the foam material, which is affected by the foam cell size, cell density, porosity, and cell-opening rate. The sound absorption performance of the material can be assessed by the resistance of airflow. When the flow resistance is low, the sound wave flows through the foam material without slowing down, the internal friction and viscous force are low, thus, the loss of sound energy would be low, which corresponds to low acoustic absorption. In the case of high flow resistance, the ability of sound waves to enter the foam material is reduced, and they are either reflected back from the material or forced into the structure [28]. Due to its great influence on sound absorption performance, to obtain foam materials with good sound absorption performance, the airflow resistance of foam materials should be selected in a reasonable range.

4.3.2.4 Tortuosity Factor

Tortuosity is a dimensionless quantity that considers the curved fluid path through the foam material, through the ratio of the actual path length of the foam material to the linear path length, ranging from 1 (low-density fiber material) to 10 (partial mesh foam), the higher the curvature, the greater the sound attenuation, the better the sound absorption performance.

Liu et al. [29] developed a semi-open cellular network by coating a melamine foam with graphene oxide (GO) and carbon nanotube (CNT). The composite foam exhibits an approximately 100% enhancement in SAC in the band gap between 250 and 1600 Hz compared to the pristine melamine foam due to the synergistic effects of optimized air-flow resistance, tortuosity against sound wave propagation, smaller viscous characteristic lengths, internal reflection, and interfacial damping. As illustrated in Figure 4.13a–c, high flow resistance, tortuosity and interfacial damping are the three major factors to enhance sound absorption performance. Generally, an open and semiopen cellular structure with a smaller cell size, lower porosity, larger sheet coverage and greater thickness induce higher air-flow resistance and vice versa. The composite foam with higher GO and CNT loading could obstruct sound wave transmission by increasing the tortuosity. In addition, the semi-open cellular structure may cause more complex sound wave reflection and greater attenuation compared with closed cells (Figure 4.13d). Thus, moderate cell-opening rate is favorable for enhancing the sound absorption performance.

4.4 Types of Acoustic Absorption Foams

4.4.1 Sound Absorption Ceramic

Ceramic foam is a class of porous inorganic materials, of which the property is related not only to the composition of the material species, but also closely to the porosity, pore structure, pore size, and pore distribution. Ceramic foams typically have the advantage of being low-cost, nontoxic, tasteless, nonflammable, waterproof, anticorrosion, high stiffness, and high strength, which makes them

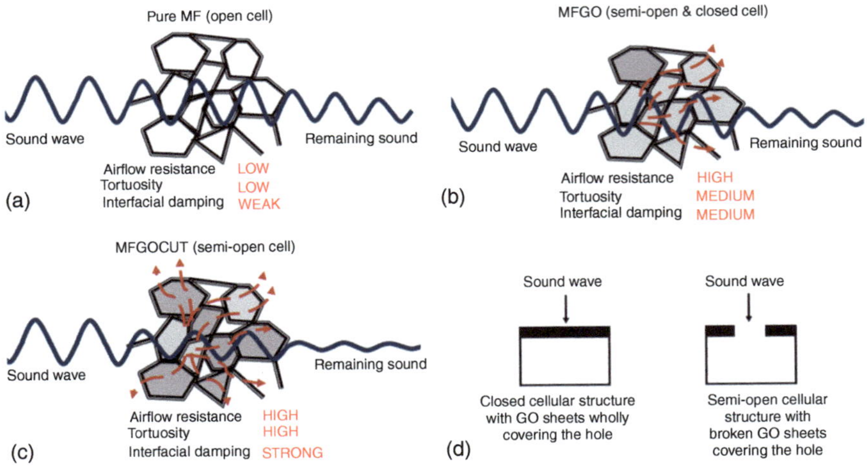

Figure 4.13 Acoustic propagation through composite foams with: (a) low airflow resistance, tortuosity, and weak interfacial damping, (b) high airflow resistance, medium tortuosity, and medium interfacial damping, and (c) high airflow resistance, high tortuosity, and strong interfacial damping, and (d) schematic of difference between the closed cellular structure and semi-open cellular structure as sound wave squeezes in the cell. Source: Liu et al. [29]/with permission of Royal Society of Chemistry.

extremely suitable for noise reduction in extreme conditions. In addition, many ceramic foams can serve as structural materials and heat-insulating materials while they are used as sound absorbers, which renders them multifunctional properties.

In the preparation of ceramic foams, a porous network is typically created in the low-temperature synthesis step, which constitutes the skeleton of the ceramic foam for the porous network generated in the calcination step. The cell size of ceramic foams could range from microporous and/or mesoporous to microporous and the gas bubbles in the ceramic precursor are usually generated using foaming agents, mechanical stirring, or porogens [30].

Liu et al. [31] mixed natural zeolite powder with binder, additive agents, and pore-forming agents to produce ceramic foam by molding, drying, and sintering. They investigated the effect of pore size on the sound absorption performance. Pore-forming agents of different sizes (e.g. 4–5 mm and 1–2 mm) were mixed with the natural zeolite powder to produce ceramic foams with relatively large pores and small pores. It was found that the absorption performance of the ceramic foam product with relatively large pores is superior to that with relatively small pores in the case of the approximately same porosity. For the ceramic foams with the same pore size, the average absorption coefficient was increased with the increase of porosity.

Du et al. [32] prepared Si_3N_4 ceramic foams with different porosity, pore size, and thickness. They investigated the effects of structural parameters on the sound absorption properties of the highly porous Si_3N_4 ceramic foams. The ceramic foam had a uniform cellular structure, and the cell surface was composed of microsized crystals. It was found that ceramic foams with relatively high porosity and small pore size possess high sound absorption in almost the entire frequency. The sound speed and acoustic impedance of the Si_3N_4 ceramic foams were decreased with the increase in porosity. In addition, the thicker samples would increase the SAC in low frequency by shifting the sound absorption peak to a lower frequency. They further inserted an air gap in the ceramic foams and realized selectively filtration of low- or medium-frequency sound waves.

4.4.2 Sound Absorption Metallic Foam

Metallic foams having open or semi-open cells are also used as sound-absorbing materials. Since they are highly conductive, metallic foams often serve as electromagnetic interference (EMI) shielding materials while they are used as a sound absorption materials. In contrast to elastic foams, the relatively large stiffness of the metal skeleton restricted the direct mechanical damping of sound waves by the metal itself. Therefore, the sound absorption mechanism for metal foam is generally realized (i) by viscous losses as the pressure wave pumps air in and out of cavities in the absorber, (ii) by thermal–elastic damping, and (iii) by Helmholtz-type resonators [33].

Yu et al. [34] fabricated functionally graded foam made of Inconel 625 superalloy using open-cell PU foams as a template through the template replication method. They fabricated alloy foams with different cell sizes and assembled these foams in different ways to investigate the effect of the arrangement of foams with different cell sizes on sound absorption properties. The fabricated metallic foam showed a highly porous structure with micro bumps presented on the foam skeleton. They found that placing the foam layer of smaller pore size nearer to the sound source is beneficial for increasing the SACs by resonance-like effects, and the sound absorption performance can be tuned by changing the arrangement of the foams with different pore sizes. With the manipulated resonance-like effects, the sound absorption performance at lower frequencies is generally boosted, and the experimental performance was accurately simulated by the mathematic models.

To enhance the mechanical damping of sound waves, Cops et al. [35] embedded PU foam into an open-cell metallic foam to construct a dual-network composite foam. As shown in Figure 4.14a, the composite foam showed a significantly increased absorption coefficient compared to pristine metallic foam since the composite foam is filled with PU foam that has a smaller pore size. Although the volume fraction of air was decreased, the dual porous network increased the boundary layer thermal and viscous losses and therefore increased the absorption coefficient. The composite foam increased the sound absorption from 0.1 to 0.6 in the low-frequency range and increased the average absorption coefficient from 0.2 to 0.4 in broadband when compared to the original metallic foam (Figure 4.14b). They also developed a lumped element model to predict and elucidate the absorption mechanisms for the composite, which revealed the physical mechanisms that control acoustic absorption, including thermo-viscous effects at pore interfaces, structural damping effects due to foam elasticity, and coupling effects.

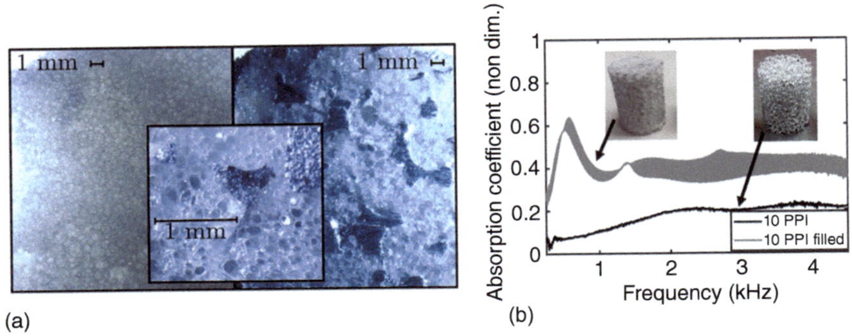

Figure 4.14 (a) Optical microscope images of pure flex foam and composite foam filled with flex foam, and (b) the measured absorption coefficient of metal foam and PU-filled composite foam. Source: Cops et al. [35]/with permission of Elsevier.

4.4.3 Sound Absorption Polymer Foam

Polymer foams are widely used in noise reduction due to their low density, low cost, controllable fine cellular structure, and elasticity. The sound absorption of polymer foam is largely related to their microstructure as we introduced in the above sections. Therefore, the majority of research is focused on manipulating the structure of polymer foams to optimize the sound absorption performance of the foams.

For example, Naguib and coworkers [36] fabricated porous PLA foams using salt particles of different sizes as porogens. The weight ratio of PLA to salt is controlled to be 1 : 9 to ensure a highly porous open-cellular structure and a constant porosity (Figures 4.15a and c). The airflow resistivity results showed that the PLA foams

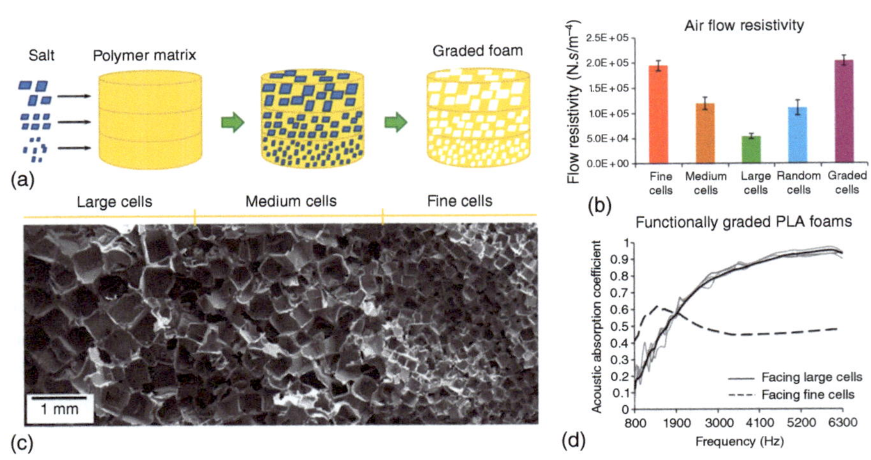

Figure 4.15 (a) Schematic of particulate leaching technique to fabricate porous structures, (b) airflow resistivity measurements of PLA open-cell foams, (c) SEM micrographs of functionally graded PLA open-cell foam, and (d) Sound absorption coefficient versus frequency for functionally graded PLA foams. Source: Mosanenzadeh et al. [36]/with permission of Springer Nature.

with fine cells and graded cells showed the highest flow resistivity while the foam with large cells had the lowest flow resistivity due to the low tortuosity of the large cell foam (Figure 4.15b). With an optimized graded microstructure, the PLA foam achieved a 20% enhancement in the maximum SAC and 9% increase in the average acoustic absorption coefficient compared to PLA foam with uniform cell size (Figure 4.15d).

Considering the low mechanical strength and shock absorption property of polymer foam based acoustic absorption foams, Oh and coworkers [37] developed a multi-purpose auxetic foam through nonsolvent induced phase separation, which has highly ordered honeycomb concave micropattern on the surface of the foam. Figure 4.16a shows the fabrication process and the corresponding microstructure of the developed auxetic foam. It was found that the dimensions of the concave micropatterns can be easily controlled by changing the nonsolvent ratio, and the mechanical strength and energy absorption were greatly enhanced by the constructed concave micropatterns. The developed foam also demonstrated significantly enhanced sound absorption properties. A 37.5% improvement in the average SAC was achieved when compared with the pristine PU foam, and the tortuosity, porosity, apparent density, and total pore area were found to be the main factors affecting the energy and sound absorption properties. The auxetic foam fabricated using 15% modification solution achieved the highest average SAC of 0.712 with the highest tortuosity of 3.72 among all foams (Figure 4.16b–d).

4.4.4 Sound Absorption Polymer Composite Foam

Constructing polymer composites is an effective approach to integrating the advantages of different components in a single composite material. In the fabrication of sound absorption polymer foams, different polymer phases or additives have been introduced into the polymer matrix to alter the microstructure of the foam and improve the sound absorption performance. For example. Yao and coworkers [38] developed a composite foam by combining PU, PVDF, and multiwalled carbon nanotube (MWCNT) . It was found that PVDF formed a separate immiscible phase and part of it was crystallized in a polar phase in the PU scaffold, which could introduce interfacial damping and local piezoelectric damping effects upon sound wave incident. The introduction of MWCNT further facilitated the dissipation of the electrical charges generated from the local piezoelectric effect. When the composition of PU/PVDF/MWCNT was 83/15/2, the composite foam reached the highest SAC of 0.85 at 1 kHz, which was significantly higher than the PU foam. Similarly, Kim and coworkers [39] found that the addition of oleic acid-treated $CaCO_3$ filler into PU foam reduced the open porosity of the foam and resulted in a high SAC due to the enhanced airflow resistance.

Low-frequency sound wave absorption is a challenge for passive acoustic absorption polymer foams because the dissipation of materials is inherently weak in this region [40]. Li et al. [41] fabricated a PP/graphite composite (PP-G) foam via supercritical CO_2 ($scCO_2$) foaming which showed enhanced sound absorption performance (Figure 4.17a). The addition of graphite nanoparticles effectively adjusted

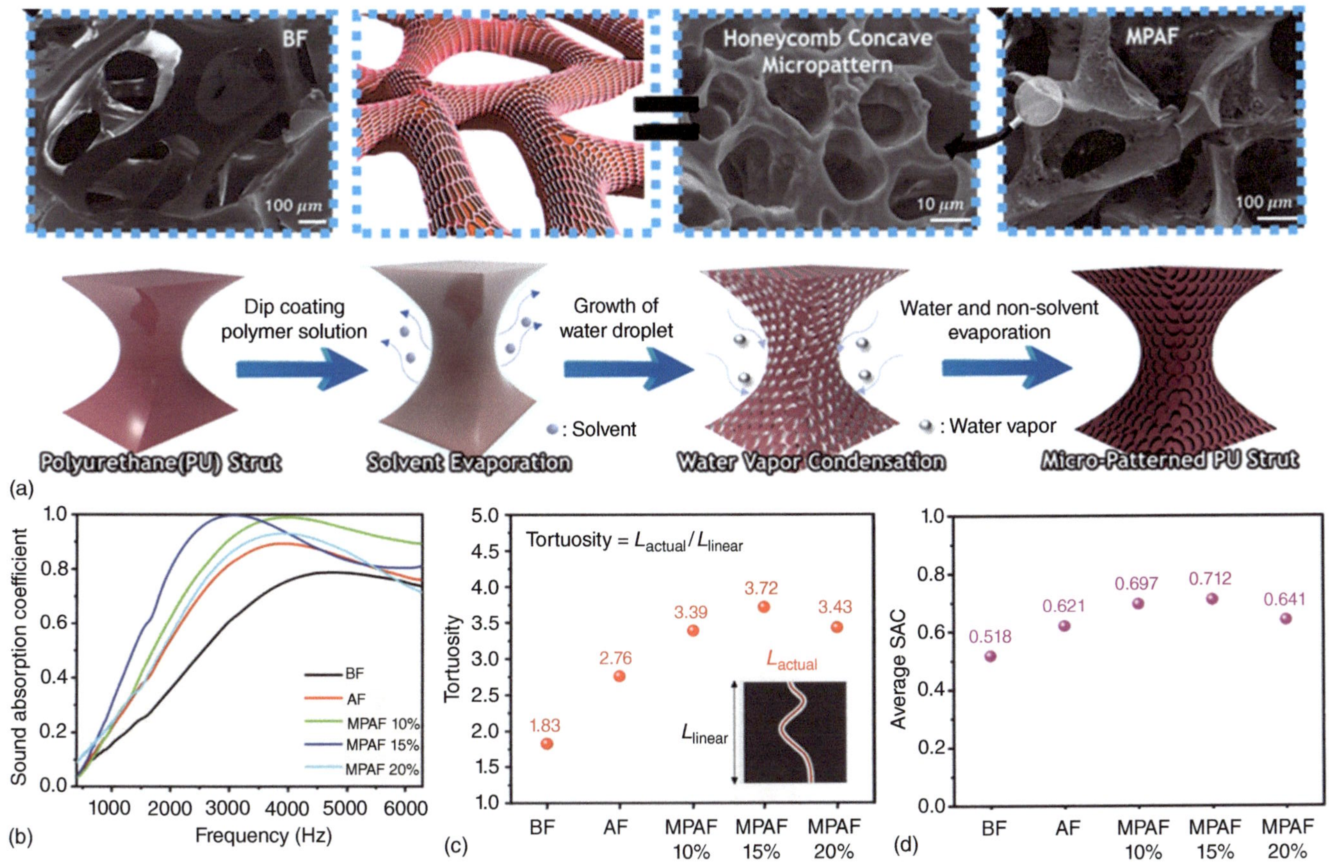

Figure 4.16 (a) Schematic diagram of fabrication steps of micro-patterned auxetic foam and enlarged SEM images. Self-assembled mechanism of nonsolvent induced phase separation method for fabricating concave honeycomb PLA micropattern on PU struts; (b) sound absorption coefficient measurements corresponding to frequencies from 125 to 6400 Hz; (c) Average tortuosity; and (d) SAC of different samples. Source: Kim et al. [37]/John Wiley & Sons/CC BY 4.0.

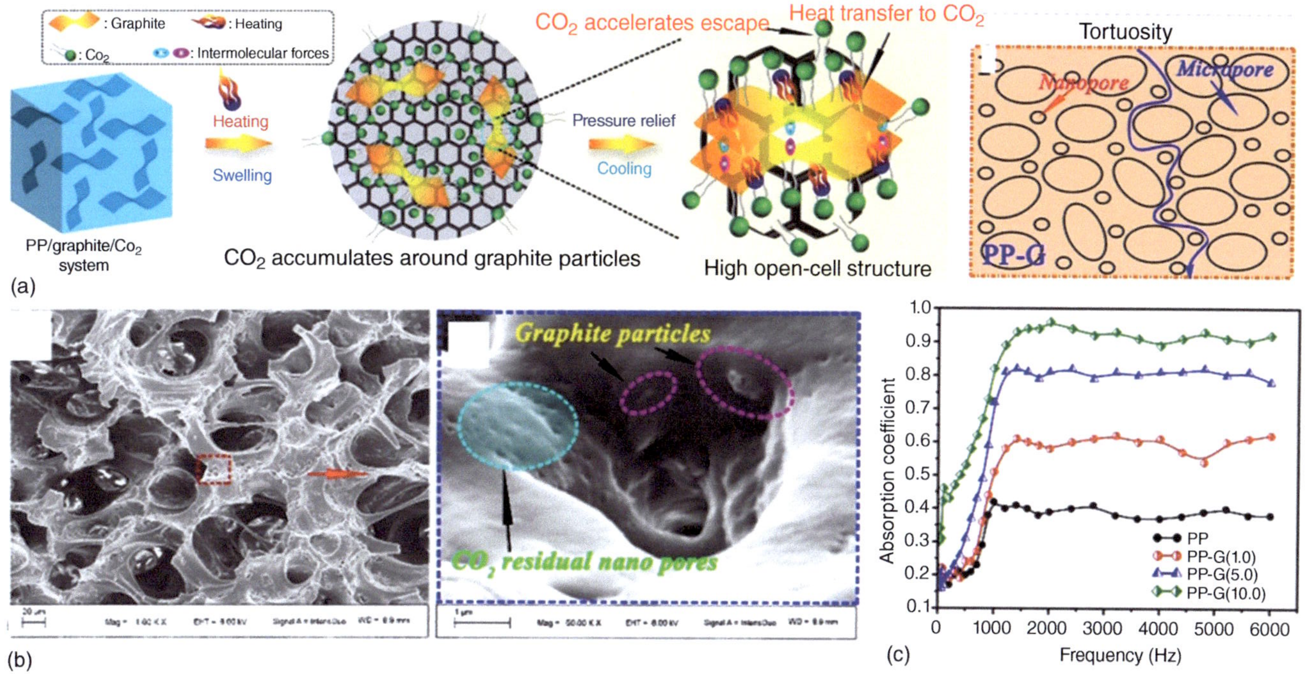

Figure 4.17 (a) Open-cell structure formation diagram and tortuosity illustration, (b) SEM images of micro/nano perforated PP-G composite foams, and (c) sound absorption coefficient results of different PP-G composite foams. Source: Li et al. [41]/with permission of Royal Society of Chemistry.

the microstructure of the PP-G foam and induce the formation of both microsized cells and nanopores inside of the foam (Figure 4.17b). The special microstructure enhanced the tortuosity and prolonged sound wave propagation length. When the amount of graphite reached 10.0 wt%, the PP-G foam exhibited optimal sound absorption performance, compression resistance, heat insulation, and hydrophobic properties. The average SAC reached ~0.9, and the internal tortuosity reached 5.27 in the frequency range of 1000–6000 Hz (Figure 4.17c) and the performance was maintained after 50 cycles of compression.

Similarly, Katbab and coworkers [42] developed an elastomeric nanocomposite foam based on ethylene–propylene–dienemonomer (EPDM) rubber filled with MWCNTs to absorb low-frequency acoustic waves. The nanocomposite foam with only 0.1% of MWCNTs regarding the rubber phase achieved a high-sound absorption coefficient with a sharp peak of 0.98 within the frequency range of 500–1000 Hz owing to the acoustic energy dissipation through the vibration and deformation of the flexible cell walls. The highly elastic matrix, high porosity and airflow resistivity of the fine partially reticulated cellular structure contributed to a sound reduction index equal to 27–33 dB. The addition of MWCNT nanoparticles increased the number of cells in the foam structure, which caused an increase in tortuosity and sound energy damping through the propagation of sound waves and consequently facilitated the transformation of sound energy to thermal energy [43]. Yu and coworkers [44] investigated the effect of silica aerogel size on the thermal insulation and acoustic absorption performance of geopolymer foam aerogel render (GFAR). They found that larger silica aerogel particles are more beneficial for acoustic absorption and thermal insulation, while smaller aerogel particles are less effective for insulation properties. With a moderate number of small cells less than 700 μm, the composite foam achieved an average acoustic absorption coefficient of 0.51 with a density of 715.2 $kg\,m^{-3}$, which contributed to the air molecules vibrating and interacting with the cell walls and facilitated the conversion of sound energy to heat.

4.5 Fabrication of Acoustic Absorption Polymer Foams

4.5.1 Chemical Foaming

Chemical foaming is a foaming process accompanied by chemical reactions to produce gas bubbles in a polymer matrix. The chemical reactions that produce gas could be the thermal decomposition of a foaming agent or the polycondensation reaction that produce gas such as the *in situ* chemical foaming of PU foams. The method of using foaming agents has advantages of board applicability, but it has disadvantages such as the narrow foaming temperature window and the emission of toxic volatiles.

In the polymerization reaction foaming method, the foaming process needs to be controlled along with the viscosity of the reaction substrate since the polymer is gradually hardened and solidified during the foaming process or curing process. To balance the polymerization and foaming processes, a small amount of catalyst and foam stabilizer (or surfactant) are generally added. Other additives such as foaming agent, water, and volatile reagents are sometimes used to assist the foaming process to form high-expansion foams or control the cell-opening rate.This method is usually used in the synthesis of PU foam, melamine foam, PI foam, silicone rubber foam, and other rubber foams. It has disadvantages of limited polymer selection, use of organic solvents, residual of catalyst and unreacted monomers, and complicated reaction process.

Nevertheless, chemical foamed polymers still occupy a large proportion of the sound absorption polymer foams and are important research topics for the development of advanced sound absorption polymer foams.

For example, Sung et al. [45] fabricated flexible PU foams by mixing high-molecular-weight copolymer polyol with polyol (CPP) through chemical foaming using CO_2 as the blowing agent. As shown in Figure 4.18a, the CO_2 is generated by the reaction of isocyanate groups and water. Cells nucleate and gradually grow as the polymerization reaction proceeded until the water is completely consumed. The development of cell morphology was controlled by the relative strength of the polymer matrix with the CO_2 pressure generated to fabricate PU foams with different porous structures and cell openness. The SAC for PU foams with different CPP contents showed an increasing trend until the CPP contents of 20 wt%, then reduced beyond this point since the sound absorption mainly results from the closed pore ratio rather than the cavity size. The highest sound absorption performance was observed at the CPP content of 20 wt% for overall frequency ranges according to the SAC at 2 kHz and the NRC at the low and middle-frequency regions (Figure 4.18b).

Baek and Kim [46] fabricated PU composite foams containing silicone–acrylic filler particles through reaction chemical foaming. The additive particles acted as nucleation sites during the foaming process and led to decreased cavity size (cell size) within 2% loading content. When the loading content exceeded 2%, the cell size was increased due to the particle agglomeration. It was found that the filler content had a great influence on the cell-opening rate and cell size. The partial open-cell rate reached the maximum level with a 2% filler content, which rendered the PU foam's excellent sound absorption performance. Both acoustic activity and NRC achieved the maximum values at this filler content due to the constructed ideal partial open porous structure.

4.5.2 Supercritical CO_2 Foaming

scCO_2 foaming is a physical foaming method that utilizes CO_2 as the blowing agent to prepare polymer foams with highly porous inner structures. Compared with other physical foaming methods such as mechanical foaming and foaming using volatile alkane gases, scCO_2 foaming has the advantages of low hazard, high repeatability, fine cell size, and superior comprehensive properties. Another inert gas, e.g. N_2, has also been widely used in the fabrication of polymer foams as a green-blowing agent. However, it usually requires higher foaming pressure and longer saturation

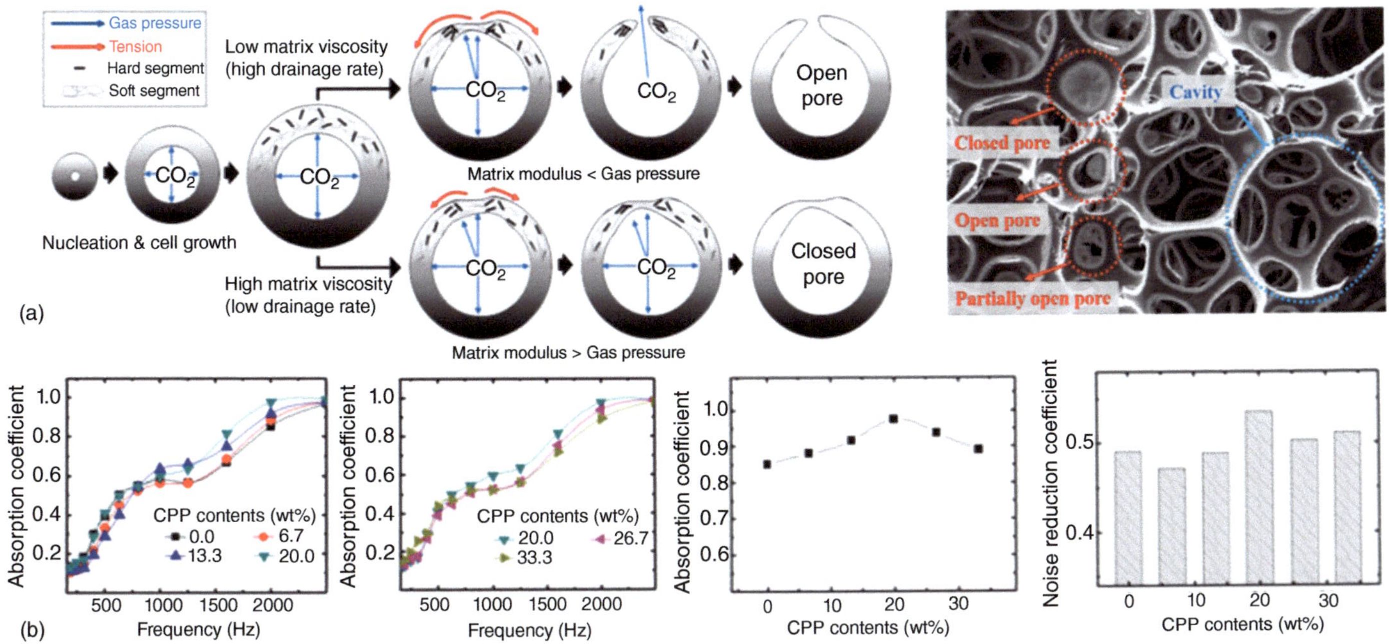

Figure 4.18 (a) Schematic demonstration of cell formation and pore-opening mechanism, and a typical cell morphology showing cavity, partially open, open, and closed pores; (b) Sound absorption coefficients at various copolymer polyol (CPP) contents, sound absorption coefficient at 2 kHz, and noise reduction coefficient as a function of CPP contents. Source: Sung et al. [45]/with permission of John Wiley & Sons.

time due to its relatively low solubility in polymer matrix. Nevertheless, supercritical N_2 (scN_2) is frequently used with $scCO_2$ as a co-blowing agent to produce polymer foams with superior properties. As introduced in Chapter 1, the $scCO_2$ foaming process mainly involves four stages including gas dissolution, cell nucleation, cell growth, and solidification. The morphology of the polymer foams generally can be controlled by adjusting the key parameters in the foaming process, including saturation temperature, foaming temperature, gas pressure, depressurization rate, cooling rate, and even mold cavity. Due to the high porosity and fine cell size, polymer foams produced via $scCO_2$ foaming have been more and more used in the application of sound absorption in recent years.

Park and coworkers [47] developed a mold opening technique accompanied by a traditional foam injection-molding machine to produce PP open-cell acoustic foams. In the process, gas counter-pressure was used to postpone the foaming of the polymer melt, which is helpful for the improvement in cell morphology uniformity and surface quality. A foamed structure with an open-cell content of 67% and an expansion ratio of 4.6 was obtained when the mold was opened by 4.5 mm. When the mold opening distance is further increased, a hollow cavity is created in the middle of the foam due to the solidification of the skin region and the crack of the cells presented in the center of the sample (Figure 4.19a). Noticeably, the foam with a cavity and a high open-cell content presented remarkable acoustic properties, and the absorption coefficient was increased by the increased cavity depth. A peak absorption coefficient of 0.65–0.95 in a frequency range of 950–1200 Hz and a maximum transmission loss of about 41 dB were achieved by the optimized foam with a cavity in the center (Figure 4.19c).

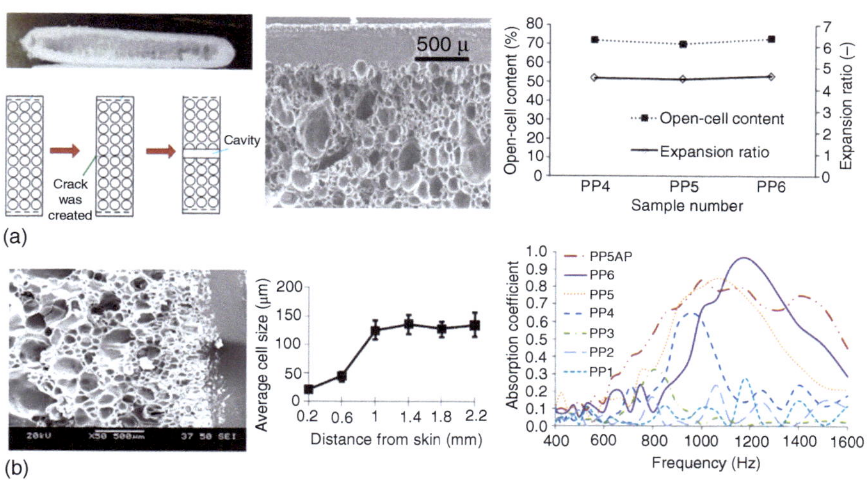

Figure 4.19 (a) Schematic of the cracking behavior during the mold open foaming process and representative sample photograph and SEM image along with the open-cell content and expansion ratio results; (b) SEM micrograph and cell size distribution across the thickness of the foamed PP sample; and (c) Absorption coefficient results for different PP foams. Source: Jahani et al. [47]/with permission of Elsevier.

Similarly, Wang and coworkers [48] fabricated PP/polytetrafluoroethylene (PTFE) composite foams by combining microcellular injection molding (MIM) using $scCO_2$ as the blowing agent and in-situ fibrillation technologies. The PTFE component formed nano-scaled networks in the PP polymer matrix during melt blending, which promoted the crystallization and melt strength of the PP matrix and enhanced the foaming ability of PP in the MIM process. The composite PP/PTFE foam with an expansion ratio of up to 20.2 and a cell density of 10^8 cells cm^{-3} showed a high open-cell content of 98.3% and contributed to a high airflow resistivity and sound insulation by reducing impedance and enhancing viscous dissipation. A SAC of over 0.5 and a transmission loss of up to 50 dB were achieved by the developed foam.

Jian et al. [49] reported a lightweight acoustic composite foam with surprisingly stable sound absorption performance in a wide bandwidth of frequency. As illustrated in Figure 4.20a, PS composite foams containing graphitic biocarbon were fabricated by sequestering recycled $scCO_2$ torsion extrusion foaming. With good mixing and thermal management instrumental design, the composite foam offered uniform cellular structures with controllable cell size and cell density (Figure 4.20b). The presence of biocarbon in the composition changed the sound insulation behavior from resonant absorber of pure PS closed cell foam to stable acoustic performance. As shown in Figure 4.20c, two additional peaks were detected at 4.3 kHz ($\alpha = 0.64$) and 6 kHz ($\alpha = 0.46$) for the composite foam and the SAC variation with frequency

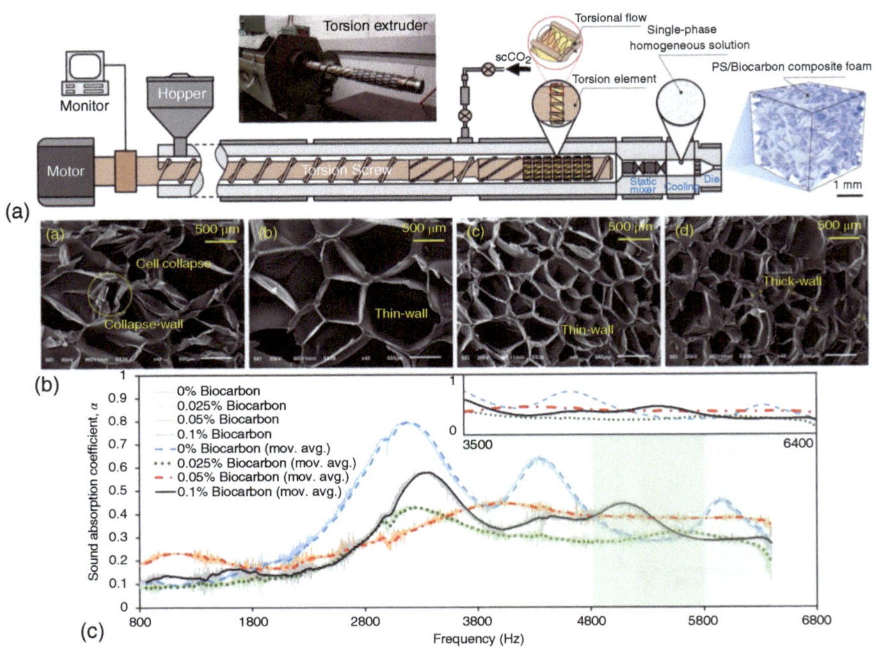

Figure 4.20 (a) Schematic of the sequestering recycled $scCO_2$ torsion extrusion foaming process, (b) SEM images of PS/biocarbon composite foams with different content of renewable carbon in PS, and (c) sound absorption coefficient of different composite foams. Source: Jian et al. [49]/with permission of John Wiley & Sons.

showed a typical behavior of a resonant absorber. The unique acoustic properties of the PS/biocarbon composite foam were attributed to the highly exfoliated nanolayered biocarbon nucleation and the closed cell microstructure.

4.5.3 Coating of Foam Skeletons

To develop high-performance sound absorption foams, researchers have developed various simple coating methods based on commercially available foams, such as PU foams and melamine foams. Through the solution-assisted coating method, the porous structure of pristine foam can easily be tuned to present desired physical and chemical properties. For instance, Jung and coworker [50] coated GO on an open-cell PU foam by a step-by-step vacuum-assisted infiltration and drying process. The sound-absorbing property of the foam was greatly enhanced by GO impregnation due to the increase in the airflow resistance and tortuosity of the foam, and the maximum absorption coefficient could be tuned to a specific frequency range by controlling GO impregnation density. The average SAC was increased more than four times by impregnating 15% of GO.

In a recent study, Gao and coworkers [51] developed a novel graphene acoustic absorber (PFGA) by integrating self-standing ultrathin graphene drums and cellular polymer microframes via a hydroplastic foaming (HPF) method. As illustrated in Figure 4.21a, the process involves coating a polymer foam with GO sheet and HPF process using N_2H_4/H_2O treatment. Atomically, the 2D graphene sheets exhibited unparalleled in-plane stiffness and large out-of-plane elasticity, which rendered them strong mechanical resonance behavior (Figure 4.21b and c). The coated graphene formed graphene drums on the foam skeleton and exhibited strong resonances and efficiently dissipated sound waves in a broad frequency range (Figure 4.21d), and the SAC was greatly enhanced with the increase of GO content especially in the high-frequency range, and a ≈320% enhancement in average absorption coefficient across wide frequencies from 200 to 6000 Hz was achieved by the composite foam. The absorption performance also showed significant improvement with the increase in foam thickness from 5 to 30 mm (Figure 4.21e and f).

4.5.4 Phase Separation and Particulate Leaching

Phase inversion or phase separation is a widely used approach to produce porous polymer foams by removing a sacrificial solvent phase from a phase-separated system. In a typical phase separation process, the polymer solute is first dissolved in a solvent or a cosolvent to form a homogenous solution; Then, the phase separation is triggered by reducing the temperature or by adding a nonsolvent. At last, the separated solvent phase is removed by means like solvent exchange, drying, freeze-drying, or $scCO_2$ drying. Through the process, porous structures could be formed in the polymer matrix. Particulate leaching refers to an approach that involves soluble particles as the porogen in the fabrication of polymer foams. In the process, the polymer matrix is filled to the space among the particles by

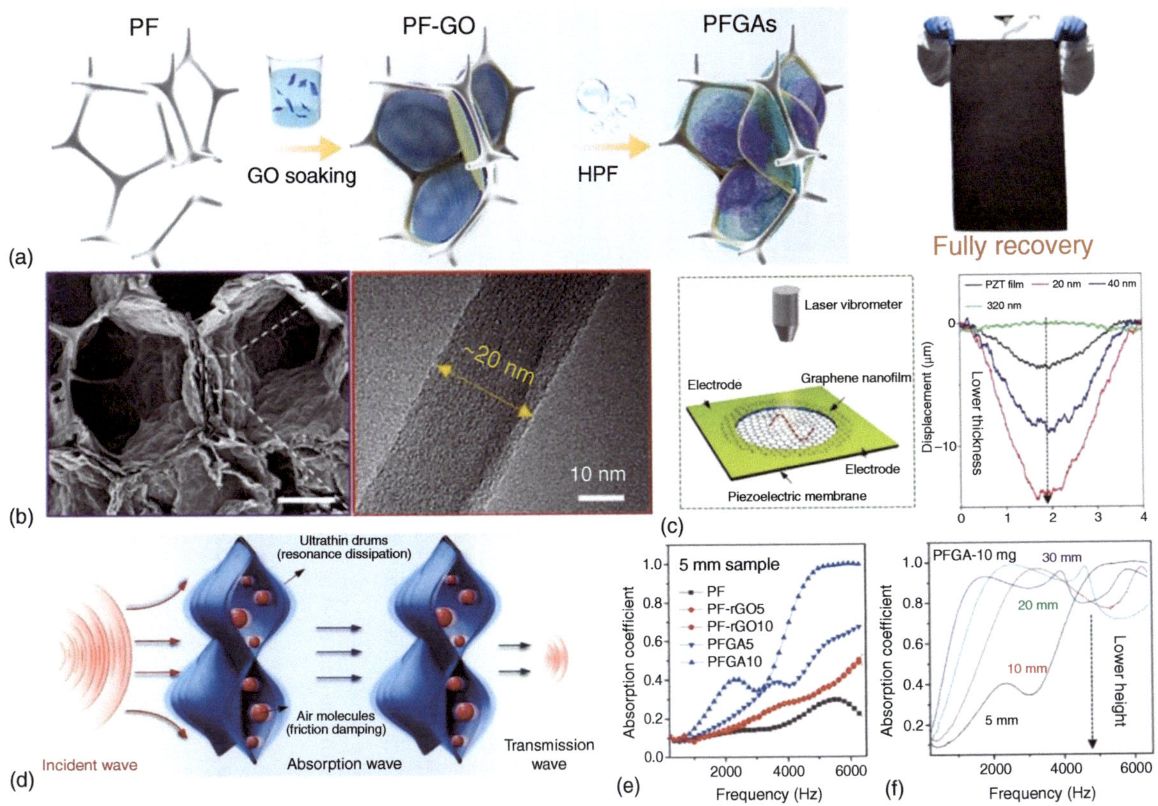

Figure 4.21 (a) Diagram of the fabrication of PFGA and a photograph of the PFGA fabricated, (b) SEM image of PFGA and the HRTEM image of graphene nanowalls, (c) The vibrational displacements of ultrathin graphene drums with 20, 40, and 320 nm thickness, (d) Resonance dissipation mechanism of the ultrathin drums, (e) The sound-absorption coefficients of PF, PF-rGO, and PFGA at 5 mm thickness, and (f) The sound absorption spectra of PFGA with different thickness. Source: Pang et al. [51]/with permission of John Wiley & Sons.

the melt process or solution process. Then, the particles were selectively leached using circulating water or a specific solvent. Commonly used porogens include salt particles, sugar particles, paraffine particles, PS spheres, etc.

It is obvious that the phase separation and particulate leaching are relatively complicated and involve a significant amount of solvent in the process. Therefore, it is not widely used in the fabrication of sound absorption foams for large-scale production. Nonetheless, these processes are promising for fabricating polymer foams using water as a solvent. For instance, Zhou and coworkers [52] combined cellulose nanofiber (CNF) with melamine foam to develop a hybrid foam with both macropores and mesopores. As illustrated in Figure 4.22a, the melamine foam was immersed in the CNF dispersion (Figure 4.22b) followed by a cyclic freezing-thawing process to allow the self-assembly of CNF on the melamine foam skeleton. Then, the hybrid foam is obtained through freeze-drying. The hierarchical pore structure (Figure 4.22c) constructed enhanced the specific surface area of the foam (Figure 4.22d) and facilitated the multiple reflections and microvibration of the sound waves when used as sound absorbers. The SAC of the hybrid foam prepared using 0.4% CNF concentration achieved an enhancement of about 107% at 500 Hz compared with the pristine melamine foam (Figure 4.22e).

Kim et al. [53] proposed a facile strategy to fabricate a composite polymer (CAP) scaffold composed of PP fiber and calcium alginate with hierarchically structured highly porous morphology and multifunctional properties. As illustrated in Figure 4.23a and b, PP fibers were dispersed in the sodium alginate solution and unidirectionally frozen via an ice template method. Then, the ice phase was

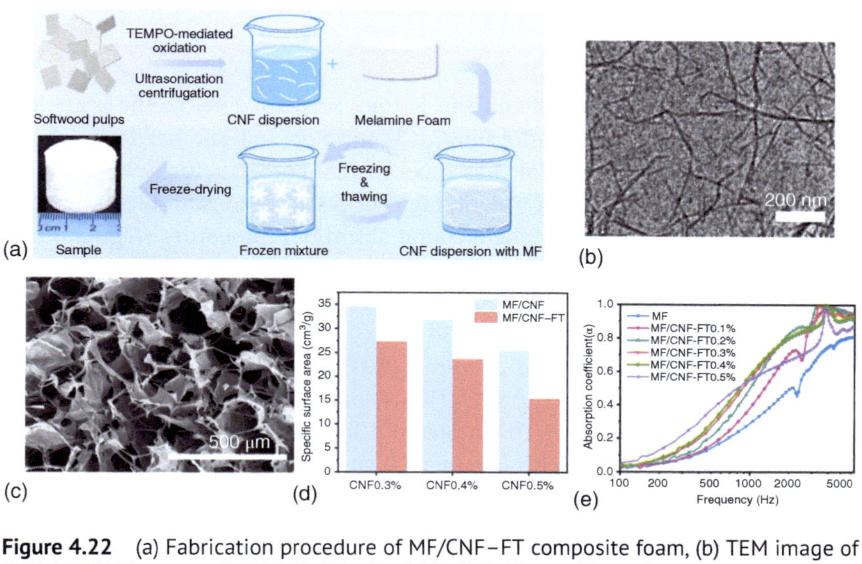

Figure 4.22 (a) Fabrication procedure of MF/CNF–FT composite foam, (b) TEM image of CNF, (c) SEM images of MF/CNF–FT with 0.4% CNF, (d) specific surface area results, and (e) sound absorption coefficient results of MF/CNF–FT with different concentrations of CNF. Source: Shen et al. [52]/with permission of Elsevier.

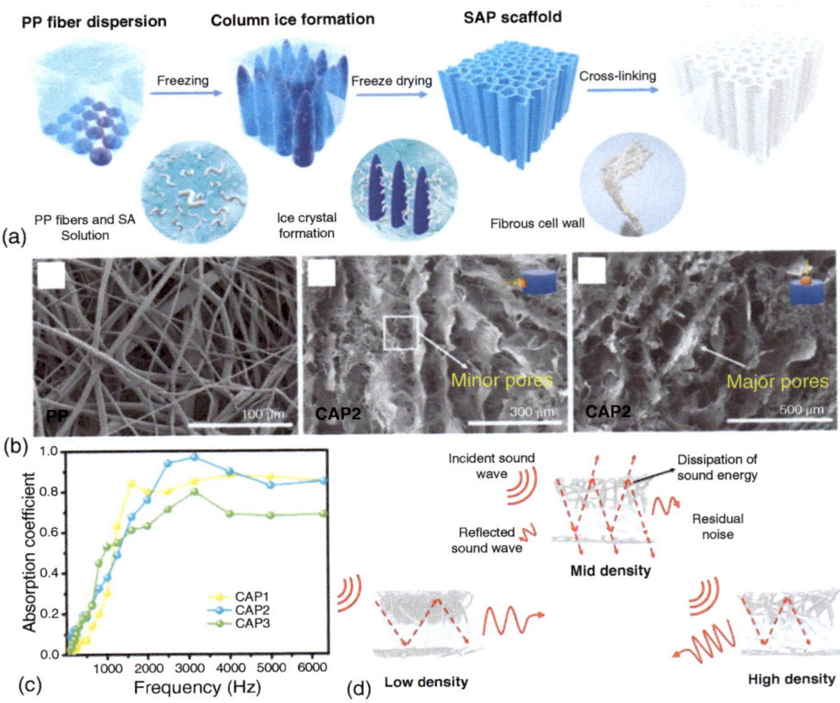

Figure 4.23 (a) synthesis route of the CAP scaffold, (b) SEM images of PP fiber network, and different positions of the CAP scaffold, (c) The acoustic absorption coefficient of different aerogels, and (d) Schematic representation of sound propagation relative to the density of the aerogel. Source: Pornea et al. [53]/with permission of American Chemical Society.

removed by freeze-drying to obtain the porous scaffold. Finally, the scaffold is cross-linked by soaking the scaffold in $CaCl_2$ solution and another freeze-drying process. The obtained CAP scaffold has a special dual-pore structure with fibrillated networks, which exhibited light weight, high surface area, anisotropic mechanical and thermal properties, and excellent sound absorption properties (Figure 4.23c). The microchannels of the dual-pore structure played a major role in the dissipation of sound waves within the interconnected fiber wall, which could resonate the sound in different directions (Figure 4.23d).

Mohamed et al. [54] developed open-cell PVDF foams by using sugar granules as the porogen. Briefly, $(K_{0.5}Na_{0.5})NbO_3$ (KNN)-nanofibers were dispersed in DMF and acetone cosolvent with PVDF to obtain a homogeneous precursor solution. Then, the solution is mixed with commercial sugar followed by sufficient drying to remove the solvents. Then, the solid samples were leached in hot water to remove the sugar porogen and obtain the porous PVDF/KNN composite foam. The airborne sound absorption performance of the foam was enhanced with the increasing KNN nanofiber content. The sound energy dissipation with the opposite charges generated through the piezoelectric effects in the two phases (e.g. the piezoelectric KNN nanofibers phase with positive piezoelectric coefficient and the PVDF matrix phased with negative piezoelectric coefficient) contributed to the significant improvement in sound absorption performance of the composite foam.

References

1 Tao, Y.P., Ren, M.S., Zhang, H., and Peijs, T. (2021). Recent progress in acoustic materials and noise control strategies-a review. *Applied Materials Today* 24: 101141.

2 Liu, X.W., Duan, M.Y., Liu, M.L. et al. (2021). Acoustic labyrinthine porous metamaterials for subwavelength low-frequency sound absorption. *Journal of Applied Physics* 129: 195103.

3 Kalauni, K. and Pawar, S.J. (2019). A review on the taxonomy, factors associated with sound absorption and theoretical modeling of porous sound absorbing materials. *Journal of Porous Materials* 26: 1795–1819.

4 Jia, Z.B., Zheng, X., Zhou, Q. et al. (2020). A hybrid active noise control system for the attenuation of road noise inside a vehicle cabin. *Sensors* 20: 7190.

5 Rajappan, S., Bhaskaran, P., and Ravindran, P. (2017). An insight into the composite materials for passive sound absorption. *Journal of Applied Sciences* 17: 339–356.

6 Rastegar, N., Ershad-Langroudi, A., Parsimehr, H., and Moradi, G. (2022). Sound-absorbing porous materials: a review on polyurethane-based foams. *Iranian Polymer Journal* 31: 83–105.

7 Rahimabady, M., Statharas, E.C., Yao, K. et al. (2017). Hybrid local piezoelectric and conductive functions for high performance airborne sound absorption. *Applied Physics Letters* 111: 241601.

8 Hutchison, T.S. (1960). Ultrasonic absorption in solids: how high-frequency sound is damped by imperfections in solids. *Science* 132: 643–652.

9 Allard, J.F. and Champoux, Y. (1992). New empirical equations for sound-propagation in rigid frame fibrous materials. *Journal of the Acoustical Society of America* 91: 3346–3353.

10 Zhang, J.Z., Chen, T.N., Zhang, B., and Zhu, J. (2023). A semi-empirical model for sound absorption by perforated plate covered open cell foam and improvements from optimising the perforated plate parameters. *Applied Sciences-Basel* 13: 78.

11 Liu, P., Yang, Y., and Yao, J. (2011). Study on absorption property of porous sound-absorbing materials. *Noise and Vibration Control* 31: 123–126.

12 Botterman, B., de la Grée, G.C.H.D., Hornikx, M.C.J. et al. (2018). Modelling and optimization of the sound absorption of wood-wool cement boards. *Applied Acoustics* 129: 144–154.

13 Oldham, D.J., Egan, C.A., and Cookson, R.D. (2011). Sustainable acoustic absorbers from the biomass. *Applied Acoustics* 72: 350–363.

14 Champoux, Y. and Allard, J.F. (1991). Dynamic tortuosity and bulk modulus in air-saturated porous-media. *Journal of Applied Physics* 70: 1975–1979.

15 Johnson, D.L., Koplik, J., and Dashen, R. (1987). Theory of dynamic permeability and tortuosity in fluid-saturated porous-media. *Journal of Fluid Mechanics* 176: 379–402.

16 Mosanenzadeh, S.G., Doutres, O., Naguib, H.E. et al. (2015). A numerical scheme for investigating the effect of bimodal structure on acoustic behavior of polylactide foams. *Applied Acoustics* 88: 75–83.

17 Chanlert, P. and Ruamcharoen, P. (2021). Sound absorption properties of rigid polyurethane foam composites with rubber-wood sawdust as a natural filler. *Journal of Physics: Conference Series* 1719: 012062.

18 ISO (1991). Acoustics-materials for acoustical applications-determination of airflow resistance. ISO 9053:1991, EN 29053:1993. Publication Date: 31 Mar 1993.

19 Seddeq, H. (2009). Factors influencing acoustic performance of sound absorptive materials. *Australian Journal of Basic and Applied Sciences* 3: 4610–4617.

20 Mosanenzadeh, S.G., Naguib, H.E., Park, C.B., and Atalla, N. (2013). Development, characterization, and modeling of environmentally friendly open-cell acoustic foams. *Polymer Engineering and Science* 53: 1979–1989.

21 Doutres, O., Atalla, N., and Dong, K. (2011). Effect of the microstructure closed pore content on the acoustic behavior of polyurethane foams. *Journal of Applied Physics* 110: 064901.

22 Chen, W.H., Lee, F.C., and Chiang, D.M. (2000). On the acoustic absorption of porous materials with different surface shapes and perforated plates. *Journal of Sound and Vibration* 237: 337–355.

23 Beranek, L.L. and Sleeper, H.P. (1946). The design and construction of anechoic sound chamber. *Journal of the Acoustical Society of America* 18: 140–150.

24 Jiang, C.Y., Zhang, S.Y., and Huang, L.X. (2016). On the acoustic wedge design and simulation of anechoic chamber. *Journal of Sound and Vibration* 381: 139–155.

25 Tsay, H.S. and Yeh, F.H. (2019). Design of acoustic foam wedge panels by finite element frequency domain acoustical analysis. *Journal of Cellular Plastics* 55: 283–295.

26 Ren, X.H., Wang, J., Sun, G.H. et al. (2020). Effects of structural design including cellular structure precision controlling and sharp holes introducing on sound absorption behavior of polyimide foam. *Polymer Testing* 84: 106393.

27 Yuan, B.G., Cheng, Y., and Liu, X.J. (2015). Conversion of sound radiation pattern via gradient acoustic metasurface with space-coiling structure. *Applied Physics Express* 8: 027301.

28 Peng, L.M., Song, B.Q., Wang, J.F., and Wang, D. (2015). Mechanic and acoustic properties of the sound-absorbing material made from natural fiber and polyester. Advances In Materials Science And. *Engineering* 2015: 274913.

29 Liu, L., Chen, Y.J., Liu, H.Z. et al. (2019). A graphene oxide and functionalized carbon nanotube based semi-open cellular network for sound absorption. *Soft Matter* 15: 2269–2276.

30 Caniato, M., D'Amore, G.K.O., Kaspar, J., and Gasparella, A. (2020). Sound absorption performance of sustainable foam materials: application of analytical

and numerical tools for the optimization of forecasting models. *Applied Acoustics* 161: 107166.

31 Chen, J.H., Liu, P.S., and Sun, J.X. (2020). Sound absorption performance of a lightweight ceramic foam. *Ceramics International* 46: 22699–22708.

32 Du, Z.P., Yao, D.X., Xia, Y.F. et al. (2020). The sound absorption properties of highly porous silicon nitride ceramic foams. *Journal of Alloys and Compounds* 820: 153067.

33 Lu, T.J., Hess, A., and Ashby, M.F. (1999). Sound absorption in metallic foams. *Journal of Applied Physics* 85: 7528–7539.

34 Chua, J.W., Li, X.W., Li, T. et al. (2022). Customisable sound absorption properties of functionally graded metallic foams. *Journal of Materials Science & Technology* 108: 196–207.

35 Cops, M.J., McDaniel, J.G., Magliula, E.A. et al. (2020). Measurement and analysis of sound absorption by a composite foam. *Applied Acoustics* 160: 107138.

36 Mosanenzadeh, S.G., Naguib, H.E., Park, C.B., and Atalla, N. (2015). Design and development of novel bio-based functionally graded foams for enhanced acoustic capabilities. *Journal of Materials Science* 50: 1248–1256.

37 Kim, J.S., Mahato, M., Oh, J.H., and Oh, I. (2023). Multi-purpose auxetic foam with honeycomb concave micropattern for sound and shock energy absorbers. *Advanced Materials Interfaces* 10: 2202092.

38 Statharas, E.C., Yao, K., Rahimabady, M. et al. (2019). Polyurethane/poly(vinylidene fluoride)/MWCNT composite foam for broadband airborne sound absorption. *Journal of Applied Polymer Science* 136: 47868.

39 Choe, H., Lee, J.H., and Kim, J.H. (2020). Polyurethane composite foams including CaCO fillers for enhanced sound absorption and compression properties. *Composites Science and Technology* 194: 108153.

40 Mei, J., Ma, G.C., Yang, M. et al. (2012). Dark acoustic metamaterials as super absorbers for low-frequency sound. *Nature Communications* 3: 756.

41 Li, Z.Y., Yang, C.G., Yan, K. et al. (2024). Rational design of a polypropylene composite foam with open-cell structure graphite conductive network for sound absorption. *Soft Matter* 20: 1089–1099.

42 Hosseinpour, A., Katbab, A.A., and Ohadi, A. (2022). Improving the sound absorption of a highly deformable nanocomposite foam based on ethylene-propylene-diene-monomer (EPDM) infused with multi-walled carbon nanotubes (MWCNTs) to absorb low-frequency waves. *European Polymer Journal* 178: 111522.

43 Basirjafari, S., Malekfar, R., and Khadem, S.E. (2012). Low loading of carbon nanotubes to enhance acoustical properties of poly(ether)urethane foams. *Journal of Applied Physics* 112: 104312.

44 Chen, Y., Klima, K.M., Brouwers, H.J.H., and Yu, Q.L. (2022). Effect of silica aerogel on thermal insulation and acoustic absorption of geopolymer foam composites: the role of aerogel particle size. *Composites Part B: Engineering* 242: 110048.

45 Sung, G., Kim, J.S., and Kim, J.H. (2018). Sound absorption behavior of flexible polyurethane foams including high molecular-weight copolymer polyol. *Polymers for Advanced Technologies* 29: 852–859.

46 Baek, S.H. and Kim, J.H. (2020). Polyurethane composite foams including silicone-acrylic particles for enhanced sound absorption via increased damping and frictions of sound waves. *Composites Science and Technology* 198: 108325.

47 Jahani, D., Ameli, A., Jung, P.U. et al. (2014). Open-cell cavity-integrated injection-molded acoustic polypropylene foams. *Materials and Design* 53: 20–28.

48 Zhao, J.C., Wang, G.L., Chen, Z.L. et al. (2021). Microcellular injection molded outstanding oleophilic and sound-insulating PP/PTFE nanocomposite foam. *Composites Part B: Engineering* 215: 108786.

49 Jian, R.R., Shahi, P., Semeniuk, M. et al. (2022). Acoustic absorption properties of polystyrene-pyrolytic pinus resinosa composite foams prepared by torsion-induced extrusion. *Macromolecular Materials and Engineering* 307: 2100622.

50 Lee, J. and Jung, I. (2019). Tuning sound absorbing properties of open cell polyurethane foam by impregnating graphene oxide. *Applied Acoustics* 151: 10–21.

51 Pang, K., Liu, X.T., Pang, J.T. et al. (2022). Highly efficient cellular acoustic absorber of graphene ultrathin drums. *Advanced Materials* 34: 2103740.

52 Shen, L., Zhang, H.R., Lei, Y.Z. et al. (2021). Hierarchical pore structure based on cellulose nanofiber/melamine composite foam with enhanced sound absorption performance. *Carbohydrate Polymers* 255: 117405.

53 Pornea, A.G.M., Puguan, J.M.C., Ruello, J.L.A., and Kim, H. (2022). Multifunctional dual-pore network aerogel composite material for broadband sound absorption, thermal insulation, and fire repellent applications. *ACS Applied Polymer Materials* 4: 2880–2895.

54 Mohamed, A.M., Yao, K., Yousry, Y.M. et al. (2020). Enhanced airborne sound absorption effect in poly(vinylidene fluoride)/(KNa)NbO-nanofiber composite foams. *Journal of Applied Polymer Science* 137: 49022.

5

Superhydrophobic Polymer Foams

5.1 Overview of Superhydrophobic Polymer Foams

In nature, certain plant stems, leaf surfaces, animal feathers, or skin exhibit remarkable superhydrophobic wetting properties that help them to survive in different ways. The most iconic example of this phenomenon is the "lotus leaf effect," which describes a self-cleaning phenomenon in which raindrops rolling down a lotus leaf surface can carry away dust, keeping the leaf clean [1]. The superhydrophobicity of lotus leaves arises from a combination of low surface energy and surface roughness. A waxy layer provides low surface energy, while micro/nanostructures reduce the contact area between water droplets and the solid surface. Leveraging advancements in bionanotechnology, researchers have reproduced this micro/nanostructured, low-surface-energy layer on the surface of polymer foam. As a result, the foam exhibits a high-water contact angle (WCA) and finds applications in antifouling, self-cleaning, selective oil absorption, and oil–water separation [2–4].

Superhydrophobic polymer foams are characterized by water droplets on their surface forming a contact angle greater than 150° and a roll angle or sliding angle (SA) less than 10° [5]. Various superhydrophobic polymer foams have been produced by means of surface modifications, such as grafting, coating, etching, etc., onto commercial foams [6, 7]. However, these methods suffer from complexities, high manufacturing costs, limited durability, difficulties in scaling production, and potential secondary pollution, which hinder their industrial applications.

An alternative approach for the preparation of superhydrophobic porous polymer foams is supercritical carbon dioxide ($scCO_2$) foaming technology, which is a low cost, environmentally friendly method that may not necessarily require a secondary modification procedure. In the $scCO_2$ foaming, CO_2 is dissolved in the polymer under high temperature and pressure conditions to form a polymer/gas homogeneous system, and micro-sized cells are generated when the pressure drops by the phase change of the CO_2 from the supercritical state to the gas state. The surface energy of the polymer foam can be tuned by combining low surface energy components in the polymer matrix, and the rough surface of the foam can be created by the micro-sized cell voids and the cell walls.

Functional Polymer Foams: Green Fabrication Methods, Performance and Applications,
First Edition. Haoyang Mi.

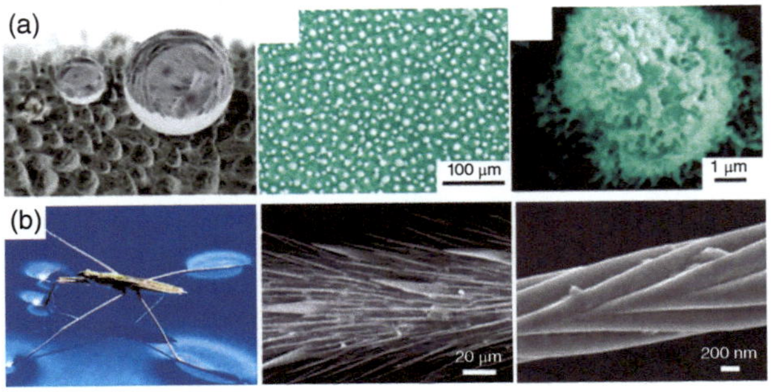

Figure 5.1 (a) Microstructure of lotus leaf in different scales, and (b) Microstructure of water strider legs in different scales. Source: (a) Feng et al. [2]/with permission of John Wiley & Sons. (b) Feng et al. [8]/with permission of American Chemical Society.

5.1.1 Superhydrophobicity in Nature

Lotus leaves exhibit a unique feature where rainwater forms spherical droplets on their surfaces without spreading, and the droplets readily roll off the lotus leaf's surface when it sways in the wind. This distinctive attribute is known as the "lotus leaf effect," manifesting as a self-cleaning and superhydrophobic phenomenon on the leaf's surface. The superhydrophobic property of the lotus leaves originates from the presence of numerous papillae structures and the secondary nanoscale structures that have wax-like low-surface-energy materials (Figure 5.1a) [2]. The water strider is another example of a superhydrophobic creature in nature, which exhibits remarkable stability and swift movement while skimming across the water surface. This special ability of the water strider is attributed to the hierarchical microstructure within the water strider's legs and the nano-groove structure presented on the needle-like bristles of its leg, as shown in Figure 5.1b [8]. This unique structure facilitates the creation of an air cushion layer within the micro- and nano-level formations, enabling water striders to glide atop the water's surface without wetting their legs.

Research into the composition and structure of various plants and animals has unveiled a common trait, which is the ability to repel water through self-cleaning and superhydrophobic properties. Scientists have devoted themselves to the fabrication of functional materials with superhydrophobic characteristics by emulating these biological attributes, combining low surface energy components with micro- and nanostructured surfaces. There exists a variety of methods for crafting hydrophobic porous materials, including surface treatment, etching, template embossing, electrostatic spinning, etc [9, 10]. These methods find extensive application in areas such as self-cleaning and oil–water separation, showcasing their versatility and utility in various fields.

5.1.2 Influencing Factors for Superhydrophobicity

The superhydrophobic properties of materials are mainly determined by a combination of two key factors including low surface free energy and multilevel surface roughness structure.

Table 5.1 Surface energy of common polymers.

Material name	γ_s	γ_s^{LW}	γ_s^+	γ_s^-
Poly(perfluoroalkylacrylate)s PFA	6.4	6.1	0.0	3.6
Poly(perfluoroalkylmethacrylate)s PFMA	38.65	35.77	0.27	7.66
Poly(methylpropenoxyfluoroalkylsiloxane)s PFE	16.0	14.1	0.5	2.0
Poly(alkylacrylate)s POA	33.1	33.0	0.01	1.5
Poly(methylpropenoxyalkylsiloxane)s PES	35.2	33.3	0.1	18.5
Poly(diperfluoroalkylitaconate)s PFI	8.3	8.0	0.3	0.1

Note: γ_s represents the surface energy of a solid, and the surface energy of a solid or liquid includes γ_s^{LW} (lifshitz van der Waals component), γ_s^+ (Lewis acid content) and γ_s^-(Lewis alkaline content). The above data are all obtained in the case of $n = 5$, and the units are mJ m^{-2}.
Source: Adapted from Tsibouklis and Nevell [11].

5.1.2.1 Surface Energy

Surface energy also referred to as solid surface free energy, signifies the specific surface Gibbs free energy of a solid material. The surface energy plays a pivotal role in determining the hydrophilic or hydrophobic nature of a material, with lower free energy indicating a higher degree of hydrophobicity. Following the principles of mechanics, a liquid with elevated surface energy cannot effectively spread on a surface with lower surface energy. Water, being a substance with high surface energy, tends to form spherical shapes on surfaces with lower surface energy. Materials containing silicon, fluorine, and similar elements, which possess lower surface energy than water, tend to exhibit hydrophobic characteristics. The surface energy values for common polymers are provided in Table 5.1 and play a significant role in material behavior and applications. Therefore, it is crucial to employ low surface energy surface treatments in various practical applications.

5.1.2.2 Surface Structure

As seen from the examples of superhydrophobic creatures in nature, the creation of a rough surface with a multiscale hierarchical surface structure is crucial for improving the hydrophobicity of the surface. Researchers have found that the incorporation of microstructures can significantly enhance the nonwetting characteristics of solid surfaces, resulting in apparent contact angles of 150° or higher. Adjusting the morphology and size of these microstructures on material surfaces to reduce droplet rolling angles or contact angle hysteresis for superhydrophobic solid surfaces holds paramount importance. Investigating techniques to engineer various micro-nanoscale structures on the material surface to render the surface with superhydrophobicity is a hot topic in the research of functional materials.

5.1.3 Methods to Engineer Hierarchical Structured Surface

5.1.3.1 Surface Treatment Method

Surface treatment is probably the most straightforward and commonly used method to produce superhydrophobic surfaces. It contains various coating approaches like spray coating, dip coating, and spin coating, and surface modification approaches, such as chemical grafting, physical deposition, and self-assembly techniques. All these methods involve introducing another component on the surface of an existing substrate, which is essentially an additive processing method. Generally, all these methods can be used for the modification of polymer foams. However, many of them can only achieve effective modification on the surface layer of the foam rather than the whole 3D foam structure. For instance, the spraying method entails applying a viscous nanoparticle suspension onto a substrate, allowing the accumulation of micro- and nanostructures that render the surface superhydrophobic. From an industrial perspective, the spraying method is straightforward, adaptable to various product sizes, and holds substantial potential for widespread application. Nevertheless, the micro/nanostructures on the coated surface are susceptible to external friction and impact, potentially compromising the coating's superhydrophobic characteristics.

To overcome this problem, Mi et al. [12] developed a superhydrophobic foam using multidimensional nanoparticle coating via chemical grafting and dip coating. The chemical linkages created between the substrate foam and different nanoparticles greatly enhanced the stability of the coating layer on the material surface. A high WCA of 153° was achieved in air, and a contact angle of 158° was reached in oil. The foam showed an absorption capacity 113 times their weight gain and a remarkable flux of 32.6 $l\,m^{-2}\,s^{-1}$ for the separation of oil from water driven by gravity (Figure 5.2).

5.1.3.2 Etching Method

Etching technology encompasses various methods, such as laser, plasma, electrochemical, and chemical techniques, to create rough microstructures on material surfaces. In these approaches, part of the material on the surface is removed by different means to leave a rough topography with micro-nanoscaled features. Therefore, these approaches are characterized as subtractive processing techniques. So far, these approaches are more frequently applied to metal surfaces than polymer surfaces for micro- and nanostructure formation since techniques like laser engraving on polymers can lead to damage in the polymer matrix due to excessive heating, making it challenging to generate dense and regular structures. Electrochemical etching or deposition requires the substrate material to be conductive, which is not applicable to most polymers. Therefore, the etching method is relatively difficult to be used for polymer foams. Kim et al. [13] engineered silicon nanowires on the surface of silicon wafer via top-down etching to create nanostructured surface structures. After coating with fluorine carbon, the surface is converted from superhydrophilic to superhydrophobic. The effect of the height of the nanoscale structures on the surface wettability was investigated, and it was found that the surface is superhydrophilic when the height is several hundred nanometers, while it changes to superhydrophobic when the height surpasses a micrometer (Figure 5.3).

5.1.3.3 Phase Separation Method

The phase separation technique [14] is an economical approach for creating superhydrophobic porous structures, which employs the solubility change of a

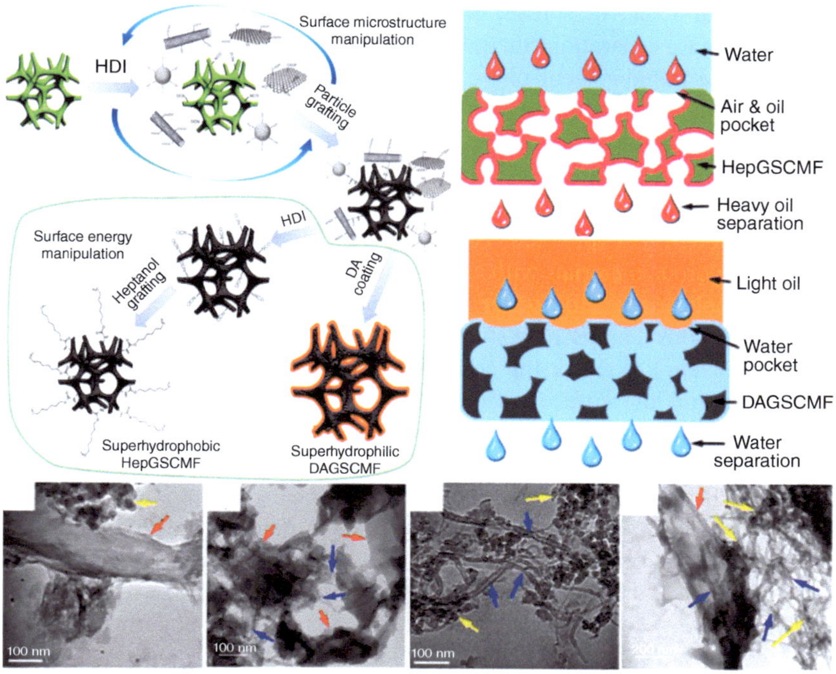

Figure 5.2 Schematic illustration of the modified foam surface microstructure and surface energy manipulation procedures to produce superhydrophobic and super hydrophilic modified foams, and the mechanism of oil–water separation; and the TEM images of multiscale nanoparticles used in the modification. Source: Mi et al. [12]/with permission of Elsevier.

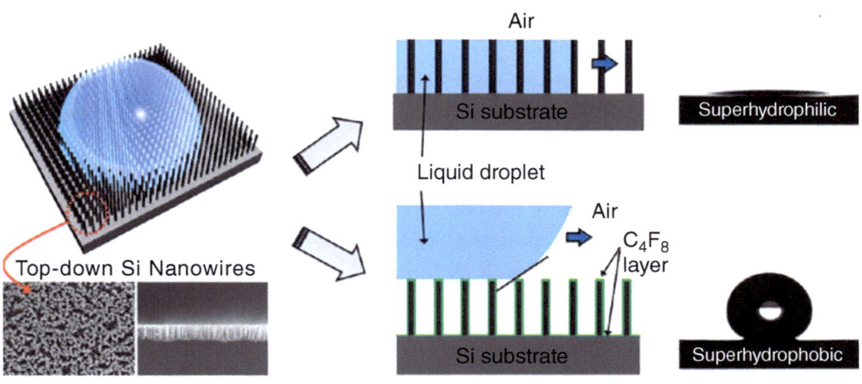

Figure 5.3 Fabrication of silicon nanowires via top-down etching and its superhydrophilicity before fluorine carbon coating and superhydrophobicity after the coating. Source: Kim et al. [13]/with permission of American Chemical Society.

polymer phase in a specific solvent phase when the temperature changes. After triggering the phase separation, the polymer phase would be separated from the solution, and the initial solvent phase forms voids in the solidified new surface, which can quickly enhance the roughness of the material surface to gain superhydrophobicity. Another similar approach, which is called phase inversion is a method that triggers the phase separation by introducing nonsolvent rather than reducing the temperature. A continuous porous structure composed of the polymer phase can be obtained through this approach. The phase separation technique is well-suited for crafting 3D porous materials with superhydrophobicity and has been frequently used in the application of oil/water separation. For example, Deng et al. [15] manufactured highly porous poly(styrene-acrylic acid 2,2,3,4,4,4-hexafluorobutyl ester-divinylbenzene) foams using the phase separation method. The resulting rough structure and low surface energy endowed the foam with superhydrophobic properties with a WCA of 150.3°, and the foam could be used to selectively separate oil from water (Figure 5.4a).

In addition, the phase separation method can also be used to modify the surface of existing foams. Ma et al. [16] developed a flame-retardant polymer foam through copolymerization of hydroxyethyl acrylate (HEA) and sodium vinyl sulfonate (VS) free radicals and induced micro/nanostructures via phase separation on rigid polyurethane (PU) foam substrates. The foam achieved superhydrophobic properties with a WCA of 151°, as shown in Figure 5.4b.

5.1.3.4 Template Replicating

Template replicating is another frequently used method to produce hierarchical structured rough surfaces. In this method, a nanostructured template is necessary to replicate a structured polymer surface with the opposite structure. One example is the replicating of the nanopillar structure using an aluminum template with a nanoporous structured surface that is prepared by the anodic aluminum oxide (AAO) method (Figure 5.5a). For instance, Xie and coworkers [17] proposed for the preparation of the bionic polystyrene surface (BNPPS) with a nanopillar structure by applying dynamic injection molding coupled with a bionic gradient template. The BNPPS film has a transmittance higher than 90% and a reflectivity lower than 5% in the visible light range, and it could prevent contamination from dyeing water, milk, juice, cola, and ink. In addition, the nanopillar structured BNPPS film can kill most of the *Escherichia coli* within 20 minutes.

To obtain a polymer surface with the same nanostructure as the template, another mold, which is often prepared by the curing of poly(-dimethylsiloxane) (PDMS) on the template is required. This approach is frequently used to replicate the nanostructures from natural creatures such as lotus leaves, shark skins, gecko feet, and insect wings. For example, Huang and coworkers [18] investigated the gradually increased hydrophilicity of cicada wings and fabricated micro-nanostructured surfaces to direct the collection of moisture and self-cleaning of the material surface (Figure 5.5b).

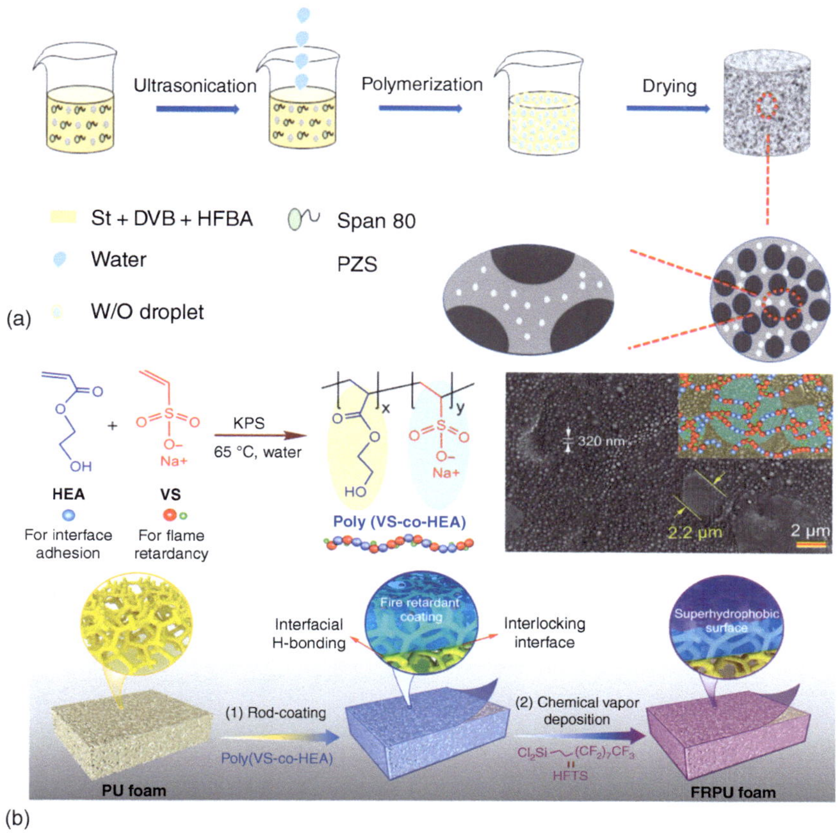

Figure 5.4 (a) Schematic illustration of the fabrication of the porous and hydrophobic foam, and (b) Schematic diagram of superhydrophobic foam prepared by coating method. Source: (a) Deng et al. [15]/with permission of John Wiley & Sons. (b) Ma et al. [16]/with permission of American Chemical Society.

5.1.3.5 Electrostatic Spinning

Electrostatic spinning is a technique that involves creating a jet of polymer solution under the influence of high-pressure electrostatic forces, ultimately yielding nanodroplets or nanofibers. When the porous nanodroplets and nanofibers are composed of low surface energy materials or further modified using low surface energy chemicals, they can exhibit superhydrophobic properties. This method is gaining traction as an emerging technology for both crafting superhydrophobic porous materials and surface modification of existing solid or foamed materials. It offers advantages such as straightforward equipment, low spinning costs, and precise process control. Lv et al. [19] prepared a spray solution by dissolving a blend of polycaprolactone (PCL) and trapezoidal fluorinated polysiloxane (F-PH-LPSQ) in chloroform.

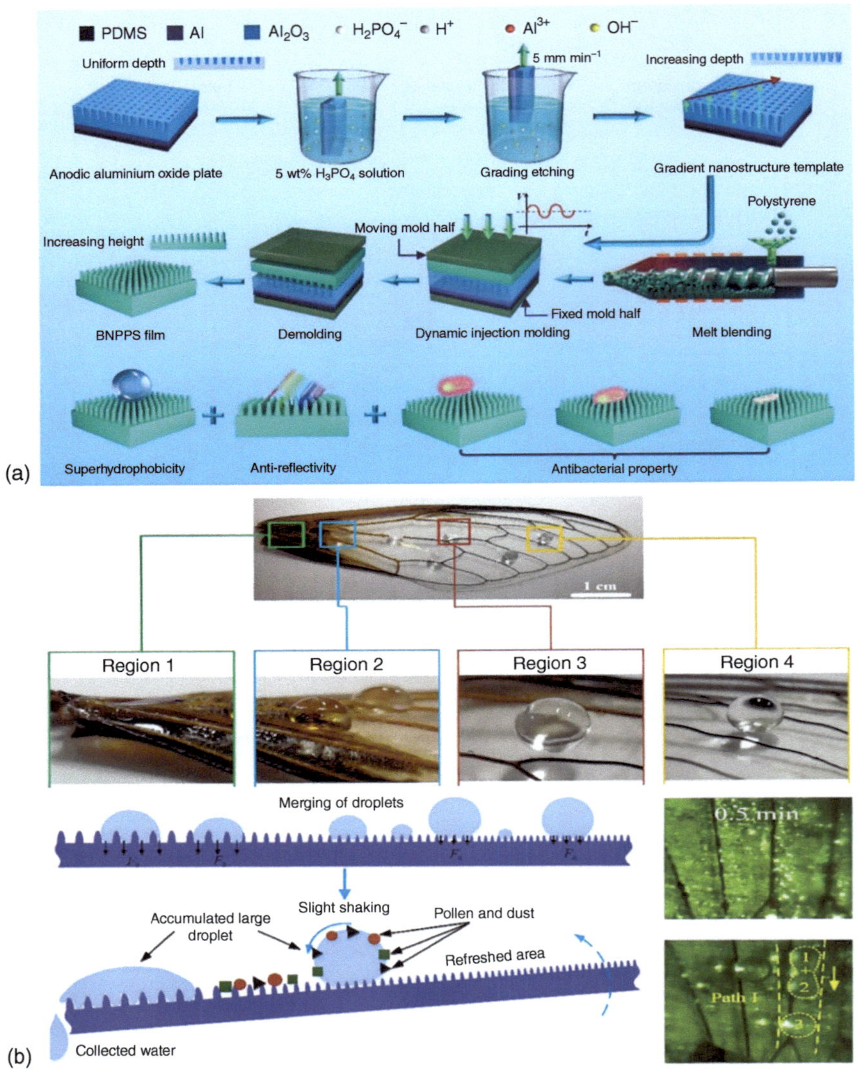

Figure 5.5 (a) Illustration of the preparation process of the bionic gradient nanostructure template and the structured BNPPS film, and (b) Photographs of different regions of a cicada wing, its wettability, and its ability to be used in droplet transportation. Source: (a) Du et al. [17]/with permission of American Chemical Society. (b) Xie et al. [18]/with permission of Elsevier.

When the spray solution is deposited on the surface of substrates under the electrostatic field, it automatically assembles into a rough surface that is composed of low surface energy microspheres with a WCA exceeding 150° and a rolling angle below 1°.

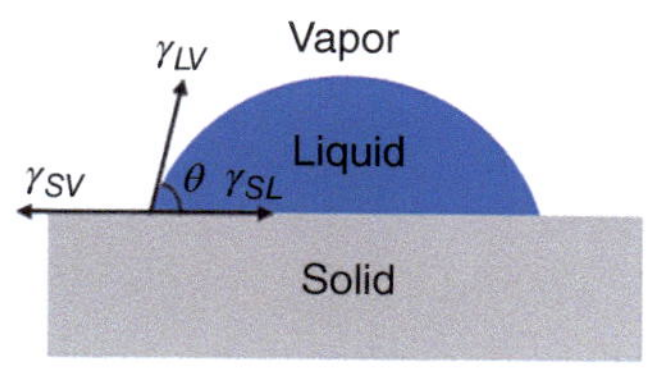

Figure 5.6 Schematic illustration of the liquid contact states described by Young's equation.

5.2 Theoretical Basis of Superhydrophobicity

5.2.1 Wetting on a Solid Surface

Wettability is the spreading of liquid on the solid surface in different degrees, which is the result of the stabilization of the interface between gas, liquid, and solid phases. When the liquid cannot be completely spread out on the solid surface and forms a certain angle with the solid surface, that is, the so-called contact angle, expressed by θ. The SA refers to the angle at which a liquid droplet starts to roll off from the material surface. A superhydrophobic surface usually exhibits a WCA greater than 150° and a SA less than 10°.

In the early nineteenth century, Thomas Young [20] proposed Young's model to explain the wettability characteristics of a material surface (Figure 5.6), which argued that the contact angle has a close relationship with the three-phase interfacial tension involved. Young's equation (Eq. (5.1)) reveals the thermodynamic equilibrium between the contact angle and the surface tension of a solid.

$$\gamma_{SV} = \gamma_{SL} + \gamma_{LV} \cos\theta \tag{5.1}$$

where γ_{SV}, γ_{SL}, and γ_{LV} are the surface tensions between solid–gas, solid–liquid, and liquid–gas, respectively.

Since this equation characterizes the contact angle of a liquid drop on a perfectly smooth ideal plane, this contact angle is called the intrinsic contact angle of the material. Young's wettability equation is the basis for studying the wettability of solids, and the contact angle can reflect the level of wettability of the solid surface.

When $\theta = 0°$, the droplet would completely spread and wet the surface of the substrate (Figure 5.7a); when $0 < \theta < 90°$, the surface of the substrate is considered to be hydrophilic, and the liquid can still wet the substrate (Figure 5.7b); when $90° < \theta < 150°$, the liquid become difficult to completely wet the substrate, the liquid starts to form clear puddles on the substrate surface, and the surface is considered to be hydrophobic; when $\theta > 150°$, the liquid cannot wet the substrate, the surface is considered to be superhydrophobic, and the liquid droplet can easily roll off from the substrate surface (Figure 5.7c). The surfaces with superhydrophobic properties normally have self-cleaning behavior. When $\theta = 180°$, the droplet does not wet the substrate at all, which means that the liquid droplet is floating in the air. It is the ideal repelling state and is difficult to achieve on Earth (Figure 5.7d).

5.2.2 Wetting on Rough Surfaces

Young's equation is applicable to smooth solid surfaces under ideal conditions, but it has limitations in explaining the special wettability on rough surfaces. In 1936,

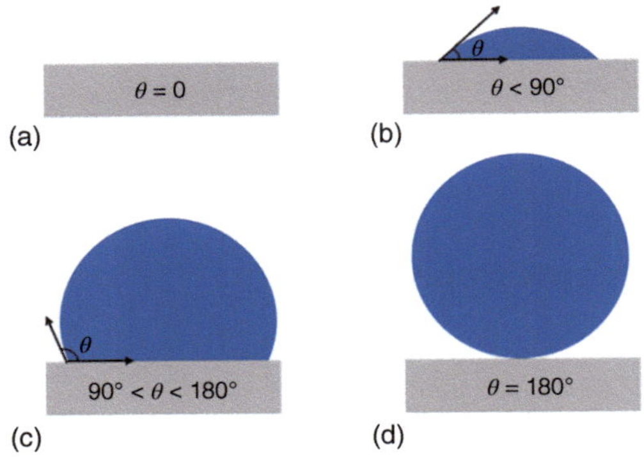

Figure 5.7 Wetting performance of water droplets on the substrate surface.

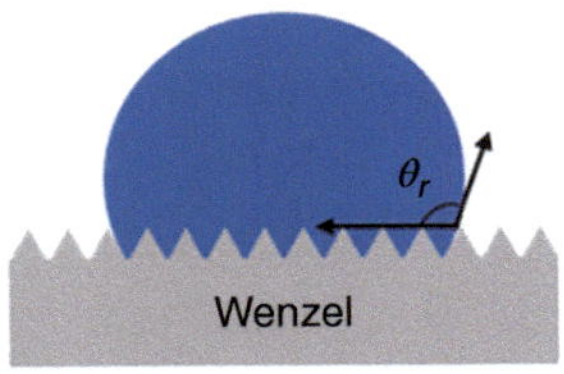

Figure 5.8 Schematic illustration of the contact state of the Wenzel model.

Wenzel [21] proposed a new theory based on the definition of surface tension pointing out that the actual surface area of a rough substrate is larger than its apparent surface area, and further elaborated the effect of nonideal surfaces on hydrophobic properties on the basis of Young's model, which is the Wenzel model (Eq. (5.2)).

$$\cos\,\theta_r = \mathrm{r}\,cos\theta \tag{5.2}$$

where θ_r is the contact angle of the nonideal surface, θ is the intrinsic contact angle of the material, and r is the roughness of the solid surface, which is the ratio of the actual solid–liquid contact area to the apparent contact area ($r > 1$). The Wenzel equation is based on the assumption that the liquid completely penetrates the roughness grooves. In the Wenzel model state, the droplet wets the surface and completely fills all the voids of the rough surface, as illustrated in Figure 5.8. When the wettability of the solid material is hydrophilic, i.e. $\theta < 90°$, the contact angle θ_r decreases as the roughness r of the solid surface increases, i.e. the solid surface exhibits more hydrophilic properties as the roughness increases. Similarly, when the wetting property of the solid material is hydrophobic, i.e. $\theta > 90°$, the contact angle θ_r increases as the roughness *r of* the solid surface increases, i.e. as the roughness increases, the solid surface exhibits more hydrophobic properties [22]. Therefore, the microstructure of the solid material surface can amplify the surface wetting properties, which means that the apparent roughness of the material can make the hydrophilic material more hydrophilic and the hydrophobic material more hydrophobic.

5.2.3 Cassie–Baxter Model

The Wenzel model investigates the wettability of a nonideal surface, but still has some limitations and cannot explain the phenomenon of a compound interface with

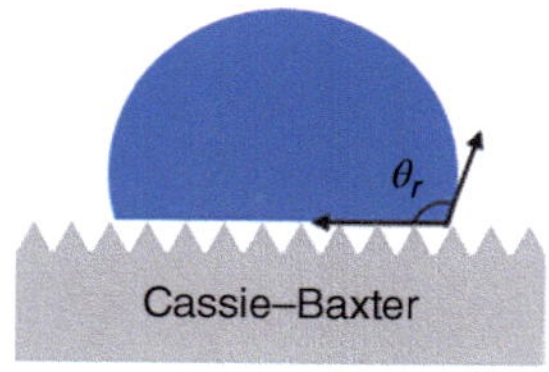

Figure 5.9 Schematic illustration of the contact state of the Cassie–Baxter model.

a gas in contact between a liquid and a rough solid surface (e.g. the superhydrophobic contact state).

When the material surface is composed of a porous structure filled with air, it is difficult for liquid droplets to wet the material surface, which means the Wenzel model is not applicable. To solve this problem, Cassie and Baxter proposed the Cassie–Baxter model in 1944 based on the Wenzel model [23], as illustrated in Figure 5.9.

When the droplet is in contact with the rough surface, the droplet is not able to penetrate into the interior of the rough structure, which leads to the formation of an "air cushion" between the droplet and the substrate surface, as shown in Eq. (5.3):

$$\cos\theta = f_{SL}\cos\theta_1 + f_{GL}\cos\theta_2 \tag{5.3}$$

where f_{SL} and θ_1 refer to the apparent area fraction of solid–liquid contact and the contact angle of solid–liquid, respectively. f_{GL} and θ_2 refer to the apparent area fraction of gas–liquid contact and the contact angle of gas–liquid, respectively. The contact angle θ is the apparent contact angle that can be measured on a rough surface. The sum of the apparent area fraction of liquid–solid contact and the apparent area fraction of liquid–gas contact is 1. Since the contact angle of water in the air is 180°, the Cassie–Baxter expression of the model can be expanded to Eq. (5.4):

$$\cos\theta = f_{SL}(\cos\theta_1 + 1) - 1 \text{ or } \cos\theta = f_{SL}\cos\theta_1 - f_{GL} \tag{5.4}$$

From the above equation, it can be seen that as the increase of solid surface roughness, the contact area between the droplet and the material surface becomes smaller, the solid–liquid contact area fraction f_{SL} becomes smaller, and the apparent contact angle between the liquid and the solid surface will be higher. Therefore, enhanced surface roughness is an effective approach to increase the apparent contact angle of a material.

5.2.4 Theoretical Basis for 3D Porous Foams

The Cassie–Baxter model finds applicability not only in characterizing the surface structure and properties of flat materials but also in porous materials with 3D structures. These porous materials can achieve remarkable superhydrophobic properties through the construction of microstructures. In the case of 3D structures, such as nanospheres presented on the foam surface, the water profile would sag along the structure, and the solid–liquid–gas contact interface would change as the sagging

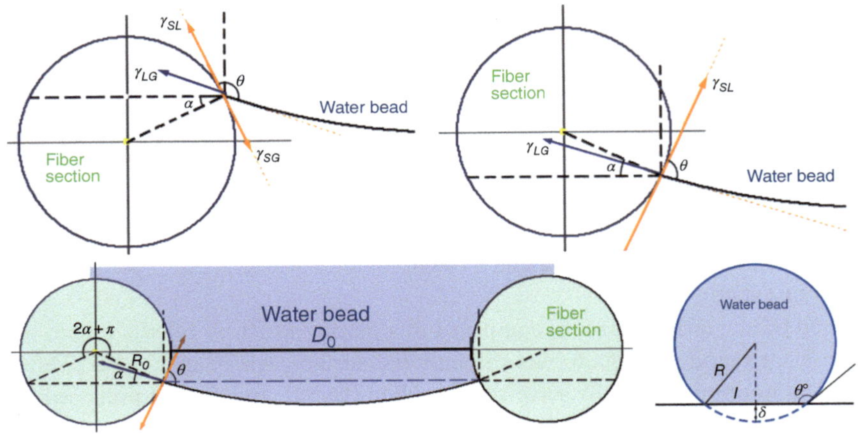

Figure 5.10 Schematic model of a water bead in contact with circular structures, and the theoretical illustration of a water bead in contact with a circular structured surface. Source: Mi et al. [24]/with permission of Elsevier.

of the water profile. As illustrated in Figure 5.10, two possible contact states can be defined by the angle α. When the composite interface is on the upper semicircle, angle α is defined to be negative, thus indicating a normal contact situation, and such structure refers to the contact state of surfaces with pyramid structure and platform structures. When the composite interface is on the lower semicircle, angle α is defined to be positive, thus representing a reentrant contact situation, since a higher α is preferred in constructing superhydrophobic surfaces. Then, the angle α equals 0, the direction of γ_{LG} is horizontal, and it represents the case of surfaces with pillar structures. The resultant interfacial tension (γ_0) was obtained from the surface tensions on solid–liquid (γ_{SL}), liquid–gas (γ_{LG}), and solid–gas (γ_{SG}) interfaces at the intersection. To achieve an equilibrium state, the combined γ_0 of all features on the material surface should cancel out the gravitational force of the water bead.

At equilibrium, if the water bead is in normal contact, the following equation, Eq. (5.5), is applicable.

$$\gamma_0 = \gamma_{SL} \cos\alpha + \gamma_{LG} \cos(\pi - \theta + \alpha) - \gamma_{SG} \cos\alpha \tag{5.5}$$

It can be simplified into Eq. (5.6), shown below:

$$\gamma_0 = (\gamma_{SL} - \gamma_{SG}) \cos\alpha - \gamma_{LG} \cos(\theta - \alpha) \tag{5.6}$$

At equilibrium, if the water bead is in reentrant contact, Eq. (5.7) applies:

$$\gamma_0 = \gamma_{SL} \cos\alpha + \gamma_{LG} \cos(\pi - \theta - \alpha) - \gamma_{SG} \cos\alpha \tag{5.7}$$

It can be simplified into Eq. (5.8):

$$\gamma_0 = (\gamma_{SL} - \gamma_{SG}) \cos\alpha - \gamma_{LG} \cos(\theta + \alpha) \tag{5.8}$$

It is clear that θ is different for these two situations at equilibrium, which implies that a larger θ is required to maintain equilibrium at normal contact. This is the reason why reducing the surface energy is important for enhancing hydrophobicity. For a substrate material with $\theta < 90°$, water would sag along the circle surface and equilibrate at a reentrant contact situation.

Combined with the above model with Yong's Equation, the θ^* of a substrate with a geometry defined by a radius, R_0, and an interstructural distance, D_0 can be obtained.

The projected length of a water arc can be expressed by Eq. (5.9).

$$L_{LG} = D_0 + 2R_0(1 - \cos\alpha) \tag{5.9}$$

Hence, the fraction of the projected length of a water arc is expressed as follows.

$$f_{LG} = \frac{D_0 + 2R_0(1 - \cos\alpha)}{2R_0 + D_0} = 1 - \frac{2R_0\cos\alpha}{2R_0 + D_0} \tag{5.10}$$

In the normal contact situation, the fraction of the projected length of an arc can be expressed as Eq. (5.11).

$$f_{SL} = \frac{2R_0\cos\alpha}{2R_0 + D_0} \tag{5.11}$$

Similarly, in the reentrant contact situation, the fraction of the projected length of an arc can be expressed as Eq. (5.12).

$$f_{SL} = \frac{4R_0 - 2R_0\cos\alpha}{2R_0 + D_0} \tag{5.12}$$

By substituting Eqs. (5.10) and (5.11), or (5.10) and (5.12), into Eq. (5.3), the apparent contact angle in both the normal contact situation (Eq. (5.13)) and the reentrant contact situation (Eq. (5.14)) can be obtained.

$$\cos\theta^* = \frac{2R_0\cos\alpha}{2R_0 + D_0}(\cos\theta + 1) - 1 \tag{5.13}$$

$$\cos\theta^* = \frac{2R_0(2 - \cos\alpha)}{2R_0 + D_0}\cos\theta + \frac{2R_0}{2R_0 + D_0}\cos\alpha - 1 \tag{5.14}$$

Note that when $\alpha = 0°$, Eqs. (5.3) and (5.5) became identical to Eq. (5.15).

$$\cos\theta^* = \frac{2R_0}{2R_0 + D_0}(\cos\theta + 1) - 1 \tag{5.15}$$

Thus, a more specific structural factor can be defined as $k = \frac{2R_0}{2R_0 + D_0}$, which is reciprocal to θ^*. It can be seen that θ is also proportional to θ^*. Therefore, increasing D_0, α, and θ, and decreasing R_0 are the suggested ways of enhancing θ^*. Similar suggestions on structured surfaces have been proposed by Tuteja and coworkers [25]. These equations provide more detailed insights into the material structure of a specific substrate. One can see that increasing α, reducing R_0, and increasing D_0 and θ are the keys to enhancing the superhydrophobicity of a surface. Therefore, it is desired to fabricate surfaces with reentrant structures with small structure size, but large gas between adjacent structures, and surfaces with low surface energy (low intrinsic contact angle) to achieve high superhydrophobicity.

However, an infinite increase in D_0 or a decrease in R_0 is irrational since the water profile would sag into the space between the structures, turn into the Wenzel contact state, and finally penetrate the 3D porous material due to the capillary effect. Thus, another criterion is required to establish a boundary condition. Tuteja and coworkers [26] introduced the breakthrough pressure as another criterion. However, it might be improper for a stationary situation or porous 3D substrate. By utilizing the Mahavedan–Pomeau model [27] and the energy equilibrium equation for water on a porous material [28], a criterion for a 3D porous material is established. Equation (5.16) suggests that the overall surface tension beneath a water bead, γ, is proportional to the third power of the size of the water bead, R^3, and reciprocal to the sagging distance, δ,

$$\gamma \propto \frac{\rho g R^3}{\delta} \tag{5.16}$$

where ρ is the density of water, g is the gravitational constant, R is the radius of the water bead, and δ is the sagging distance assuming that the water bead is spherical.

Substituting γ with the individual surface tension of each circular structure, $\gamma_{0,}$ and the number of contacting structures, n, Eq. (5.17) is obtained:

$$n \propto \frac{\pi l \rho g R^3}{\gamma_0 \delta (D_0 + R_0)} \tag{5.17}$$

Where l is the radius of the contact area of the water bead with the substrate.

It is clear from Eq. (5.17) that the number of contacting structures needs to increase as the water bead volume (R) increases and as the structure diameter (R_0) decreases to maintain a stationary state. In other words, a critical number of contacting structures (n_c) is required to achieve a stationary state.

For a surface with micro-nanostructures in multiscales, the water repellency behavior will usually be enlarged. Tuteja and coworkers [29, 30] proposed Eq. (5.18) to account for this phenomenon. The equation evidently showed that θ_n^* increases with the increase of n.

$$\cos\theta_n^* = (1 - f_{LG,n})\cos\theta_{n-1}^* - f_{LG,n} \tag{5.18}$$

where $n = 1, 2, 3$, and so on, and is the number of layers of hierarchical structures, f_{LG} is the projected area fraction of the liquid–gas interface for the nth layer.

θ_n^* and θ_{n-1}^* are the apparent contact angles on the hierarchically structured substrate with n and $n-1$ layers of nanostructures.

For example, Mi et al. [31] fabricated superhydrophobic 3D fibrous foam composed of silica fibers decorated with hierarchical Co nanoparticles. They verified that the "solid–liquid–gas" interfaces generated by the hierarchical nanoparticles could enhance the amount of "air pockets" presented on the material surface and significantly reduce the sagging distance needed for the water to achieve equilibrium. On the contrary, the fibrous foam without hierarchical Co nanoparticle decoration showed a much lower WCA and a large water profile sagging distance, as illustrated in Figure 5.11.

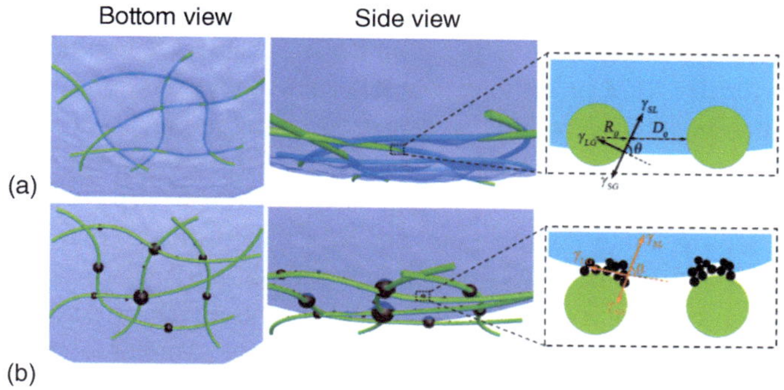

Figure 5.11 Schematic illustration of liquid droplets in contact with (a) smooth fibers, and (b) hierarchically structured fibers. The enlarged diagrams show the interfacial contacting surfaces at equilibrium. Source: Mi et al. [31]/with permission of Elsevier.

5.3 Characterizations of Superhydrophobic Foams

5.3.1 Morphological Characterization

Microstructure is a critical factor that determines the wettability of a material with a rough surface. It is important to characterize the morphology of the foam material to reveal the underlying mechanism of its surface wettability.

5.3.1.1 Scanning Electron Microscope (SEM)

Most nonconductive polymer foams need to be sprayed with a gold film using a coating instrument before observation. The scanning electron microscope (SEM) images can be further analyzed by the contrast of solid and void fractions. For polymer foams, the average cell size and cell density are often obtained from the SEM images using various image analyzing software.

In addition, with the help of high-resolution SEM, researchers are able to observe the hierarchical structures presented on the material surface with size down to a few nanometers. Thus, SEM is the most useful tool for analyzing the surface morphology of superhydrophobic materials.

5.3.1.2 Atomic Force Microscope (AFM)

Although SEM can view the surface morphology of materials in detail, it cannot detect the height of the nanostructures on the material surface. Atomic force microscope (AFM) is another method that can map the surface topography in 3D, which is widely used for nanoscale imaging and force spectroscopy. In AFM, an optical microscope is equipped with a cantilever that interacts with the sample surface to obtain the height profile of the sample surface. Typically, tapping and noncontact modes can be selected in AFM measurement, and the researchers can obtain 3D topography images with a proper AFM scan, which can be further used to analyze the height and size details of the features on the material surface.

5.3.1.3 White-Light Interferometry (WLI)

In the case of polymer foam characterization, the material surface would often be too rough or too soft for the AFM probe to detect. Alternatively, white-light interferometry (WLI), which is an effective and widely used technique for structured surface measurement, can be used in these cases. In the WLI measurement, a high-precision laser interferometer and a 3D motion carrier platform with an angle adjustment actuator are the main components. Samples plated on the platform would be scanned and the height difference on the sample surface would be reflected by the interference mapping image. Although WLI has a lower resolution than AFM, it is powerful in quickly analyzing the surface of samples without contact.

5.3.2 Surface Chemistry and Wettability Characterization

5.3.2.1 Surface Chemistry Characterization

As mentioned earlier, surface energy is an important factor affecting the wettability of material surfaces. The surface energy is highly related to the functional chemical bonds on the material surface. Therefore, it is important to evaluate the surface chemistry of a material surface.

Fourier transform infrared (FTIR) spectra are frequently used to characterize the surface chemistry of material by detecting specific frequencies of light that correspond to the vibrational bond energies within molecules. Usually, the material is scanned in the wavelength range of 4000–600 cm^{-1} using the FTIR spectrometer to analyze the molecular structure of the material.

In addition, X-ray photoelectron spectroscopy (XPS) is a powerful tool to detect the elements presented on the material surface. The measurements are often performed on an X-ray photoelectron spectrometer with a focused, monochromatic K-alpha X-ray source and a monoatomic/cluster ion gun. The Survey scan in XPS can reveal the elements that exist on the material surface, and the core-level signal spectra of each element can be Gaussian fitted and the proportion of each bond can be determined from the peak area ratios.

5.3.2.2 Surface Wettability Characterization

The surface wettability of a material can be analyzed using WCA measurement. The measurement is typically carried out on a static contact angle tester using deionized water as the target liquid. The contact angle can be measured by attaching the droplet to the surface of the sample. The rolling angle can be measured by first dropping a 5 μl droplet on the flat sample surface and then slowly increasing the angle of the platform, and the angle when the droplet rolled on the platform is recognized as the SA.

5.3.3 Selective Oil Adsorption Experiments

In the application of selective oil adsorption or oil–water separation, it is critical to measure the ability of the functional foam material to selectively absorb oil from water. In a typical process, organic solvents with lighter density than water (i.e. Cyclohexane) and organic solvents with heavier density than water (i.e. Carbon tetrachloride) were stained with Sudan III dyes or oil O Red, and the two organic solvents are dropped into water separately, and the superhydrophobic polymer foams are used to selectively adsorb the organic solvents on and below the water surface.

5.3.4 Absorption Capacity of Superhydrophobic Foams

In the application of oil absorption, it is necessary to assess the capacity of the foam toward various organic solvents. In the absorption capacity experiments, different kinds of organic solvents (e.g. toluene, anhydrous ethanol, carbon tetrachloride, chloroform, xylene, cyclohexane, etc.) are used to measure the adsorption and storage capacity of the developed superhydrophobic porous foam. At room temperature, the mass of the dried porous foam is weighed first, then the porous foam is immersed in a beaker containing organic solvents. After saturating the porous foam with organic solvents, the foam is removed, and the residual organic solvents on the foam surface are wiped off with weighing paper. The absorption capacity (g/g) is calculated using Eq. (5.19):

$$Q = \frac{m_a - m_0}{m_0} \tag{5.19}$$

where m_0 and m_a are the masses of the porous foam before and after the adsorption of organic solvents, respectively.

To evaluate the reusability of the absorption foam, the adsorption and desorption cycling capacity is often characterized. In the process, porous foam was placed in beakers containing different solvents, such as cyclohexane and carbon tetrachloride, respectively. After adsorption, the adsorbed organic solvent is removed in a vacuum oven. The mass of porous foam before and after each cycle is recorded to calculate the adsorption and desorption cycling capacity of the porous foam for the adsorbed and desorbed oil.

5.4 Superhydrophobic Foam Preparation Technology

5.4.1 Nanoparticle/Porous Material Complexes

Incorporating nanoparticles into porous materials represents a commonly employed technique for fabricating oil–water separation materials with enhanced adsorption properties. The introduction of nanoparticles into porous materials serves several purposes, including the enhancement of surface roughness, the augmentation of hydrophobicity to prevent water infiltration, and the promotion of oleophilicity required for efficient oil absorption. Various nanoparticles have been used in the preparation of superhydrophobic foams, such as graphene, carbon nanotube (CNT), carbon black, and silica nanoparticles. In most cases, nanoparticles are decorated

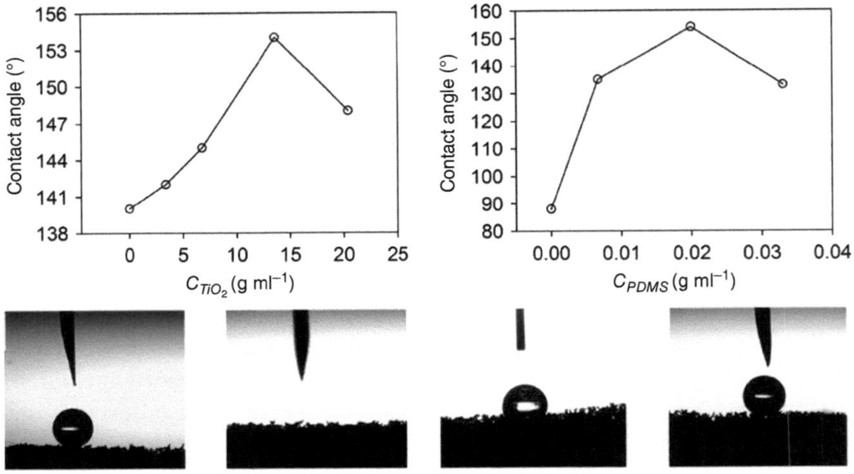

Figure 5.12 Effects of the concentrations of TiO_2 sol gel and PDMS solution on the hydrophobic properties of the prepared sponges. Source: Shuai et al. [33]/with permission of Elsevier.

on the surface of porous foams via chemical grafting/deposition, physical coating, *in situ* growth, etc.

Tamsilian et al. [32] developed a superhydrophobic nanocomposite sponge using a zeolite imidazole framework (ZIF8) metal-organic framework as a low-density crystal. While obtaining superhydrophobicity, it exhibits a WCA of up to 156° and can maintain a high WCA even in acidic and alkaline environments. Shuai et al. [33] fabricated a novel superhydrophobic PDMS-TiO_2 coated PU sponge by sol-gel growth of TiO_2 nanoparticles on the surface of the PU sponge, followed by in situ polymerization of PDMS. The as-prepared PDMS-TiO_2-PU sponge shows a superhydrophobic surface with a WCA of 154° and exhibits high sorption capacities of more than 16.7 g g^{-1} for the studied oils and organic compounds (Figure 5.12).

5.4.2 Superhydrophobic Foams Prepared by Phase Separation

Phase separation offers an economical method for fabricating superhydrophobic surfaces through the controlled thermal- or solvent-induced phase separation of a stable polymer solution. A continuous micro- and nanostructured superhydrophobic surface can be obtained after the phase separation and solvent removal process [34]. The microstructure of the foam prepared is mainly determined by the phase separation behavior during the formation of the polymer-rich and solvent-rich phases. Therefore, the morphology of the foam can be regulated by controlling the phase separation process.

For instance, Mi et al. [35] fabricated anisotropic foams (MCGA) composed of cellulose nanofibril (CNF) and graphene oxide (GO) via a special bidirectional freeze-drying process (Figure 5.13). In the process, the solution is frozen in two directions and the CNF and GO are assembled parallel to form a bidirectional

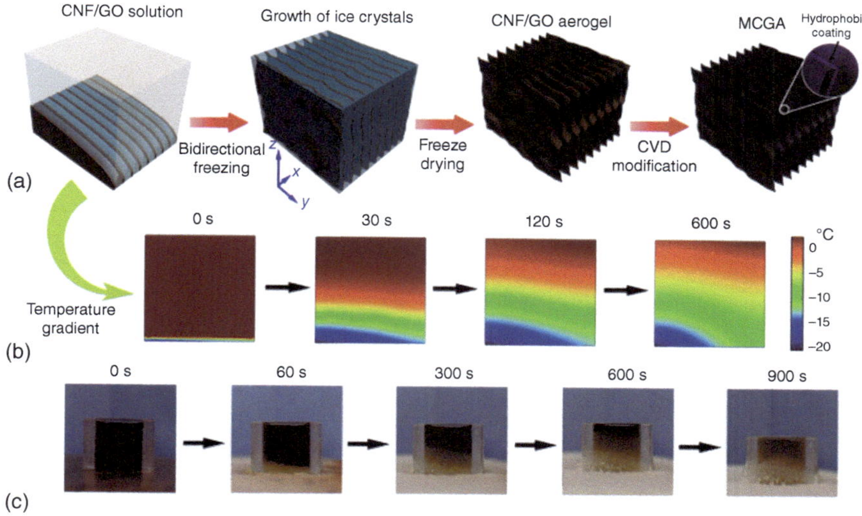

Figure 5.13 (a) Schematic illustration of the bidirectional freeze-drying fabrication process of MCGA, (b) Temperature gradient simulation in the freezing process using the COMSOL Multiphysics software, and (c) The solution freezing process via bidirectional freezing. Source: Mi et al. [35]/with permission of Elsevier.

aligned porous structure. The fabricated foam gained superhydrophobicity after a hydrophobic coating and can be directly used in selective oil absorption from water. The anisotropic structure of the hybrid foam rendered it excellent compressibility and recoverability.

In addition, Wang et al. [36] fabricated a superhydrophobic polylactic acid (PLA) foam using a dioxane solvent combined with water-assisted phase separation. This foam boasts a distinctive micro/nanostructure, featuring a remarkable WCA of 151° and an impressive maximum adsorption capacity of $34.5\,g\,g^{-1}$ for organic solvents, rendering it highly suitable for practical applications. Kato and Sato [37] developed a unique UV light-triggered micro/nanofabrication system by utilizing the polymerization-induced phase separation (PIPS) behavior. They first coated a precursor mixture for PIPS on a substrate, then UV cured to induce phase separation between the growing polymers and solvents, finally the solvents were flushed using ethanol to remove solvents and leave a specific texture at the material surface which possesses a remarkable superhydrophobic surface with a WCA of 163° (Figure 5.14).

5.4.3 Superhydrophobic Aerogels

Sol-gel synthesis represents another highly effective approach for the production of 3D porous materials, as these materials readily develop porous structures after specific reactions or self-assembly processes and the removal of the solvent phase. The obtained porous material is often called aerogel since the solvent phase in the pristine gel is replaced with air in the final product. Mu et al. [38] employed a one-pot reaction method utilizing methyltrimethoxysilane (MTMS),

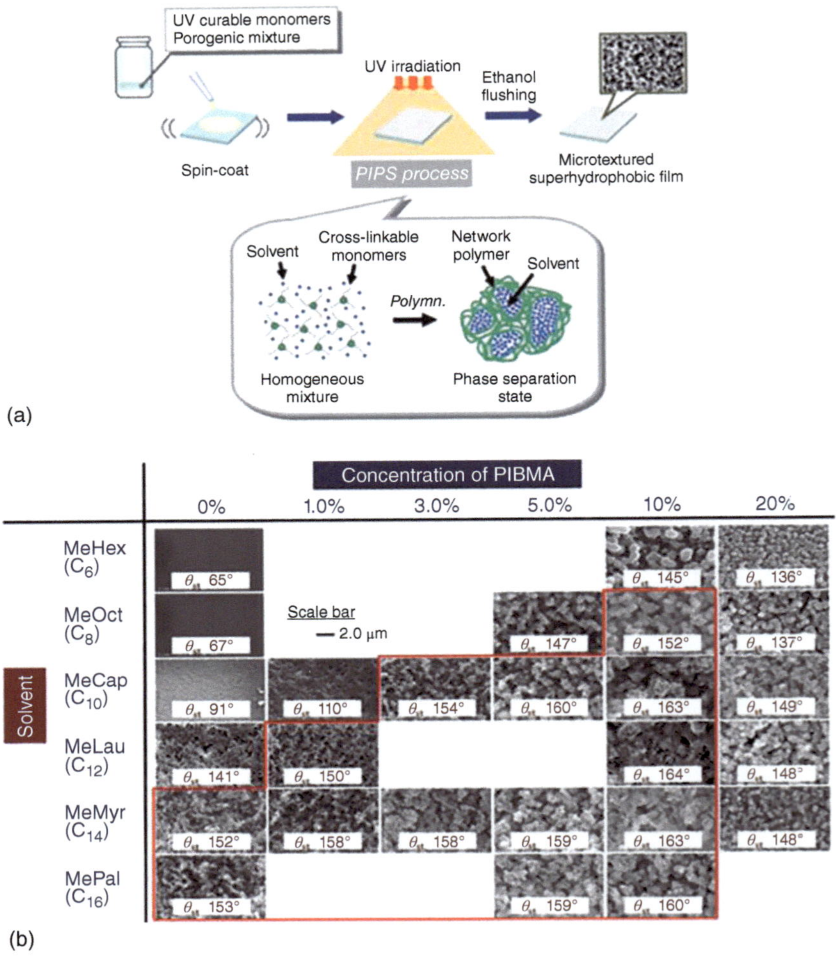

Figure 5.14 Schematic diagram (a) and morphology, and (b) of the micro-nanostructure prepared by UV-induced phase separation. Source: (b) Kato and Sato [37]/with permission of Royal Society of Chemistry.

dimethyldimethoxysilane (DMDMS), and tetraethoxysilane (TEOS) to create remarkably porous superhydrophobic silicone aerogels composed of spherical particles. The fabricated sponges exhibit an oil absorption capacity ranging from 6 to 14 g g^{-1}, and they can retain 95–98% of their adsorption capacity after undergoing 50 cycles with various organic liquids. In addition, the aerogels demonstrated excellent resilience even when exposed to temperatures as high as 200 °C without significant deterioration of their functionality (Figure 5.15a). Mi et al. [39] used a similar one-pot hydrothermal synthesis and subsequent freeze-drying and chemical vapor deposition modification process to fabricate a fluorinated hybrid aerogel (FHA) consisting of GO, CNFs, and silica nanoparticles. The CNF phase was used as the linker phase, and the GO and silica components were composed of the

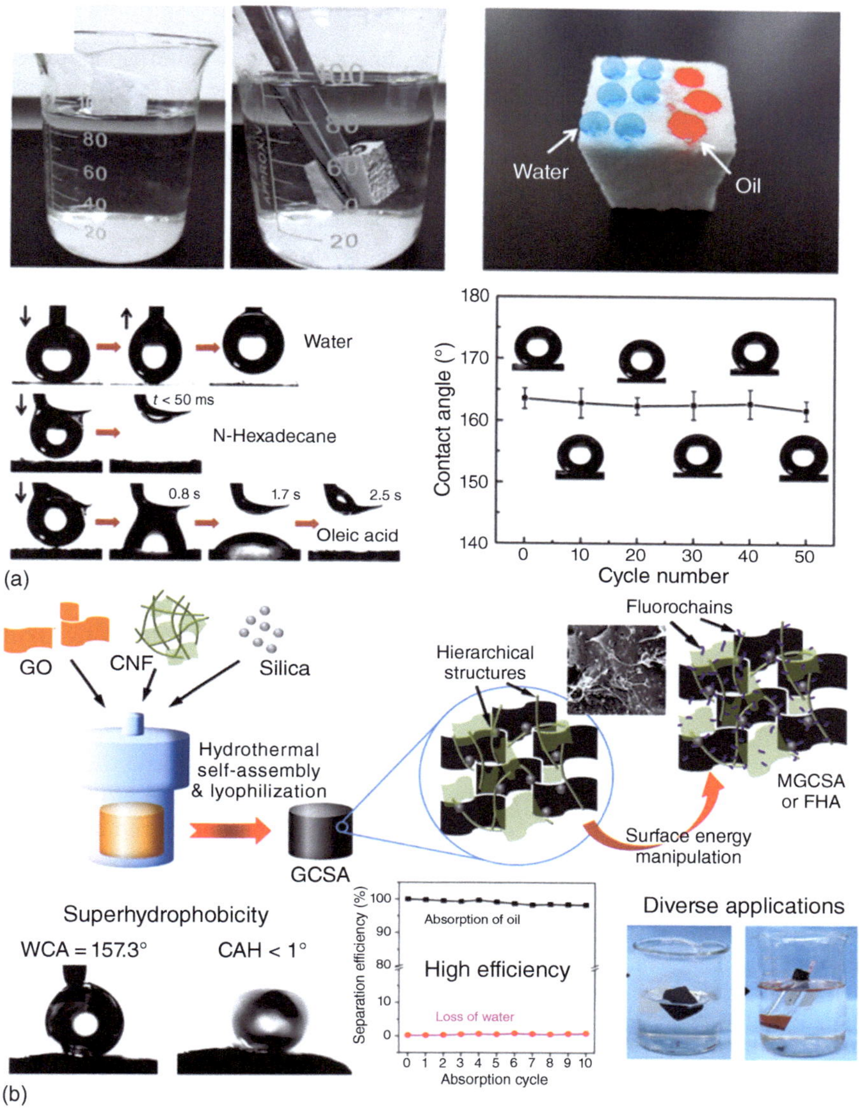

Figure 5.15 (a) Schematic diagram of hydrophobicity and lipophilicity of the sponge, and (b) Schematic diagram of the FHA-manufacturing process and the results of superhydrophobic performance and selective oil absorption testing. Source: (a) Mu et al. [38]/with permission of Royal Society of Chemistry. (b) Mi et al. [39]/with permission of American Chemical Society.

hierarchical structure of the aerogel. The grafting of fluoro chains rendered the aerogel excellent supherhydrophobicity with a high WCA of 157° and a water SA of less than 1°. In addition, the aerogel demonstrated superoleophilicity, extraordinary absorption efficiency (~100%), and high absorption capacity (39–68 times weight gain) to various oils and chemical solvents (Figure 5.15b).

5.4.4 Superhydrophobic Fibrous Sponge Prepared by Electrospinning

Electrospinning is a technique that produces fibers in nano to submicro diameter through a polymer solution with the presence of a high electric field. The polymer solution with a desired viscosity is positively charged and stretched in the electric field, and the polymer fiber is gradually solidified with the evaporation of the solvent phase during the fiber collecting processing. Electrospinning has been used to fabricate membranes with superhydrophobic properties by using low-surface-energy polymers. For instance, Su et al. [40] fabricated a superhydrophobic electrospun polyvinylidene fluoride (PVDF) membrane via electrospinning and solvent modification. The developed PVDF membrane showed a high WCA of 152.4°, fouling resistance, and ability for seawater desalination.

However, it is challenging to use electrospinning to produce a fibrous sponge with a 3D structure due to the electrostatic attraction of the fibers toward the negatively charged collector. To overcome this problem, Mi et al. [41] developed a self-assembly approach to fabricate 3D silica fiber sponges by controlling the viscosity, the gelation speed, and the direct deposition process of the fibers. After grafting the 3D silica sponge using glutaraldehyde and surface modification using perfluorodecyltriethoxysilane, the 3D sponge achieved superhydrophobicity, and the WCA of the sponge could be tunned to allow the material to exhibit the "rose peddle" effect. Hence, the developed sponge could be used for selective oil absorption in harsh environments (Figure 5.16a) [24]. In addition, they further combined these 3D silica fibers with CNF and Fe_3O_4 nanoparticles to produce hybrid fibrous aerogel via freeze drying. As shown in Figure 5.16b, the hybrid fibrous aerogel achieved a superhydrophobic surface after surface modification with florin chains, and the fibrous aerogel could be attracted by the magnetic field due to the presence of the Fe_3O_4 nanoparticles in it, which rendered the fibrous aerogel ability to be remotely controlled to absorb oil from harsh environments [42].

5.4.5 Superhydrophobic Foams Fabricated via Supercritical CO_2 Foaming

$scCO_2$ foaming is an environmentally friendly method to produce polymer foams using CO_2 as the physical blowing agent. Polymer foams with open-cell structure and closed-cell structure can be produced by controlling the foaming processing parameters [43, 44]. To achieve the superhydrophobic property of the foam, it is necessary to expose the highly porous structures on the surface of the foam, and various nanoparticles are also frequently used to enhance the complexity of the surface roughness and alter the surface energy. Highly porous polymer foams with open cell structures have been used in industrial applications such as antifouling and self-cleaning, solvent absorption, and oil–water separation. Li et al. [45] prepared PLA/polybutylene succinate (PBS) foam using $scCO_2$ foaming. The foam has a highly interconnected lace-like open cell structure with 98.2% openness, and the adsorption capacity for various organic solvents and grease reached 7.9–21.9 $g\,g^{-1}$. However, the hydrophobicity did not reach the superhydrophobicity region due to the relatively high surface energy of PLA and PBS.

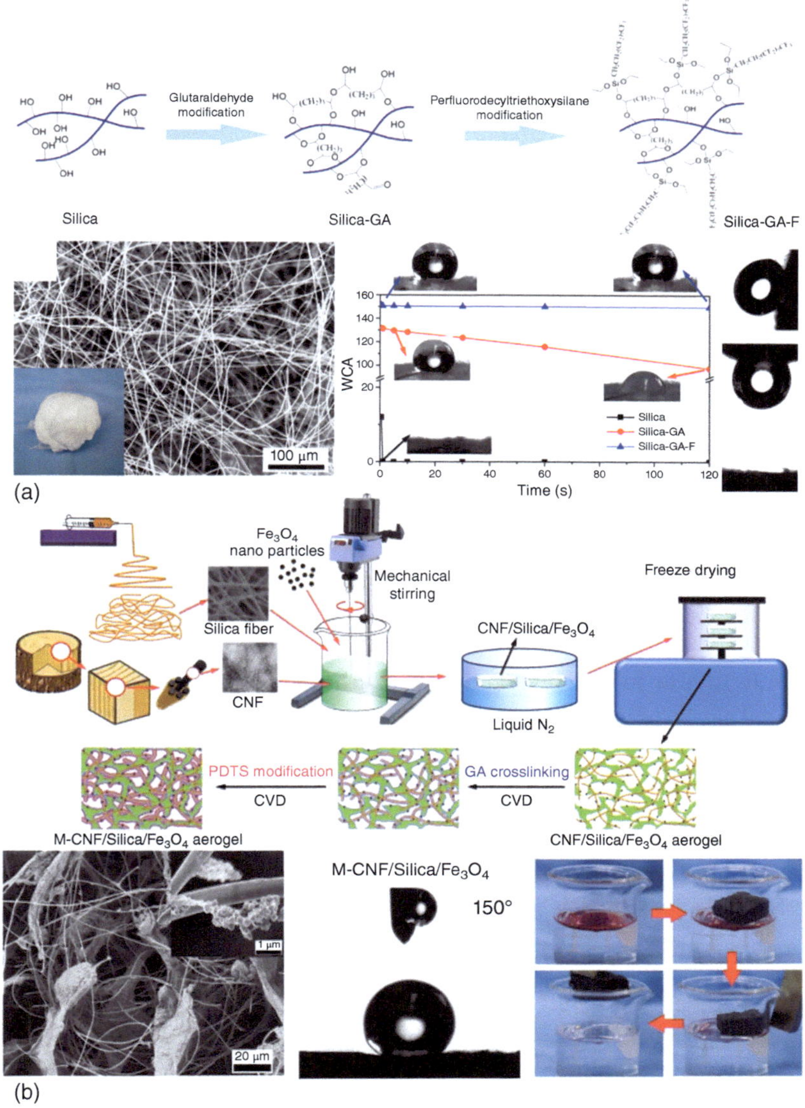

Figure 5.16 (a) Schematic diagram of as-spun self-assembled 3D TEOS/PVA fibers, and (b) Fabrication process of CNF/silica/Fe_3O_4 aerogel, and modification illustration. Source: (a) Mi et al. [24]/with permission of Elsevier. (b) Mi et al. [42]/with permission of Springer Nature.

In oil–water separation applications, it is essential to involve a selective absorption process for liquids with varying surface tensions. Liquids with low surface tension assume a Wenzel state on the material's surface, entering the 3D structure through capillary action. In contrast, high surface tension liquids exhibit a Cassie–Baxter state on the material's surface, making it challenging for them to infiltrate the internal structure, as is the case with water. The predominant method for producing superhydrophobic foam involves surface modification. However, once the surface structure is even slightly damaged, liquids can penetrate the foam's interior, significantly reducing its hydrophobic performance. Hence, the construction of a durable surface structure is of paramount importance for foam materials.

Mi and coworkers [46] prepared a superhydrophobic foam using ultra-high molecular weight polyethylene (UHMWPE) via $scCO_2$ foaming and subsequent surface modification. As shown in Figure 5.17a, $scCO_2$ foamed UHMWPE showed a highly porous structure, and the foam was decorated with modified silica nanoparticles via a solvent-assisted coating process. It was found that the WCA was enhanced from 144° to 158° with the increase of fluorine chain-modified silica nanoparticle content to 1 wt%, and the developed foams showed excellent performance in selective oil absorption. Moreover, Mi et al. [43] further developed a selective etching approach to generate UHMWPE crystal lamellar on the UHMWPE foam surface by immersing the foam into a lean solution containing UHMWPE and toluene. The partial dissolving of the amorphous phase and the crystallization of UHMWPE lamellar occurred on the struts of the UHMWPE foam, which resulted in a nanostructured surface on the foam skeleton. The developed foam with an optimized structure achieved an exceptional WCA of 162° and a SA of 1°, and the foam demonstrated remarkable stability in repetitive use and tolerance to mechanical damage (Figure 5.17b).

To avoid the additional coating and modification approaches and eliminate organic solvent use in the preparation of superhydrophobic foams via $scCO_2$ foaming, Mi et al. [44] developed a polypropylene (PP)/polytetrafluoroethylene (PTFE))composite foam using twin-screw extrusion and $scCO_2$ foaming process. As illustrated in Figure 5.18, micro-sized PTFE particles were *in situ* fibrillated into PTFE nanofibers and uniformly distributed in the PP matrix with the PTFE nanoparticles introduced. The presence of low surface energy PTFE nanofiber networks and nanoparticles contributed to the excellent superhydrophobicity of the developed composite foam. The PTFE component not only reduced the surface energy of the foam but also created hierarchical structures on the foam surface, which resulted in a high WCA of 156.8° and a SA of 1.9°. Noticeably, the foam processes exceptional durability and abrasion resistance since the porous structure and PTFE components were distributed within the whole foam, which allows the explosion of new superhydrophobic surfaces after the existing one is damaged.

Another issue faced with the $scCO_2$ foamed polymer foams is the presence of solid skin layers. The skin layers were caused by the rapid escaping of foaming agents (e.g. CO_2 gas) in the region close to the surface of the material due to the high-pressure difference and rapid gas diffusion. Therefore, an additional cutting, etching, or modification process is often needed for the fabrication of superhydrophobic foams.

To solve this problem, Mi and coworkers [47] introduced a confinement layer in the fabrication of PVDF foam. They coated the solid PVDF sheets with silicon rubber layers and natural rubber layers, separately, to serve as the gas barrier to reduce the CO_2 escaping rate. As shown in Figure 5.19, with the surface confinement layer, the

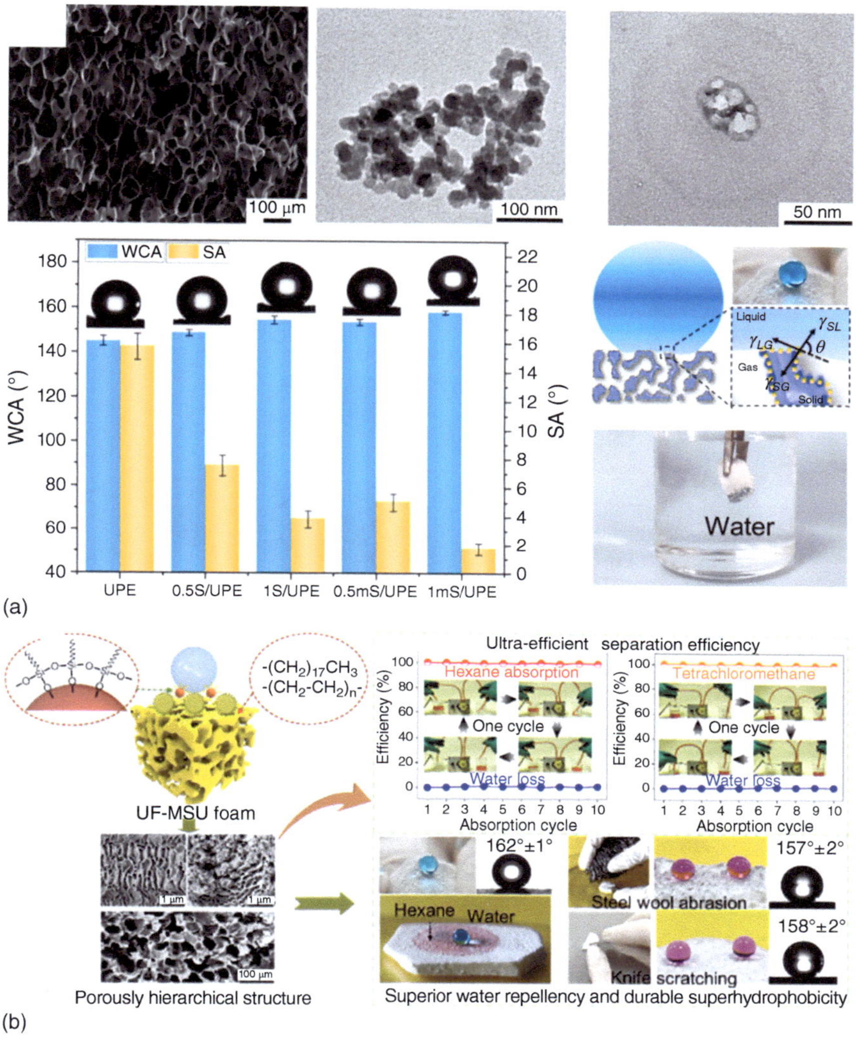

Figure 5.17 (a) SEM images of the UHMWPE nanocomposite foam and the modified silica nanoparticles, as well as the superhydrophobic properties of the UHMWPE nanocomposite foam, and (b) schematic diagram of the preparation process, hydrophobic performance, and the mechanical durability of the developed modified UHMWPE foams. Source: (a) Ren et al. [46]/with permission of Elsevier; (b) Sun et al. [43]/with permission of American Chemical Society.

PVDF foam demonstrated a highly porous structure on its surface, and the cells of the whole foam were mostly open cells. The skinless open-cellular structure of the fabricated PVDF foam is recognized as a promising material for oil absorption and oil–water separation.

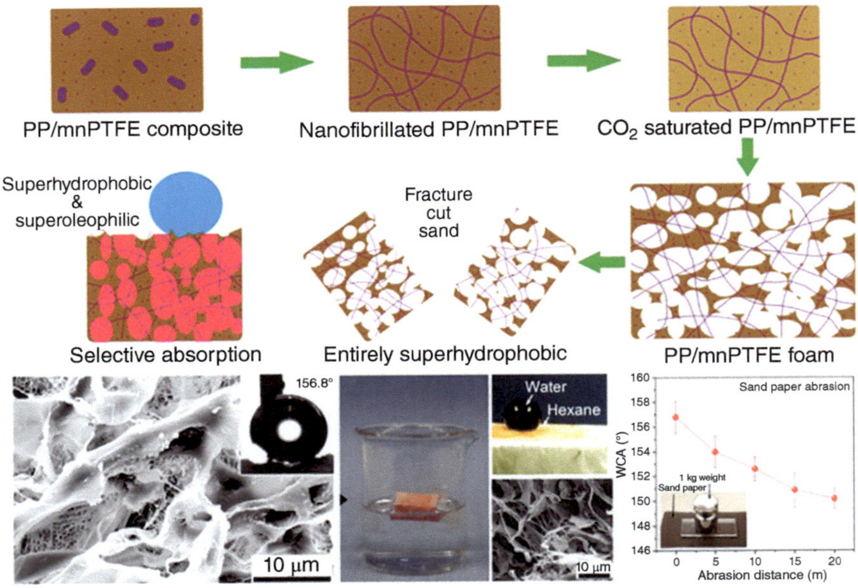

Figure 5.18 Illustration of the fabrication process of entirely superhydrophobic PP/mnPTFE foam and hydrophobic performance and mechanical durability results. Source: Mi et al. [44]/with permission of American Chemical Society.

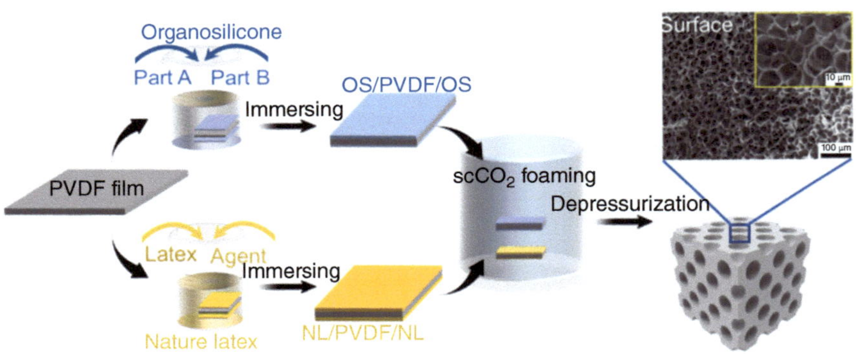

Figure 5.19 Schematic illustration of the fabrication process of skinless cellular PVDF films using PI, organosilicon, and natural latex as surface-constrained materials. Source: Wang et al. [47]/with permission of Elsevier.

5.5 Advanced Application of Superhydrophobic Polymer Foams

5.5.1 Self-Cleaning and Antifouling

Self-cleaning is an attractive feature for human beings, which can provide unique contamination-free surfaces spontaneously driven by nature. The self-cleaning behavior is originally inspired by the "Lotus effect," namely, solid particles can be easily carried away when water droplets roll off the leaves, resulting in a cleaned surface [48]. With significant benefits to the environment, energy, and labor-saving, self-cleaning surfaces have gained not only considerable scientific interest but also extensive industrial applications. Researchers have found that surfaces with self-cleaning properties usually possess considerable antifouling properties, which means that the surface can effectively slow growth and facilitate the detachment of subaquatic organisms [49].

Mi and coworkers [50] developed a porous polymer coating using UHMWPE via a controlled phase separation and crystallization process. When UHMWPE is dissolved in ortho-xylene and coated on the substrate surface, the surface forms a lamellar crystalline structure and results in the cassie-impregnating contact state with water, which essentially shows the "rose petal effect." When cyclohexane was introduced into the UHWMPE solution system, the resultant coating film showed a hierarchical micro-nanostructured surface, which demonstrated a Cassie–Baxter contact state with water and resulted in a special "lotus leaf effect" (Figure 5.20). The produced surfaces can be used for droplet transportation, self-cleaning, and antifouling applications. For 3D porous materials, Gao et al. [51] constructed a robust graphene-based superhydrophobic foam with graphene as a raw material built on a 2D copper foil to obtain a 3D porous foam with ordered vertical honeycomb nanostructures, which was further modified with PDMS to obtain a superhydrophobic foam with self-cleaning and water repellence properties.

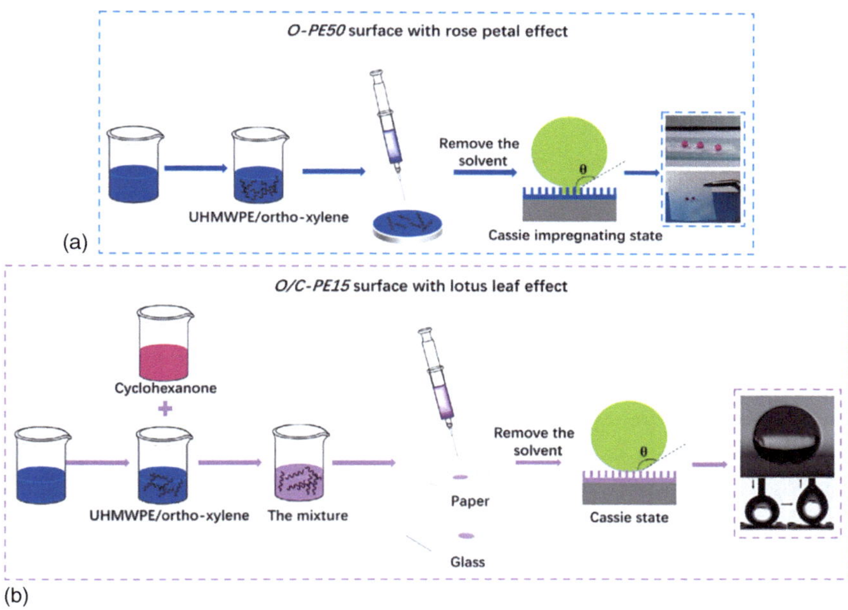

Figure 5.20 Schematic illustrations of the fabrication processes and properties of (a) high-adhesion surface with "rose petal effect," and (b) low-adhesion surface with "lotus leaf effect.". Source: Sun et al. [50]/with permission of Elsevier.

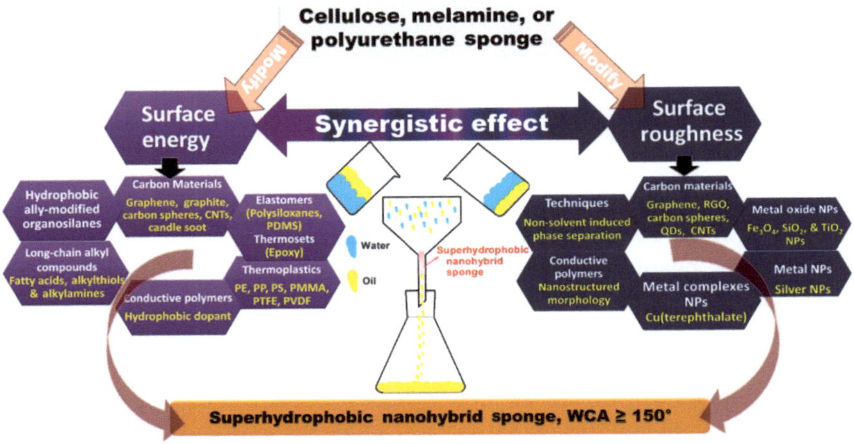

Figure 5.21 Illustration for the fabrication of superhydrophobic nanohybrid sponges via a synergistic effect between surface energy and surface roughness properties. Source: Abu-Thabit et al. [52]/with permission of Elsevier.

5.5.2 Oil–Water Separation and Oil Absorption

Foam materials have received a lot of attention because of their low price and good oil absorption properties. To achieve selective absorption of oil from water, it is essential to render the foam superhydrophobicity and superoleophilicity at the same time. Thus, it is important to simultaneously control the surface energy and surface roughness of a foam material. Abu-Thabit et al. [52] reviewed prevalent approaches to manipulate the surface energy and the surface roughness of porous foams such as cellulose, melamine, and PU. As illustrated in Figure 5.21, the surface energy manipulation methods include using carbon material, elastomers, thermosets, and thermoplastics with low surface energy, and the addition of hydrophobic dopant, long-chain alkyl compounds, and hydrophobically modified organosilanes. The surface roughness control can be realized by creating nanostructured surface morphology via the incorporation of carbon materials, metal oxide nanoparticles, metal nanoparticles, and metal complexes nanoparticles, and using other techniques such as phase separation to generate rough surfaces on the pristine foam.

Mi and coworkers [53] developed a novel approach to produce a hierarchically nanofibrous structured UHMWPE film on a commercial melamine foam. The film was prepared using the rapid flow-induced crystallization behavior of UHMWPE molecules in the xylene solution. As shown in Figure 5.22, the fabricated membrane is composed of UHMWPE shish and parallel aligned lamellae crystals perpendicular to the shish direction. The special crystalline structure is called the "shish–kebab" structure. With the shish–kebab structured UHMWPE film, the coated foam demonstrated an excellent WCA of 157°, a low contact angle hysteresis of 4.5°, and ultra-high capacity (70–191 $g\,g^{-1}$) in selective oil absorption toward various organic solvents.

In addition, Lv et al. [54] introduced a straightforward 3D printing technique for fabricating superhydrophobic surfaces. They utilized hydrophobic nano silica-filled

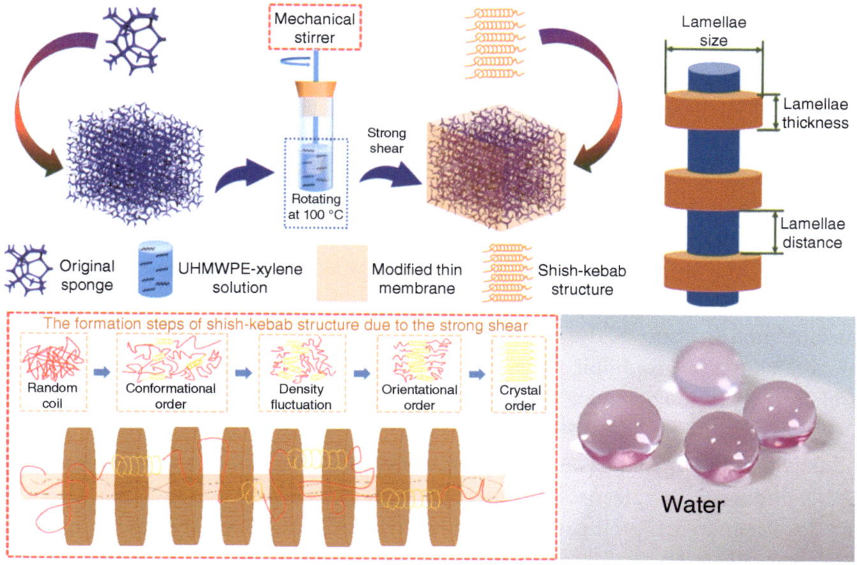

Figure 5.22 Schematic diagram of the preparation process of UHMWPE coated sponge and illustration of structural parameters of the shish–kebab crystallite. Source: Dong et al. [53]/with permission of American Chemical Society.

PDMS in the process, yielding superhydrophobic porous film coatings characterized by enhanced superhydrophobicity and mechanical robustness. Zhang et al. [55] fabricated superhydrophobic porous material F-rGO@TPU by coating a fluoroalkyl silane functionalized rGO layer on thermoplastic Polyurethane (TPU) foam. The prepared foam showed a WCA of 153° and a porosity of 87.4% and exhibited outstanding photothermal properties for efficiently cleaning up solar-driven crude oil spills. The foam also displayed high saturation adsorption capacities for various low-viscosity oils and organic solvents (ranging from 6.5 to 53.7 $g\,g^{-1}$) and excellent oil–water separation efficiency (exceeding 99.3%), demonstrating great potential for oil spill remediation.

5.5.3 Integration with Other Functions

5.5.3.1 Superhydrophobic Foams for Electromagnetic Interference (EMI) Shielding

Lightweight electromagnetic interference (EMI) shielding materials are desired for electromagnetic wave absorption. Featuring the EMI shielding foam with superhydrophobicity could greatly enhance the durability and lifespan of the material via the self-cleaning and antifouling attributes. Therefore, researchers have devoted themselves to fabricating EMI shielding foams integrated with superhydrophobic properties.

Zhang et al. [56] used a tannin-assisted silver-nanoparticle deposition method to fabricate a degradable cellulose paper followed by assembling with an electroless copper-plated paper (ECP) as the conductive layer. After the spray coating a mixture of PDMS and clustered Fe_3O_4 microspheres layer, a micro-nano hierarchical structured superhydrophobic surface was constructed. As shown in Figure 5.23a, the material surface demonstrated a high WCA of 153.5° and a SA of 3.2°, and the synergistic effect of the conductive and magnetic layers of the ECP resulted in an excellent absorption-dominated EMI shielding effectiveness of >50 dB with absorption to reflection ratio of ~95 : 5. Li and coworkers [57] developed an MXene and AgNW-coated hybrid foam using the same dip-coating approach. The fabricated foam showed a high absolute EMI shielding effectiveness (SSE/t) of 10,128 dB $cm^2\ g^{-1}$ with an ultra-low density of 11.98 mg cm^{-3} (Figure 5.23b). After a surface modification of the foam using perfluorodecyltriethoxysilane, the foam achieved a superhydrophobic property with a WCA of 159°, which demonstrated a self-cleaning surface behavior.

5.5.3.2 Superhydrophobic Foams for Piezoresistive Sensors

Flexible and wearable piezoresistive sensors have been raised as a hot spot with the rapid development of intelligent wearable devices. Conductive polymer foams have been frequently used as piezoresistive pressure sensors for the detection of tapping signals and monitoring human motion behavior. Piezoresistive pressure sensors rely on the resistance change when the material is subjected to deformation. Rendering the sensing material with superhydrophobicity is expected to be able to enhance the sensors' durability and prevent the sensors from contamination.

Wang et al. [58] prepared a flexible conductive polymer foam composite (CPFC) by anchoring CNTs onto the skeleton of the polymer foam with the assistance of ultrasonication and simultaneous nonsolvent-induced phase separation (NIPS). The Rubber sponge is composed of 40% natural rubber (NR) and 60% styrene butadiene rubber (SBR), which could be swollen and form hemisphere arrays on the sponge skeleton. The CNTs were decorated on the surface of the sponge and formed a special biomimetic lotus leaf-like microstructure. The obtained superhydrophobic CPFC exhibits excellent anticorrosive, photothermal conversion performance, and piezoresistive sensitivity when used as a pressure sensor.

5.5.3.3 Superhydrophobic Foams for Radiative Cooling

Due to climate change and global warming, novel cooling methods with low energy consumption are highly desired. Passive radiative cooling (PRC) technology provides a sustainable way of cooling by the combination of emitting heat to the cold universe and reflecting solar light without any energy input. Highly porous polymer with high sunlight reflectance can be used in the application of the PRC process. Integrating the superhydrophobic self-cleaning property to the PRC polymer foams could render the material longer in service life and reduce the efforts needed to clean the material in practical application. Liu et al. [59] fabricated a PLA-based PRC aerogel by the combination of biomimetic hierarchical structure design and stereo-complex crystals. As illustrated in Figure 5.24, poly-L-lactic acid (PLLA) and poly-D-lactic acid (PDLA) was dissolved in dioxane and water co-solvent and freeze-dried to obtain the PLA aerogel. The prepared aerogel showed a low

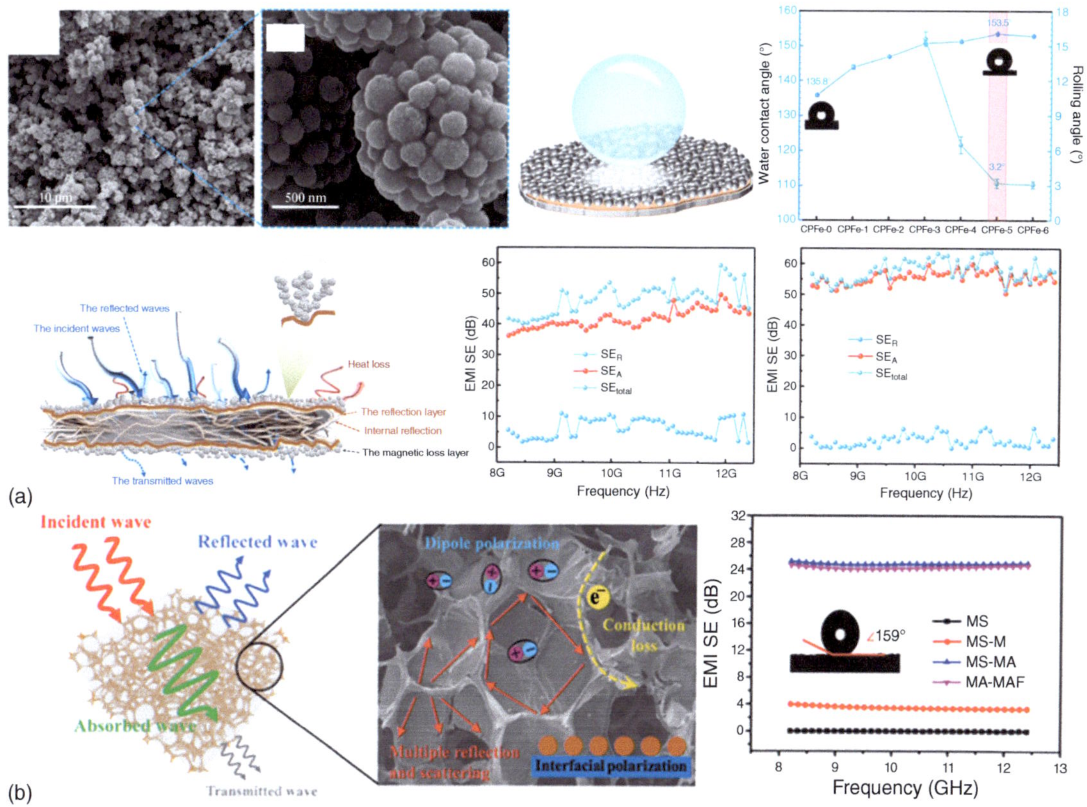

Figure 5.23 (a) SEM images of Fe_3O_4 microspheres, water contact angle and sliding angle of ECPs with different Fe_3O_4 loading, schematic diagram of the EMI shielding mechanism, and the EMI SE of ECPs sprayed with 3 and 4 times of clustered Fe_3O_4 microspheres and PDMS, and (b) MXene and AgNW-coated melamine foam prepared by dip coating and perfluorodecyltriethoxysilane modification for EMI shielding and self-cleaning. Source: (a) Zhang et al. [56]/with permission of American Chemical Society. (b) Wang et al. [57]/with permission of Royal Society of Chemistry.

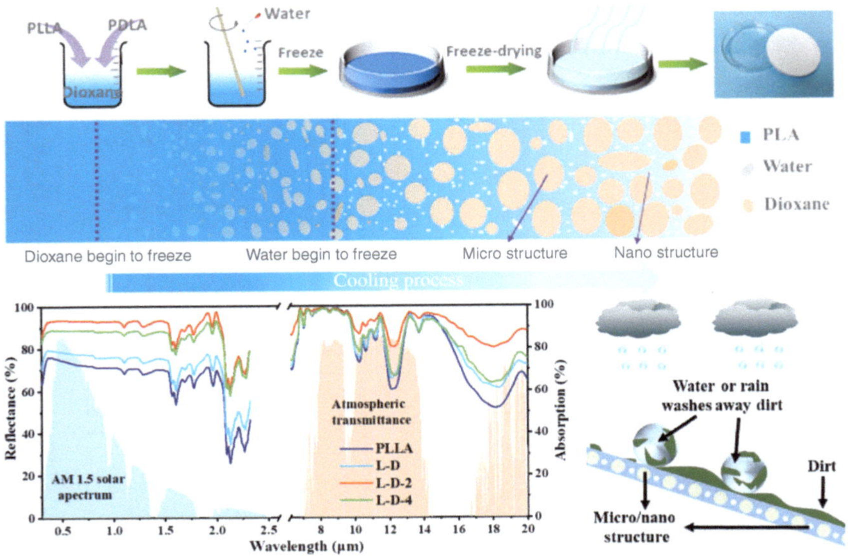

Figure 5.24 Schematic illustration of the preparation of PLA aerogels and their applications in PRC and self-cleaning. Source: Liu et al. [59]/with permission of John Wiley & Sons.

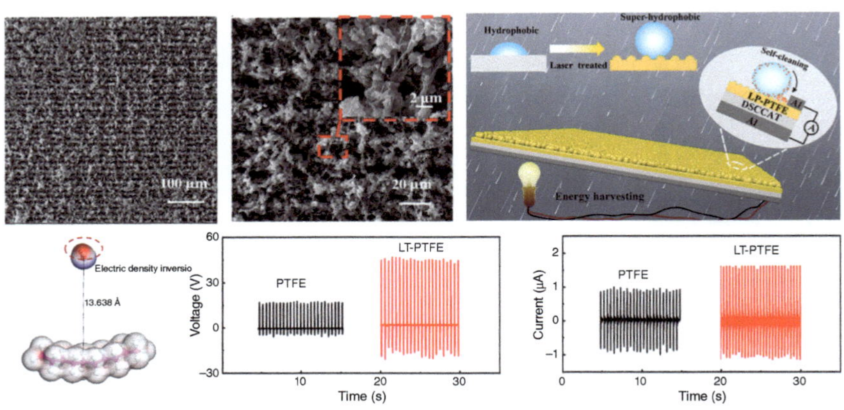

Figure 5.25 The porous micro-nanostructured PTFE with superhydrophobicity and self-cleaning ability fabricated by femtosecond laser direct processing and its application in triboelectric nanogenerator. Source: Zhang et al. [60]/with permission of American Chemical Society.

thermal conductivity of 37 mW $(m\,K)^{-1}$, high compression strength of 0.1 MPa, and self-cleaning ability with a WCA of 152°. The optimized PLA aerogel exhibited a high sunlight reflection of 89% and a strong infrared emissivity of 93%, which realized a temperature drop of 3.5 °C during the daytime and 5.8 °C during the nighttime.

5.5.3.4 Superhydrophobic Surfaces for Nanogenerators

The superhydrophobic surface can be combined with porous materials in the application of energy harvesting. Li and coworkers [60] developed a porous micro-nanostructured PTFE with superhydrophobicity and self-cleaning ability by femtosecond laser direct processing. As shown in Figure 5.25, the PTFE surface engineered via femtosecond laser demonstrated a porous morphology with multiscale hierarchical structures, which rendered the material with excellent superhydrophobicity with a WCA of 161° and outstanding self-cleaning properties. When the rough surface PTFE film is used as a dielectric layer to assemble a triboelectric nanogenerator, the device is capable of generating energy from the falling water droplet due to the triboelectric difference between water and PTFE molecules. The output voltage and current were greatly enhanced with the use of superhydrophobic PTFE rough film.

References

1 Koch, K., Bhushan, B., and Barthlott, W. (2009). Multifunctional surface structures of plants: an inspiration for biomimetics. *Progress in Materials Science* 54: 137–178.

2 Feng, L., Li, S., Li, Y. et al. (2002). Super-hydrophobic surfaces: from natural to artificial. *Advanced Materials* 14: 1857–1860.

3 Wang, G.Y., Guo, Z.G., and Liu, W.M. (2014). Interfacial effects of superhydrophobic plant surfaces: a review. *Journal of Bionic Engineering* 11: 325–345.

4 Bhushan, B. and Jung, Y.C. (2011). Natural and biomimetic artificial surfaces for superhydrophobicity, self-cleaning, low adhesion, and drag reduction. *Progress in Materials Science* 56: 1–108.

5 Barthlott, W. and Neinhuis, C. (1997). Purity of the sacred lotus, or escape from contamination in biological surfaces. *Planta (Germany)* 202: 1–8.

6 Zhang, L., Dong, D., Shao, L. et al. (2019). Cost-effective one-pot surface modified method to engineer a green superhydrophobic sponge for efficient oil/water mixtures as well as emulsions separation. *Colloids and Surfaces A: Physicochemical and Engineering Aspects* 576: 43–54.

7 Han, N., Zhang, Z., Gao, H. et al. (2020). Superhydrophobic covalent organic frameworks prepared via pore-surface modifications for functional coatings under harsh conditions. *ACS Applied Materials & Interfaces* 12: 2926–2934.

8 Feng, X.Q., Gao, X.F., Wu, Z.N. et al. (2007). Superior water repellency of water strider legs with hierarchical structures: experiments and analysis. *Langmuir* 23: 4892–4896.

9 Jie, L., Wentao, Y., Deyin, Z. et al. (2018). Hot embossing for whole teflon superhydrophobic surfaces. *Coatings* 8: 227.

10 Jucius, D., Grigaliūnas, V., Mikolajūnas, M. et al. (2011). Hot embossing of PTFE: towards superhydrophobic surfaces. *Applied Surface Science* 257: 2353–2360.

11 Tsibouklis, J. and Nevell, T.G. (2003). Ultra-low surface energy polymers: the molecular design requirements. *Advanced Materials* 15: 647–650.

12 Mi, H.-Y., Jing, X., Huang, H.-X., and Turng, L.-S. (2017). Controlling superwettability by microstructure and surface energy manipulation on three-dimensional substrates for versatile gravity-driven oil/water separation. *ACS Applied Materials & Interfaces* 9: 37529–37535.

13 Kim, B.S., Shin, S., Shin, S.J. et al. (2011). Control of superhydrophilicity/superhydrophobicity using silicon nanowires via electroless etching method and fluorine carbon coatings. *Langmuir* 27: 10148–10156.

14 Wang, Y., Yang, H., Chen, Z. et al. (2018). Recyclable oil-absorption foams via secondary phase separation. *ACS Sustainable Chemistry & Engineering* 6: 13834–13843.

15 Deng, S., Zhao, C., Zhu, L. et al. (2021). Thermally stable and superhydrophobic foam prepared by high-internal phase emulsion for oil/water separation with good recycling capacity. *Journal of Applied Polymer Science* 139: 51521.

16 Ma, Z., Liu, X., Xu, X. et al. (2021). Bioinspired, highly adhesive, nanostructured polymeric coatings for superhydrophobic fire-extinguishing thermal insulation foam. *ACS Nano* 15: 11667–11680.

17 Du, Y., Wu, T., Zhou, W. et al. (2023). Cicada wing-inspired transparent polystyrene film integrating self-cleaning, antifogging, and antibacterial properties. *ACS Applied Materials & Interfaces* 15: 46538–46549.

18 Xie, H., Huang, H.-X., and Mi, H.-Y. (2018). Gradient wetting state for droplet transportation and efficient fog harvest on nanopillared cicada wing surface. *Materials Letters* 221: 123–127.

19 Lv, C., Wang, F., He, C. et al. (2022). Preparation of silicone-PCL composite particles with hierarchical structure and the super-hydrophobic fabrics via directly electrostatic spraying. *Surface and Coatings Technology* 449: 128933.

20 Young, T. (1805). An essay on the cohesion of fluids. *Philosophical Transactions of the Royal Society of London* 95: 65–87.

21 Wenzel, R.N. (1936). Resistance of solid surfaces to wetting by water. *Transactions of the Faraday Society* 28: 988–994.

22 Herminghaus, S. (2007). Roughness-induced non-wetting. *Europhysics Letters (EPL)* 79: 59901.

23 Cassie, A.B.D. (1944). Wettability of porous surfaces. *Transactions of the Faraday Society* 40: 546–551.

24 Mi, H.-Y., Li, H., Jing, X. et al. (2020). Robust superhydrophobic fluorinated fibrous silica sponge with fire retardancy for selective oil absorption in harsh environment. *Separation and Purification Technology* 241: 116700.

25 Kota, A.K., Kwon, G., and Tuteja, A. (2014). The design and applications of superomniphobic surfaces. *NPG Asia Materials* 6: e109.

26 Kota, A.K., Li, Y., Mabry, J.M., and Tuteja, A. (2012). Superoleophobic surfaces: hierarchically structured superoleophobic surfaces with ultralow contact angle hysteresis (Adv. Mater. 43/2012). *Advanced Materials* 24: 5837.

27 Mahadevan, L. and Pomeau, Y. (1999). Rolling droplets. *Physics of Fluids* 11: 2449–2453.

28 Gennes, P.G.D., Brochard-Wyart, F., Quéré, D. et al. (2004). Capillarity and wetting phenomena: drops, bubbles, pearls, waves. *Physics Today* 57: 66–67.

29 Kota, A.K., Choi, W., and Tuteja, A. (2013). Superomniphobic surfaces: design and durability. *MRS Bulletin* 38: 383–390.

30 Tuteja, A., Choi, W., Ma, M. et al. (2007). Designing superoleophobic surfaces. *Science* 318: 1618–1622.

31 Mi, H.-Y., Jing, X., Xie, H. et al. (2018). Magnetically driven superhydrophobic silica sponge decorated with hierarchical cobalt nanoparticles for selective oil absorption and oil/water separation. *Chemical Engineering Journal* 337: 541–551.

32 Tamsilian, Y., Ansari-Asl, Z., Maghsoudian, A. et al. (2021). Superhydrophobic ZIF8/PDMS-coated polyurethane nanocomposite sponge: Synthesis, characterization and evaluation of organic pollutants continuous separation. *Journal of the Taiwan Institute of Chemical Engineers* 125: 204–214.

33 Shuai, Q., Yang, X., Luo, Y. et al. (2015). A superhydrophobic poly(dimethylsiloxane)-TiO_2 coated polyurethane sponge for selective absorption of oil from water. *Materials Chemistry and Physics* 162: 94–99.

34 Wang, X., Pan, Y., Liu, X. et al. (2019). Facile fabrication of superhydrophobic and eco-friendly poly(lactic acid) foam for oil-water separation via skin peeling. *ACS Applied Materials & Interfaces* 11: 14362–14367.

35 Mi, H.Y., Jing, X., Politowicz, A.L. et al. (2018). Highly compressible ultra-light anisotropic cellulose/graphene aerogel fabricated by bidirectional freeze drying for selective oil absorption. *Carbon* 132: 199–209. S000862231830157X.

36 Wang, X., Pan, Y., Yuan, H. et al. (2020). Simple fabrication of superhydrophobic PLA with honeycomb-like structures for high-efficiency oil-water separation. *Chinese Chemical Letters* 31: 365–368.

37 Kato, S. and Sato, A. (2012). Micro/nanotextured polymer coatings fabricated by UV curing-induced phase separation: creation of superhydrophobic surfaces. *Journal of Materials Chemistry* 22: 8613–8621.

38 Mu, L., Yang, S., Hao, B., and Ma, P.-C. (2015). Ternary silicone sponge with enhanced mechanical properties for oil–water separation. *Polymer Chemistry* 6: 5869–5875.

39 Mi, H.Y., Jing, X., Huang, H.X. et al. (2018). Superhydrophobic graphene/cellulose/silica aerogel with hierarchical structure as superabsorbers for high efficiency selective oil absorption and recovery. *Industrial & Engineering Chemistry Research* 57: https://doi.org/10.1021/acs.iecr.7b04388.

40 Su, Q., Zhang, J., and Zhang, L.Z. (2020). Fouling resistance improvement with a new superhydrophobic electrospun PVDF membrane for seawater desalination. *Desalination* 476: 114246.

41 Mi, H.Y., Jing, X., Huang, H.X., and Turng, L.S. (2017). Instantaneous self-assembly of three-dimensional silica fibers in electrospinning: Insights into fiber deposition behavior - ScienceDirect. *Materials Letters* 204: 45–48.

42 Mi, H.-Y., Li, H., Jing, X. et al. (2020). Superhydrophobic cellulose nanofibril/silica fiber/Fe_3O_4 nanocomposite aerogel for magnetically driven selective oil absorption. *Cellulose* 27: 8909–8922.

43 Sun, B., Li, J., Guo, Y. et al. (2021). Superhydrophobic UHMWPE foams with high mechanical robustness and durability fabricated by supercritical CO_2 foaming. *ACS Sustainable Chemistry & Engineering* 9: 12663–12673.

44 Mi, H.Y., Jing, X., Liu, Y. et al. (2019). Highly durable superhydrophobic polymer foams fabricated by extrusion and supercritical CO foaming for selective oil absorption. *ACS Applied Materials & Interfaces* 11: 7479–7487.

45 Li, B., Zhao, G., Wang, G. et al. (2020). Biodegradable PLA/PBS open-cell foam fabricated by supercritical CO_2 foaming for selective oil-adsorption. *Separation and Purification Technology* 257: 117949.

46 Ren, X.Y., Li, H., Mi, H.Y. et al. (2022). Superhydrophobic ultra-high molecular weight polyethylene nanocomposite foams fabricated by supercritical CO_2 foaming for selective oil absorption. *Applied Surface Science* 602: 154344.

47 Wang, L., Cui, W., Mi, H.Y. et al. (2022). Fabrication of skinless cellular poly (vinylidene fluoride) films by surface-constrained supercritical CO_2 foaming using elastic gas barrier layers. *The Journal of Supercritical Fluids* 184: 105562.

48 Frank, M.A. (2017). Functionalization of steels and other metallic materials with hydrophobic layers - influence on wetting and corrosion resistance. Doctoral thesis. University of Erlangen-Nuremberg.

49 Kuzmyn, A.R. (2021). The light way towards bioactive and antifouling coatings. Internal PhD thesis. Wageningen University.

50 Sun, S., Li, H., Guo, Y. et al. (2021). Superefficient and robust polymer coating for bionic manufacturing of superwetting surfaces with “rose petal effect” and “lotus leaf effect”. *Progress in Organic Coatings* 151: 106090.

51 Gao, X., Zhou, J., Du, R. et al. (2015). Robust superhydrophobic foam: a graphdiyne-based hierarchical architecture for oil/water separation. *Advanced Materials* 28: 168–173.

52 Abu-Thabit, N.Y., Uwaezuoke, O.J., and Abu Elella, M.H. (2022). Superhydrophobic nanohybrid sponges for separation of oil/water mixtures. *Chemosphere* 294: 133644.

53 Dong, B., Guo, Y., Sun, S. et al. (2020). Shish–kebab-structured UHMWPE coating for efficient and cost-effective oil–water separation. *ACS Applied Materials & Interfaces* 12: 58252–58262.

54 Lv, J., Gong, Z., He, Z. et al. (2000). 3D printing of a mechanically durable superhydrophobic porous membrane for oil-water separation. *Journal of Materials Chemistry A* 5: 12435–12444.

55 Zhang, Y., Wang, B., Wang, B. et al. (2022). Super-hydrophobic graphene-coated thermoplastic polyurethane porous monolith with superior photothermal effect for solar-assisted efficient cleanup of crude oil spill. *Applied Surface Science* 605: 154701.

56 Zhang, T.C., Lv, D., Liu, R.T. et al. (2021). Developing a superhydrophobic absorption-dominated electromagnetic shielding material by building clustered FeO nanoparticles on the copper-coated cellulose paper. *ACS Sustainable Chemistry & Engineering* 9: 6574–6585.

57 Wang, S., Li, D., Meng, W. et al. (2022). Scalable, superelastic, and superhydrophobic MXene/silver nanowire/melamine hybrid sponges for

high-performance electromagnetic interference shielding. *Journal of Materials Chemistry C* 10: 5336–5344.

58 Wang, L., Huang, X., Wang, D. et al. (2021). Lotus leaf inspired superhydrophobic rubber composites for temperature stable piezoresistive sensors with ultrahigh compressibility and linear working range. *Chemical Engineering Journal* 405: 127025.

59 Liu, X., Zhang, M., Hou, Y. et al. (2022). Hierarchically superhydrophobic stereo-complex poly (lactic acid) aerogel for daytime radiative cooling. *Advanced Functional Materials* 22: 2207414.

60 Zhang, H., Yin, K., Wang, L. et al. (2023). A robust droplet triboelectric nanogenerator with self-cleaning ability achieved by femtosecond laser. *ACS Applied Materials & Interfaces* 15: 30902–30912.

6

Electromagnetic Shielding Polymer Foams

6.1 Electromagnetic Pollution and Electromagnetic Interference Shielding

6.1.1 The Cause of Electromagnetic Radiation and Its Harm

With the development and application of EM science and related technologies, EM waves are almost everywhere in the current human society, and the EM pollution problem becomes more and more severe. When electric and magnetic fields that are perpendicular to each other oscillate in the same direction in space, they transfer energy and momentum through waves, forming EM radiation. Superstrength disordered EM waves will form EMI pollution, which brings various potential negative effects [1].

There are typically two major sources of EM pollution. The first is natural EMI, such as thunderstorms, sandstorms, and other climate and EM radiation from the universe. The second source of EMI is brought by human activities [2]. At present, EM technology is used in tremendous industries and human daily life, making the EM environment very complex. Disturbed and high-intensity EM waves will not only affect information security and cause EM information leakage but also reduce the stability and service life of precision instruments and are more likely to cause irreversible harm to human health. Therefore, the shielding of EMI is becoming a more and more prominent demand for informative security and human health. Due to the highly porous structure, foam materials or composite foams have been employed more and more to block EM wave conduction.

6.1.2 Electromagnetic Interference Shielding Mechanism

The basic idea of EMI shielding is to use conductive or magnetic materials to prevent the coupling of EM fields. When the EM wave passes through the conductor, it generates an induced current inside the conductor so that the EM field can be converted into the internal energy of the conductor to limit or weaken the intensity of the EM wave and control the long-distance transmission of high-intensity EM wave [3, 4].

Functional Polymer Foams: Green Fabrication Methods, Performance and Applications,
First Edition. Haoyang Mi.

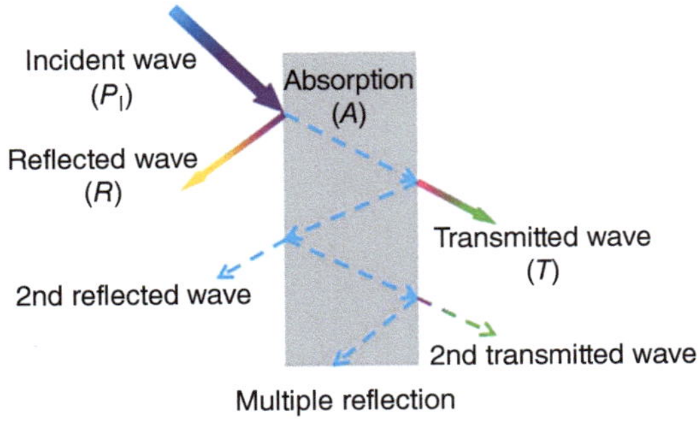

Figure 6.1 Illustration of electromagnetic shielding mechanism. Source: Zhao et al. [8]/with permission of Royal Society of Chemistry.

According to EMI shielding theory, low-frequency and high-frequency alternating EM fields have different EM shielding means. In the case of low-frequency alternating EM fields, a good conductor can enclose the target body and ground the electric field to stop the conductor surface to prevent interference caused by electrostatic coupling, which is referred to as electric field shielding. The means of magnetic field shielding typically uses a low-reluctance material to shunt magnetic fields by sealing the magnetic field line of the target body in a low-frequency alternating magnetic field to prevent internal magnetic field diffusion and external magnetic field entry [5]. In the case of high-frequency alternating EM field, which is the most common case in people's daily lives, EM shielding can be conducted in the following three ways to block the mutual induction between EM fields: (i) Reflection loss (RL) of incident surface; (ii) Absorption loss that is not reflected into the shield; (iii) Multiple RL inside the shield [6].

Various shielding mechanisms, such as the eddy current effect theory, EM field theory, and transmission line theory have been proposed. Among them, the transmission line theory is the most widely recognized, in which the EM wave passing through the shielding material is affected by three different shielding mechanisms: reflection, absorption, and multiple reflections [7]. As illustrated in Figure 6.1, the EM wave incident on the surface of the shielding material will be reflected due to the mismatch between the impedance of the shielding material and the external free space and may cause secondary EM radiation pollution. The unreflected EM wave enters the interior of the shielding material and attenuates continuously under the action of dielectric and magnetic losses. Besides, EM waves can be dissipated at the interface inside the material by repeated reflections, and the remaining EM waves will pass through the shielding material and become transmitted waves.

The attenuation or loss of EM waves is one of the key indexes to evaluate the EM shielding capability of materials. EM shielding efficiency (EMI SE, unit: dB) is used to quantitatively evaluate the ability of materials to shield EM waves. The *SE* is defined as the ratio of electric field intensity (E), magnetic field intensity (H), or power (P) before the EM wave enters the shielding material and after it passes through the shielding material [9], and it is expressed as [10]:

$$SE = 20\lg\left(\frac{E_t}{E_i}\right) = 20\lg\left(\frac{H_t}{H_i}\right) = 20\lg\left(\frac{P_t}{P_i}\right) \tag{6.1}$$

The subscripts i and t represent incident and transmitted EM waves, respectively. E_i, H_i, and P_i represent the incident electric field, magnetic field, and power intensity, respectively; E_t, H_t, and P_t represent the transmitted electric field, magnetic field, and power intensity, respectively.

According to the Schelkunoff theory, EMI SE is defined as the sum of three shielding mechanisms: reflection (SE_R), absorption (SE_A), and multiple reflection (SE_M) of EM wave attenuation by shielding materials.

The total EM loss is calculated as follows:

$$SE_T = SE_R + SE_A + SE_M \tag{6.2}$$

where, SE_T is the total EM loss, SE_R is the RL, SE_A is the absorption loss, and SE_M is the multiple RL. They are defined by the following equations [11]:

$$SE_R = 39.5 + 10\ \log(\sigma/2\pi f\mu_r) \tag{6.3}$$

$$SE_A = 8.7\, d\sqrt{\pi f \mu_r \sigma_r} \tag{6.4}$$

$$SE_M = 20\,\mathrm{lg}(1 - e^{-2d/\delta}) \tag{6.5}$$

In the formula, μ_r is the vacuum permeability ($\mathrm{H\,m^{-1}}$), σ_r is the relative conductivity of the material ($\mathrm{S\,m^{-1}}$), f is the EM frequency (Hz), d is the thickness of the shielding material (m), and δ is the skin depth of the material for a specific wavelength of EM wave (m).

At the same time, if $SE_R > SE_A$, it can be considered that the main EM shielding mechanism of this material is reflection. On the contrary, if $SE_R < SE_A$, it can be considered that the main EM shielding mechanism of the material is absorption. From Eqs. (6.3) and (6.4), it is seen that the SE_R is negatively correlated with f and μ_r, and positively correlated with σ_r, while the SE_A is positively correlated with t, f, μ_r, and σ_r of the shielding material. Thus, the SE_A of shielding material can be improved by increasing the magnetic properties, which refers to the improvement in the EMI wave absorption. In addition, the EMI shielding performance can be greatly enhanced by elevating the electrical conductivity of the material since SE_R and SE_A would be enhanced simultaneously. The SE_M can normally be ignored for a thick shielding material when the EM wave reaches the second boundary of the shielding material. In other words, SE_M can be ignored when the SE_A of the shielding material exceeds 10 dB [12].

6.1.3 EMI Shielding Mechanism for Porous Materials

Porous materials, especially porous polymer composites, are considered to be the most promising materials for EMI shielding. According to Sherkunov's EM wave interface conduction theory, the EM shielding ability of porous materials comes mainly from EM reflection and EM absorption. Appropriate material composition and structural design of CPFs can endow them with excellent electrical conductivity and hysteresis loss and effective multi-interface reflection and absorption. When the composite material has a large number of conductive interfaces, EM waves can

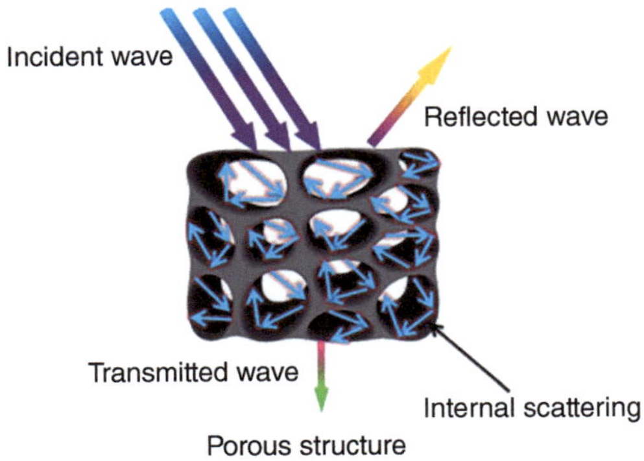

Figure 6.2 The interaction between incident EM waves and porous structure. Source: Zhao et al. [8]/with permission of Royal Society of Chemistry.

be continuously reflected between these interfaces and attenuated and absorbed in the reflection process, which can improve the EMI shielding performance by the absorption mechanism.

To realize the EM wave absorption rather than reflection is the main reason for the employment of EMI shielding porous materials. For the EM wave absorption mechanism, the shielding material is supposed to have electric and/or magnetic dipoles for interacting with the EM radiation so as to absorb and dissipate the EM waves. The electric dipoles may be provided by dielectric materials, and the magnetic dipoles are usually provided by magnetic constituents or those with moderate complex permeability. In addition, the multiple reflections mechanism is another major EM wave absorption mechanism for porous materials. As illustrated in Figure 6.2, porous materials can increase the multi-interface reflection and scattering loss of EM waves and prolong the transmission path of EM waves due to their unique microporous structure, thus effectively adjusting impedance matching and enhancing EM wave attenuation [8].

6.2 Conventional EMI Shielding Materials and Conductive Polymer Foams

6.2.1 Conventional EMI Shielding Materials

(1) Ferromagnetic materials, such as ferroferric oxide and nickel. This kind of material has high permeability and achieves EM shielding effect by attracting magnetic inductance lines and reducing magnetic flux density.
(2) Metal conductors, such as copper, silver, platinum, steel, and aluminum. These materials are rich in free electrons, which could generate reverse EM fields that counteract incident EM waves when EM waves pass through metal conductor materials, according to Lenz's law. They mainly shield the EM waves by counteracting incident EM waves. The limitation of metal-based shielding materials is that the method of EMI shielding is highly reflection-dominated, which means

the most portion of the EM waves are reflected by the conductors. Thus, it may cause secondary EM wave pollution.

(3) Intrinsic conducting polymer is a kind of polymer material with conductive ability itself. Conductive polymers have conductive channels between insulating polymer molecular chains with π conjugated bonds to transfer carriers. They work in the same way as the conductive metals with EM wave attenuation functions. At present, commonly used intrinsic conductive polymer shielding materials include polyaniline (PANI), polypyrene (PPY), polythiophene (PTH), etc.

    Compared with conductive metals, conductive polymer materials have the advantages of low density, corrosion resistance, and high strength, which make them promising materials for the EMI shielding field. However, due to their brittleness and difficulty in processing, conductive polymers often need to be doped in practical applications.

(4) Conductive polymer composites (CPCs) refer to the composites formed after adding conductive materials into insulating polymers and engendering conductive properties through melt blending, solution blending, *in situ* polymerization or coprecipitation, etc. [13]. The types of conductive fillers added to the insulating polymer matrix mainly include carbon base, metal base, polymer base, and hybrid conductive particles. Carbon base is mainly carbon black, graphene (GR), carbon nanotubes (CNTs), carbon fiber, etc. Metal fillers are mainly silver, copper, nickel, and aluminum particles that are in different dimensions, including spheres, rods, wires, and flakes. Polymer-based CPCs typically refer to composites that combine intrinsic conducting polymers with insulating polymers. The hybrid fillers for CPCs are usually materials like silver-coated microspheres, nickel-modified GR, nanoparticle-decorated carbon fiber, etc. Since CPCs combine the properties of polymer matrix and filler and have the advantages of large-scale production, they can overcome the defects of metal shielding materials and coating shielding materials and are expected to replace metal materials as the main EM shielding materials in the future. However, the shielding mechanism of these materials for EM waves is mostly reflection, which is easy to be accompanied by serious secondary reflection pollution, and it is difficult to meet the shielding requirements of various fields.

6.2.2 Conductive Polymer Foams

The development of EMI shielding composites with low-EM wave reflection and high-absorption characteristics has become a research focus of CPCs. Polymer foams naturally have the advantage of rich internal porous structure [14]. Polymer foams developed using CPCs as the matrix are referred to as CPFs, which usually have enhanced conductive networks due to the principle of volume repulsion during the cell formation, which promotes the formation of an efficient conductive path of conductive fillers [15]. The highly porous conductive structure of CPFs is beneficial for significantly enhancing the EM wave absorption performance since the cells reduce impedance mismatch, allowing EM waves to enter the material and reducing interfacial reflection. The multiple RLs of EM waves on the multiple interfaces of the conductive cell walls are the main cause of the EMI shielding performance improvement. In addition to the high EMI shielding ability, the CPFs also have the advantages of low density, absorbing impact load, absorbing sound, heat insulation, good chemical resistance, and high-temperature resistance [16].

In addition, the volume repulsion effect during the CPF formation process would promote the formation of an efficient conductive path of conductive fillers by the voids inside of the materials occupied by the microcells, which is beneficial to increase the compactness of the conductive fillers in the CPC matrix. GR, CNTs, silver nanowires (AgNWs), carbon black, and MXene are commonly used conductive nanofillers in the fabrication of CPFs. The compact of these fillers in the foam matrix would result in a significant increase in the EMI shielding performance in the means of EM wave absorption compared with their solid counterparts [17].

Therefore, introducing cellular structure to CPCs can significantly reduce the material density, which will allow the application of these materials to expand to a point where weight reduction is essential and improve the EM SE by selectively locating the conductive nanoparticles on the cell wall, which reduces their average degree of separation and optimizes the formation of an efficient conductive network.

6.3 Characterization of EMI Shielding Polymer Foams

6.3.1 EMI Shielding Effectiveness Measurement

Vector network analyzer (VNA) is generally used to test the EMI SE of a material. The main test methods include the shielding chamber test, the coaxial transmission line method, and the coaxial transmission line method. These methods are widely used in the laboratory. As shown in Figure 6.3, S_{11} and S_{21} represent the reflection parameters obtained through VNA port 1 and the transmission coefficient from port 1 to port 2. S_{22} and S_{12} represent the reflection parameter obtained from port 2 and the transmission coefficient from port 2 to port 1. These S parameters correlate with

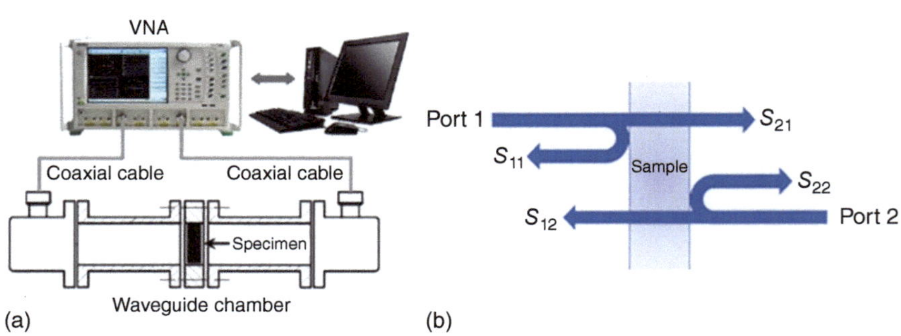

Figure 6.3 (a) VNA pictures. (b) The measurement principle of VNA scattering parameters. Source: Sun et al. [1]/with permission of Elsevier.

the reflection and transmission coefficients. Especially when the measured material is uniform EM shielding material, $S_{11} = S_{22}$, $S_{21} = S_{12}$.

The calculation of EMI SE uses the following formula to obtain the scattering parameters from two ports.

$$T = |S_{21}|^2 \tag{6.6}$$

$$R = |S_{11}|^2 \tag{6.7}$$

$$1 = A + R + T \tag{6.8}$$

$$SE_A = -10\lg\frac{T}{1-R} \tag{6.9}$$

$$SE_R = -10\lg(1-R) \tag{6.10}$$

$$SE_T = SE_R + SE_A + SE_M \tag{6.11}$$

where A, R, and T are absorption, reflection, and transmission power coefficients, respectively.

6.3.2 EM Wave Absorption

Based on the fundamental knowledge of EM wave absorption, the reflection coefficient could be expressed as $R = \frac{Z-Z_0}{Z+Z_0} = \frac{Z_{in}-1}{Z_{in}+1}$, where Z and Z_0 are the transmission impedance in absorption medium and air, respectively; $Z_{in} = Z/Z_0$ is the transmission impedance of the material interface. The RL value is used to intuitively evaluate the EM wave absorption performance of the absorbent. According to transmission line theory, the Z_{in} and RL can be expressed as follows [18]:

$$Z_{in} = \sqrt{\frac{\mu_r}{\varepsilon_r}}\tan h\left[j\frac{2\pi fd}{c}\sqrt{\mu_r\varepsilon_r}\right] \tag{6.12}$$

$$RL(dB) = 20\lg\left|\frac{Z_{in}-1}{Z_{in}+1}\right| \tag{6.13}$$

where, d is the absorption thickness, f is the EM wave frequency, Z_{in} and Z_0 are the input impedance and impedance in free space, and c is the velocity of microwave in free space. It is generally considered that the frequency band with $RL < 10$ dB (90% EM wave absorption) is the effective frequency band and the target frequency band for basic engineering applications.

6.3.3 Electrical Conductivity

The electrical conductivity is an important factor for EMI shielding materials. Generally, the conductivity of the specimen can be measured using a four-point probe meter or a digital precision multimeter. To enhance the accuracy of the measurement and reduce the contact resistance, the ends of the sample need to be coated with highly conductive silver paint. The conductivity (σ) in the unit of S m^{-1} can be calculated using the following equation,

$$\sigma = \frac{l}{RS} \tag{6.14}$$

where, l is the length of the sample (cm), R is the resistance (Ω), and S is the cross-sectional area of the sample (cm^2).

6.3.4 Magnetic Property

The magnetic properties of magnetic nanomaterials, conductive composites, and CPFs is usually measured using a vibrating sample magnetometer (VSM) instrument. VSM measures magnetic properties based on Faraday's Law of Induction. In the measurement, the sample is placed in a constant magnetic field, and the sample will align its magnetization with the external field if it is magnetic. The magnetic field of the sample would change when it is moved up and down in the coils of the VSM instrument. The current is proportional to the magnetization of the sample, thus, the magnetic property is deduced by the hysteresis curve measured [19].

6.4 Preparation of EMI Shielding Polymer Foams

6.4.1 Composition of EMI Shielding Polymer Foams

CPFs are typically composed of CPCs and micro-sized voids within the matrix. The matrix material has a profound influence on the conductivity and EMI shielding performance of CPFs. In the design and preparation process of polymer composite materials, commonly used polymer substrates include high-density polyethylene (HDPE), polystyrene (PS), polyvinylidene fluoride (PVDF), polypropylene (PP), thermoplastic polyurethane (TPU), polymethyl methacrylate (PMMA), silicone rubber, etc.

To render the material with high conductivity and high-EMI shielding performance. Additive fillers are necessary components for the polymer matrix. Commonly used conductive fillers include carbon fiber, CNTs, GR, MXene, carbon black, AgNWs, slive flakes, other metallic nanowires or particles, etc. The dispersion and distribution of conductive fillers in the polymer matrix would greatly affect the conductivity and the EMI shielding properties of the CPCs and the resultant CPFs. With the creation of porous structures, the conductive fillers would be located in a specific microregion of the polymer matrix due to the volume repulsion effect, which would generally lead to a more efficient conductive network and reduced percolation threshold.

Besides the conductive filler, magnetic fillers are also incorporated in the CPFs to boost the magnetic loss when EM waves interact with the shielding materials. According to classical EM theory, magnetic materials can shield EM waves mainly through hysteresis loss, eddy current loss, domain wall resonance, ferromagnetic resonance, and natural resonance mechanisms [20, 21]. Common magnetic shielding materials are mainly divided into two categories, namely, the metal alloy materials based on iron, cobalt, nickel, etc., and the metal oxide materials such as carbonyl iron, carbonyl nickel, ferrite, and garnet [22, 23]. In many cases, magnetic fillers are combined with conductive fillers in the development of EMI shielding CPFs to boost the EM wave absorption properties.

For example, Mi and coworkers [1] developed a novel asymmetric magnetic-electric dual-functional composite foam for ultra-efficient EMI shielding. Iron oxide decorated reduced GR oxide (Fe_3O_4@rGO) and CNT were added to the TPU matrix to render the material with both conductivity and magnetic properties. After vacuum-assisted hot-compressing and scCO_2 foaming, a highly conductive silver layer is pasted on the other side of the composite foam, which is named T/C/MG/Ag composite foam (Figure 6.4a and b). The prepared composite foam with a thickness of ~2 mm reached a high EMI SE of 89.2–103.5 dB with an ultralow low SE_R of 0.24–0.83 dB. Due to the magnetic–electric dual-functional network constructed by Fe_3O_4@rGO and CNT nanofillers contributed to the high-EM wave absorption and low-EM wave reflection with a minimum reflectivity (R) value of 0.001, which holds the lowest record among EMI shielding materials with comparable SE (Figure 6.4c and d).

6.4.2 Factors Influence EMI Shielding Performance of Polymer Foams

The EMI shielding and EM wave absorption performance of a polymer foam is highly related to the properties of the matrix material and the microcellular structure of the foam material. The critical factors mainly include the conductivity, magnetic property, cell size, cell density, porosity, material thickness, and the configuration of the shielding materials.

6.4.2.1 Conductivity and Magnetic Properties

EM wave attenuation is generally regulated by the dielectric loss and magnetic loss of EMI shielding materials, which is the key to obtaining desirable shields [24]. EMI SE is related to the ratio between electrical conductivity (σ) and magnetic permeability (μ) of the shielding material. According to the EMI shielding theory, SE_R decreases with the increase in frequency and SE_A is proportional to $(\sigma\mu)^{1/2}$, thus, high-electrical conductivity and permeability are desired for EMI shielding materials to realize EMI shielding and EM wave absorption simultaneously.

For porous CPFs, the dielectric loss is mainly determined by conductivity, polarization relaxation, and multiple scattering. Higher electrical conductivity would produce a large number of free charges, which leads to more reflection as a result of the impedance mismatch, thus leading to serious secondary EM pollution. Compared with solid CPC, CPF usually possesses higher conductivity when normalized by the bulk density, which means that the cell walls of CPF would be more conductive than a CPC film with the same thickness as the cell wall. This is particularly important for understanding the superior EMI shielding performance of the CPFs since the volume resistance of CPF foamed using CPC is usually higher than that of the CPC with the same size due to the high unit volume filler content in CPC. This is beneficial

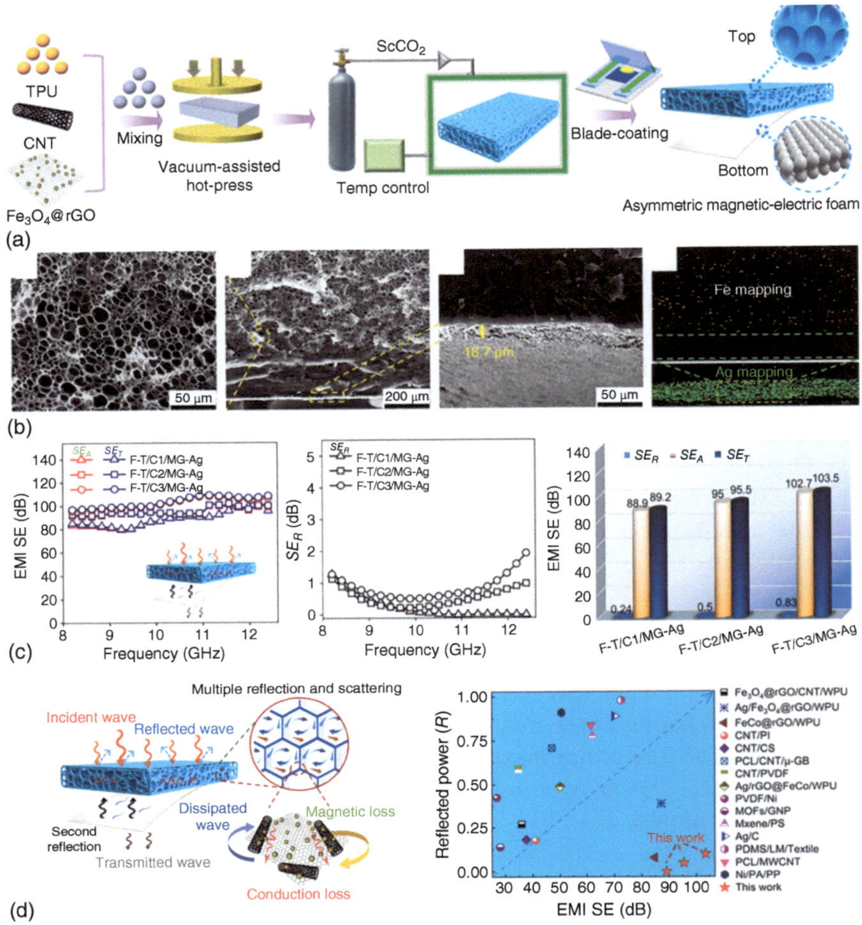

Figure 6.4 (a) Illustration of the fabrication process of T/C/MG/Ag composite foams; (b) SEM images and element distribution of T/C/MG/Ag composite foam; (c) EMI SE and SE_R results of different T/C/MG/Ag composite foams; (d) Scheme of EM wave dissipation mechanism and the comparison of EMI SE and relative R values among reference. Source: Sun et al. [1]/with permission of Elsevier.

for reducing the impedance mismatch of the CPF, which allows more EM waves to enter the interior cells of the CPFs. At the same time, the well-established conductive network on the cell wall of CPF would trigger multiple reflections of the EM wave in the CPF, which causes significant attenuation of the EM waves.

Generally speaking, the more the content of conductive filler, the higher the density of the conductive network, the longer the transmission path of the EM wave in the shielding material, the greater the dielectric loss would be caused, and the higher the total EMI SE would be. However, the excessive conductive filler is not obvious in enhancing the performance of EM shielding when the loading content exceeds the percolation threshold, which greatly increases the material cost and may lead to difficulties in material processing and foaming. Therefore, the content of the conductive filler needs to be optimized when developing a CPC for the preparation of EMI shielding CPF.

Combining magnetic materials into the CPF matrix is a favorable approach to enhance the magnetic loss of the EMI shields by increasing the EM wave absorption. As shown in Figure 6.5a, EM waves include two basic components magnetic field (H) and electric field (E). The ratio of E to H is defined as wave impedance (Z) by the following equation,

$$Z = \frac{E}{H} = \sqrt{\frac{\mu}{\varepsilon}} \tag{6.15}$$

Magnetic field and electric field are perpendicular to each other, and their direction of propagation is at right angles to the plane containing these two components. The desirable impedance matching requires the complex permeability (μ) close to complex permittivity (ε). To achieve high μ, magnetic fillers with high permeability are often introduced into the matrix of CPFs to enhance the ferromagnetic resonance and the eddy current. The "exchange coupling behavior" of hard/soft ferrite nanocomposites might be an effective approach to enhance the permeability of magnetic absorbers. As shown in Figure 6.5b, the hysteresis loop of exchange coupled magnet showed superior comprehensive magnetic permeability than the hard magnetic phase and soft magnetic phase [25].

6.4.2.2 Material Thickness and Resonance Behavior

Thickness is another vital factor to obtain the desirable skin depths (δ), which was defined as the thickness when the strength of the EM field is reduced to $1/e$ from the initial value. The relationship between δ and the material can be evaluated by the following equation:

$$\delta = 1/\sqrt{\pi f \mu \sigma} \tag{6.16}$$

where μ is the permeability, σ is the electrical conductivity, and f is the frequency. According to the equation, δ would decrease with the increase of conductivity/permeability of material and frequency inversely. The small skin depth would lead to the reflection from the material surface, and result in a low absorption rate, and the field strength of the EM wave will be reduced to 37% of the incident field strength at the skin depth [26].

In terms of the EMI shields' thickness (t_m), the EMI SE would generally increase with the increase of t since the EM waves have to propagate a longer distance in a shield with higher thickness. In the case of EM wave absorption, the absorption peak would shift to the lower frequency band with the increase of matching thickness, which can be explained by the 1/4 wavelength ($\lambda/4$) matching theory:

$$t_m = \frac{n\lambda}{4} = \frac{nc}{4f_m\sqrt{|\mu_r \varepsilon_r|}}, n = (1, 3, 5 \ldots) \tag{6.17}$$

When the t_m and EM frequency (f) meet the model, destructive interference will be generated, which would cancel the reflected EM waves at the absorption–air

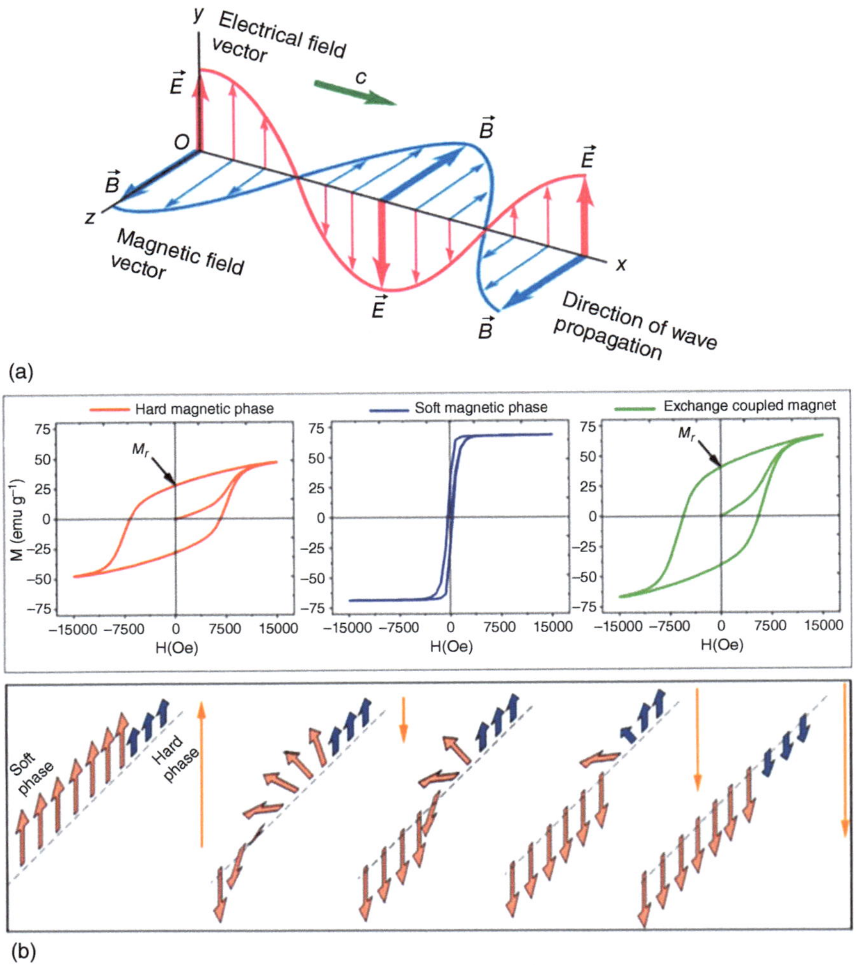

Figure 6.5 (a) Illustration of the electric and magnetic waves and their direction of propagation. (b) Hysteresis loop representation of hard phase, soft phase, and exchange-coupled hard/soft phases of the nanocomposite along with the portrait of the shift of dipole moments of soft and hard phases with the reversal of magnetic field reversal. Source: Mathews and Babu [25]/with permission of Elsevier.

interface, thus consuming the EM waves caused by the 180° phase difference between the reflected and incident microwaves.

To utilize the interference effect of internally reflected EM waves, materials with resonant cavity (RC) structures have been developed. For example, Mi and coworkers [27] fabricated a PLA/MXene multilayered EMI shielding material containing an RC structure. By introducing the double-layered structure and increasing the RC distance, a 48.5% and 74.8% improvement in SE_T and SE_A, and a 24.6% reduction in SE_R were achieved without varying the material composition and cost. A high SE of 92.3 dB was obtained at an RC distance of 6 mm owing to the synergetic effects

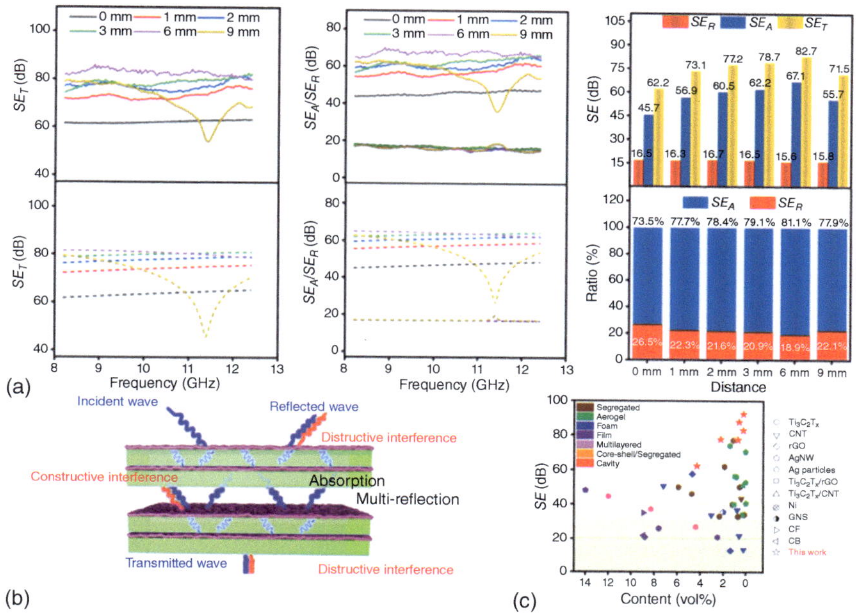

Figure 6.6 (a) Measured and calculated SE_T, SE_A, and SE_R of PLA/MXene multilayered shields with various RC distances. (b) Schematic illustration of the EMI shielding mechanism, and (c) the performance comparison among references of the PLA/MXene multilayered shields. Source: Du et al. [27]/with permission of American Chemical Society.

of multiple reflections and destructive EM wave interference. The EMI shields developed reached a remarkable EMI SE with an ultralow content of conductive components (Figure 6.6).

6.4.2.3 Cell Size, Cell Density, and Porosity

The cellular structure of CPFs has a significant influence on their EMI shielding properties. The cell size, cell density, and porosity are factors that are frequently manipulated in the fabrication of EMI shielding CPFs. As illustrated in Figure 6.2, the cell walls of the CPFs play a major role in triggering the multiple reflections of the EM waves inside of the foam body, which is the main source for EM wave attenuation. In this respect, the cell morphology of CPFs could govern the transmission and propagation route of the EM waves, thus, determining the amount of EM wave that enters the foam body, attenuated inside of the foam body, and transmitted through the foam body.

Generally, the foam with smaller cell size and higher cell density possesses superior EM wave absorption properties due to the higher surface area and greater amount of cell walls of such foam. The porosity of CPFs has a complex impact on the EMI shielding performance, especially the EM wave absorption properties. When the porosity is too low, a large portion of the foam is the thick solid cell walls and the amount of cells is limited, which would cause limited EM wave reflection and attenuation within the foam body. If the porosity is too high, the cell size has to

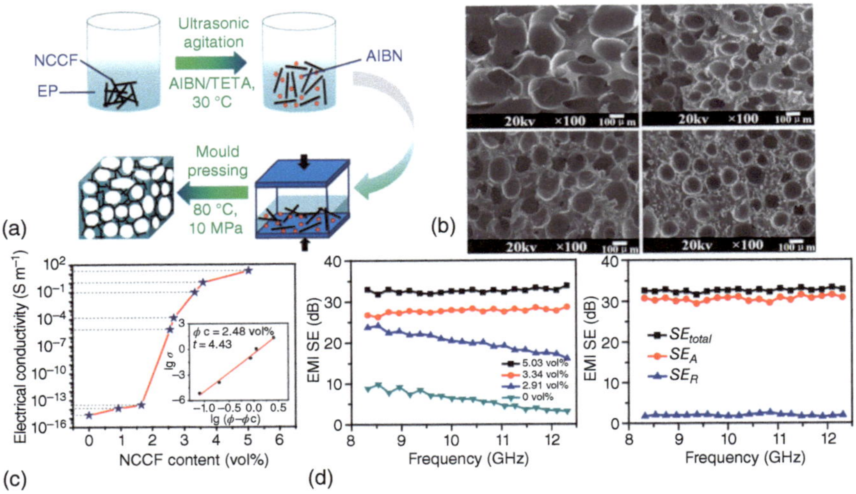

Figure 6.7 (a) Schematic for the fabrication of epoxy/NCCFs composite foams; (b) Morphology of composite foams filled with 0, 1.65, 2.91, and 5.03 vol% of NCCFs; (c) Electrical conductivity as a function of NCCFs content; (d) EMI SE of composite foams with various NCCFs loadings, and the SE_{Total}, SE_A, and SE_R of foam with 5.03 vol% of NCCFs.

be taken into account to understand whether the foam is composed of large cells or fine cells since both cases could lead to high porosity, while their specific surface area would be significantly different. In the case of high porosity with large cells, the EM waves are easy to enter and pass through the foam body. However, in the case of high porosity with fine cells, the EM waves could be difficult to enter the foam body, and the entered EM waves could be effectively attenuated by the cell walls.

Park and coworkers investigated the effect of cell size, void fraction, and thickness on the EMI shielding properties using PVDF/CNT/SiC nanowire (SiCnw) composite foams [28]. The PVDF/CNT/SiCnw composite foams with different void fractions and thicknesses are prepared by $scCO_2$ foaming. When the thickness is varied in 1.4–2.4 mm, the foam demonstrated an absorption-dominant EMI shielding mechanism. As shown in Figure 6.7b and c, it was found that the *R*-value would be reduced as the increase in the void fraction of the foam, which is realized by enhancing the porosity and expansion ratio of the foam. For a with an 85% void fraction, the EM wave attenuation is dominated by the absorption mechanism, and the effect attributed to EM wave absorption would be more significant with the reduction of the cell size from 36 to 10 μm (Figure 6.7d). Therefore, CPFs with high porosity and fine cell size are generally favorable for EM wave absorption.

6.4.3 Foaming Methods for Conductive Polymer Foams

Various approaches have been developed for the fabrication of EMI shielding CPFs, such as chemical foaming, supercritical fluid foaming, phase-separation/freeze drying, electrospinning, template sacrifice, pyrolysis, hydrothermal reaction, supercritical CO_2 drying, 3D printing, bottom-up self-assembly, foam template coating, etc. In this section, we will mainly focus on the chemical foaming and physical foaming approaches to fabricate EMI shielding CPFs since these methods possess the advantages of low cost, high productivity, and scale-up capability [29].

In chemical foaming, the gas formation path is a chemical decomposition process, in which gas is released through chemical reactions or heat-induced decomposition. Then, the molten or viscous polymer matrix is foamed by the released gas. In chemical foaming, the chemical blowing agent (CBA) usually needs to be mixed with the CPC matrix first before the foaming process. The mixing of CBA can be done by twin screw extrusion, melt blending, or solution mixing.

In physical foaming, the cells are formed by the phase change of the physical blowing agent (PBA). When using supercritical fluid as the blowing agent, compressed physical gases like N_2 and CO_2 in the supercritical state are mixed with the polymer matrix by screw mixing or saturation first. Then, the polymer is foamed by the expansion of gas during the phase change of gases from the supercritical state to the gas state.

6.4.3.1 Chemical Foaming of EMI Shielding CPFs

CBA is widely used in the chemical foaming of thermoplastic and thermosetting polymers. Various CBAs have been used in the development of CPFs with high conductivity and EMI shielding properties. Azodiisobutyronitrile (AIBN) is one of the most frequently used CBAs in the fabrication of CPFs. For example, Gupta and coworkers [30] developed a PS/CNT composite foam using AIBN as the foaming agent in 2005. They dispersed CNT into a PS/toluene solution containing 5 wt% AIBN through ultrasonication followed by spray coating and sufficient drying. The dried PS/CNT film is hot compressed at 80 °C to produce the final PS/CNT composite foam. The decomposition of AIBN facilitated the formation of microsized cells in the PS/CNT matrix, and the EMI SE of the foam was enhanced with the increase of CNT content. With a CNT content of 7%, an EMI SE of 18.2–19.3 dB was achieved. Similarly, Ramaprabhu and coworkers [31] prepared functionalized GR (f-G)/PVDF composite foam using AIBN as the CBA. They found that when the f-G content was only 0.5 wt%, the SE_T of the f-G/PVDF CPF was able to reach 20 dB at the frequency of 8–12 GHz.

Besides thermoplastic polymers, CBAs are also used in the fabrication of thermoset conductive foams. Duan and coworkers [32] developed a conductive epoxy foam containing nickel-coated carbon fiber (NCCFs) as the conductive filler using AIBN as the foaming agent (Figure 6.7a). As shown in Figure 6.7b and c, the cell size was reduced, and the conductivity was enhanced with the increase of NCCF loading content. The EMI SE of the composite foam reached about 33 dB when the NCCF loading content was 5 vol%, and the foam demonstrated an ultralow SE_R with a high SE_A suggesting absorption-dominated EMI shielding properties.

Hu and coworkers [33] developed a carbonized phthalonitrile (PN) polymer foam by chemical foaming using azodicarbonamide as the CBA following the same approach. After the foaming process, they carbonized the foam at 1000 °C to obtain a porous foam with a carbon skeleton. The developed carbon foam showed a high compressive strength of ~6.0 MPa cm^3 g^{-1} and a high EMI SE of ~51.2 dB, with absorption being the primary shielding mechanism.

6.4.3.2 Supercritical CO_2 Foaming of EMI Shielding CPFs

In physical foaming, microcells are generated through the cell nucleation and cell growth processes that are realized by the saturation and phase change of physical foaming agents. In the fabrication of EMI shielding CPFs, supercritical CO_2 (scCO_2) foaming is the most frequently used PBA due to its high solubility, diffusivity, volatility, and low cost. In a typical scCO_2 foaming process, CO_2 is dissolved into the polymer matrix under high pressure to produce a single-phase polymer/gas mixture. Then, the cell nucleation is triggered by rapid depressurization at high temperatures or temperature rise under ambient pressure. The pressure-drop-induced cell nucleation refers to the one-step foaming approach, and the temperature-increase-induced cell nucleation refers to the two-step foaming approach. After cell nucleation, the cell nucleus would grow due to the dissolution and concentration of gas molecules, during which the matrix polymer is expanded. After a certain cell growth procedure, the foamed samples are normally transferred to a low-temperature medium for solidification to maintain the cellular structure [34].

In the scCO_2 foaming, the microstructure of polymer foams can be controlled by the foaming parameters, including saturation temperature, saturation pressure, saturation time, depressurization rate, and cooling temperature. In the case of CPC foaming, the presence of conductive fillers would greatly affect the foaming behavior and the final morphology of the CPF due to the heterogeneous cell nucleation caused by the fillers. The geometry, composition, and content of the filler in the CPC matrix would all affect the micropore structure, as well as the conductivity and EMI shielding performance of the CPFs [35].

Ozisik and coworkers [36] studied the relationship between the EMI shielding properties of scCO_2 foamed polycarbonate (PC)/GR composites and the cell morphology and dispersion of GR particles. Using a two-step foaming method, the dispersion of GR was improved, and different cell structures were formed. A maximum normalized EMI SE of ~78 dB cm^3 g^{-1} was obtained by optimizing the composition and microstructure of the foam. Liu and coworkers [37] prepared a multifunctional composite foam with high-efficiency EMI shielding and excellent photoelectrothermal properties by melt extrusion of PVDF/Ni composite and PE/CNTs composite and following scCO_2 batch foaming. The foam showed a highly porous structure that contains CNT conductive network and Ni magnetic components, which contributed to a high electrical conductivity of up to 1.08×10^{-2} S cm^{-1} and a high EMI SE of approximately 27.8 dB in the K band when the content of CNT reached 8% (Figure 6.8). In addition, the porous co-continuous structure, surface plasmonic resonance effect, and high electrical conductivity give the composite foam hydrophobicity, significant photothermal effect, and voltage-driven joule heating characteristics, which can ensure normal operation under wet and extremely cold conditions.

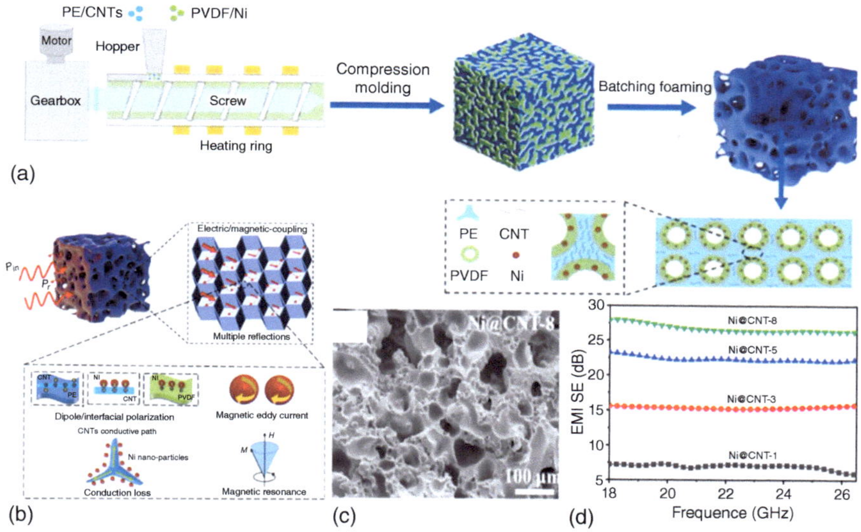

Figure 6.8 (a) illustration of the fabrication process of PVDF/PE composite foams. (b) Schematic illustration of EMI shielding mechanisms; (c) SEM image of a PVDF/PE foam; (d) EMI SE of PVDF-Ni/PE-CNTs composite foams at K band. Source: Cheng et al. [37]/with permission of Engineered Science Publisher.

6.4.4 EMI Shielding CPFs with Special Microstructures

The microstructure of CPFs is a key factor determining the EMI shielding performance. The arrangement, alignment, cell size distribution, and cell aggregation state affect the proportion of EM waves entering the foam body and the multi-reflection behavior of the EM waves inside the foam body. Conventional polymer foams prepared by chemical foaming or physical foaming possess a porous structure with spherical or hexagonal cell morphology. To alter the EM wave interaction with the CPFs, researchers have developed various methods to fabricate CPFs with special microstructures such as reentrant cell structure, aligned cell structure, segregated cell structure, and gradient cell structure.

6.4.4.1 Reentrant Cell Structure

Increasing the amount of EM wave reflection surfaces in a unit volume is a theoretical means to enhance the EM wave reflection inside of the CPF body. Polymer foams with reentrant cell structure or concave cell structure are recognized as auxetic foams since they usually possess a negative Poisson's ratio, which means that these foams would be expanded upon stretching while contracted upon compressing.

In the field of EMI shielding, auxetic foams with reentrant cell structures have more bending and folding on the wall of the cell wall, which would increase complexity and the amount of reflection cell walls in a unit volume, thereby potentially enhancing the EMI shielding performance. In addition, as we have introduced in Chapter 2, auxetic foams also possess superior properties in shear modulus, fracture toughness, yield strength, notch resistance, resilience, and energy absorption. Although several approaches have been developed to fabricate auxetic foams via $scCO_2$ foaming, the production of auxetic foams with high conductivity is still challenging due to the high filler content in the polymer matrix, which limits the reentrant conversion of the foamed cells.

So far, the most prevalent approach to produce conductive auxetic foams with reentrant cell structure is still the traditional triaxial compression and thermal annealing method. For example, Shen and coworkers [38] developed an MXene/auxetic PU composite foam by coating an MXene layer on commercial PU foam followed by triaxial compression to convert the original foam into an auxetic foam. The developed MX/PU composite foam presents a superior EMI SE of 76.2 dB, which is 31.2% higher than its nonauxetic counterpart at the same density owing to the larger specific surface area arising from the auxetic structure. Tang and coworkers [39] prepared an auxetic composite foam by coating the commercial PU foam with a polydopamine (PDA) layer and a GR layer using a dip-coating and ultrasonic immersing followed by a triaxial compression-heating process (Figure 6.9a). The developed PU/PDA/GR auxetic composite foam had a Poisson's ratio of −2.36 and an open cellular structure (Figure 6.9b). The foam showed a high EMI SE of 59.75 dB, which is 8.9 times the positive Poisson's ratio composite foam with the same GR content (Figure 6.9c). They found that the reentrant cell structure of the auxetic foam with a smaller Poisson's ratio contributed to more effective multireflection and the attenuation of the EM waves (Figure 6.9d).

6.4.4.2 Oriented Cell Structure

Polymer foams with anisotropic porous structure have special mechanical properties and unidirectional thermal and electrical properties. The honeycomb structure is one of the highly oriented cellular structures that possess unique anisotropic properties and super lightweight. When a CPF is processed with a honeycomb structure, it may have various advantages, including the high length-diameter ratio and large specific surface area of conductive filler network are conducive to the establishment of efficient conductive path and the formation of multiple interfaces with polymer matrix would enhance the interfacial polarization and improve the EM shielding performance. So far, it is difficult to produce a highly oriented honeycomb porous structure through gas foaming. The most effective and widely used approach is the ice template unidirectional freezing method.

For example, Siqueira and coworkers [40] prepared a conductive composite foam composed of CNT and nanofibrillate cellulose (NFC) with a micro-honeycomb porous structure through a freezing–thawing–drying approach. As illustrated in Figure 6.10a, the CNT/NFC solution was unidirectional frozen to allow the growth of vertically aligned ice crystals. The ice phase is removed by the following solvent exchange, thawing, and drying procedures. The obtained anisotropic CNT/NFC foam showed a highly oriented cellular structure and exhibited excellent stiffness and conductivity in the cell alignment direction (Figure 6.10b). The oriented microcell channels allowed the EM waves to enter the foam body and facilitated the attenuation of EM waves through effective multiple reflections (Figure 6.10c),

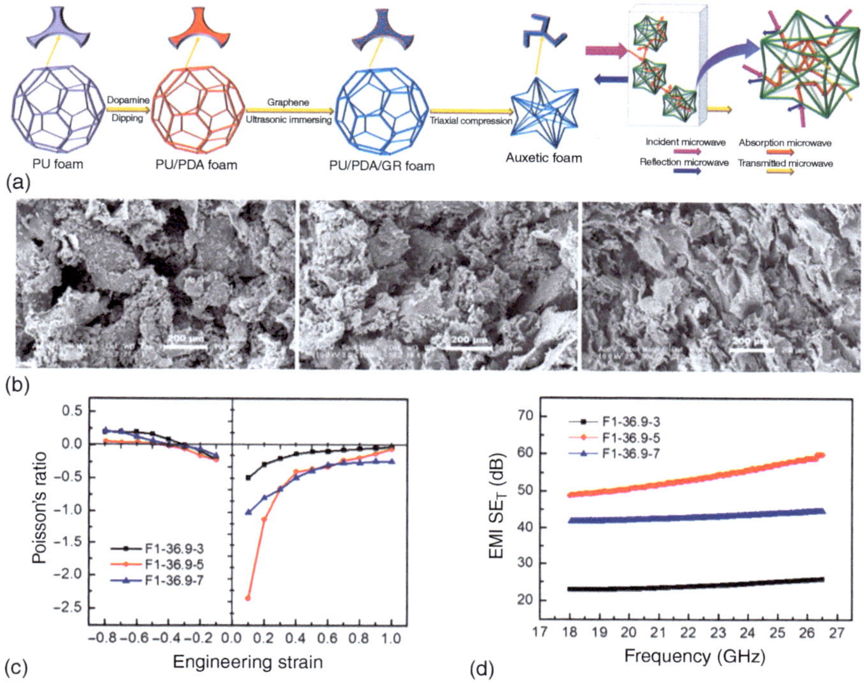

Figure 6.9 (a) Experimental flow chart of preparation of auxetic composite foam and the EMI shielding mechanism illustration; (b) SEM images of composite foams with compression factors of 3, 5, and 7; (c) The Poisson's ratio change with the engineering strain and the EMI SE results of auxetic composite foams with different compression factors; (d) Schematic illustration for EMI shielding mechanism of the auxetic composite foam. Source: (b) Fan et al. [39]/with permission of Elsevier.

which resulted in a high EMI SE of more than 70 dB and a specific SE of up to 6111 dB cm^2 g^{-1}.

6.4.4.3 Segregated Cell Structure

The construction of an effective conductive network is essential for realizing high EMI shielding performance. In the conventional foaming process, the compaction of conductive fillers in the cell walls is generally improved by the volume exclusion effect of the cells during the cell growth stage. To further enhance the compactness of the conductive fillers, researchers have developed a special segregated microstructure. The segregated structure is a commonly used material design in the fabrication of EMI shielding composite materials. In a segregated structure, the conductive phase in the composite foam forms a continuous phase to achieve high conductivity. The conductive network is segregated by dielectric phases, which provide the mechanical strength of the composite. CPFs with segregated cell structures could more effectively utilize the conductive fillers, and the multiple reflections of EM waves inside the foam body could also be emphasized.

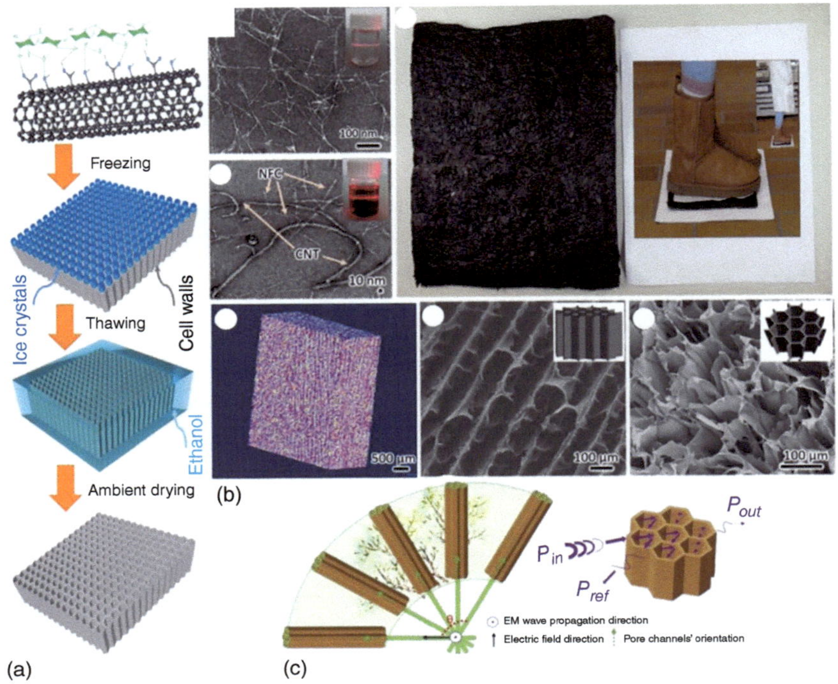

Figure 6.10 (a) Preparation process for the oriented CNT/NFC foams; (b) TEM images of NFC and NFC attached to CNTs, SEM images of the CNT/NFC foams in the aligned direction and cross-section, and a large-size CNT/NFC foam with a person standing on it; (c) Illustration of the EMI shielding mechanism for the CNT/NFC foam with oriented cellular structure. Source: Zeng et al. [40]/with permission of Royal Society of Chemistry.

Liao and coworkers [41] fabricated thermoplastic polyurethane/reduced GR oxide (TPU/RGO) composite foams with segregated cell structures for efficient EMI shielding. As shown in Figure 6.11a, conductive GO sheets were coated on the surface of TPU pellets, followed by *in situ* reduction using L-ascorbic acid, hot compression, and $scCO_2$ foaming to prepare composite foams with different RGO loading. The RGO formed a continuous network in the matrix (Figure 6.11b), and the micro-sized cells were formed in the TPU phase after foaming (Figure 6.11c). The construction of the segregated structure enhanced the conductivity and the EMI shielding performance of the composite foam. With a 3.17 vol% of RGO loading, the foam achieved an EMI SE of 21.8 dB owing to the enhanced multiple reflections of EM waves in the TPU/RGO composite foam (Figure 6.11d).

Wang et al. [42] developed an innovative approach to fabricating poly(ether-block-amide) (Pebax) composite foams with segregated structures. Foamed Pebax beads were first prepared by $scCO_2$ batch foaming. Then, the foamed beads were placed into a mode and filled with a solution containing Pebax and multiwalled CNT (MWCNT) to fill the space among the foamed Pebax beads. After a hot molding process, the segregated composite foam is obtained and the conductivity of the foam could be enhanced by simply increasing the content of MWCNT. The composite

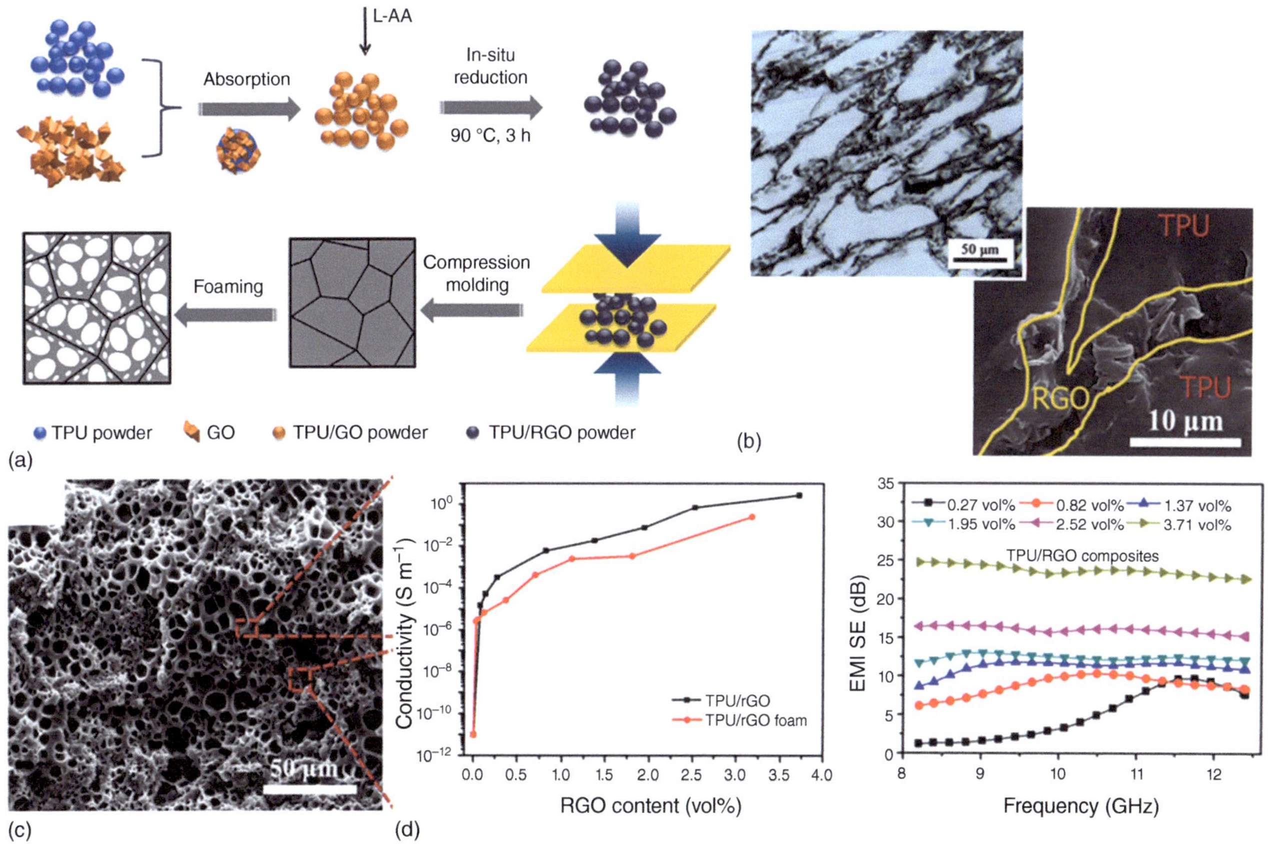

Figure 6.11 (a) Schematic showing the fabrication process of segregated TPU/RGO composite foams, (b) segregated structure of the solid composite, (c) SEM images of the foamed TPU/RGO composite, and (d) conductivity and EMI SE results of segregated TPU/RGO composite foams with different RGO loading. Source: Jiang et al. [41]/with permission of Elsevier.

foam showed ultralow density, excellent tensile and compression properties, and achieved an EMI SE of 16 dB in the X-band with the MWCNT content of 1.12%. Owing to the segregated structure, the foam attenuates EM waves mainly by the absorption mechanism.

6.4.4.4 Gradient Cell Structure

Although highly conductive network is desired to achieve high EMI SE, the EM wave reflection would be high when the material has a high conductivity on the surface due to the impedance mismatch. Therefore, to achieve high EM wave absorption, the CPF shields are desired to possess a low conductive surface layer, which allows the EM wave to enter the foam body, and a high conductive bottom layer, which facilitates EM wave attenuation. Based on this concept, CPFs with gradient cell structure and asymmetric structure have been developed to simultaneously increase the EMI SE and the EM wave absorption. Filleter et al. [43] fabricated a high-density polyethylene (HDPE)/Graphene nanoplatelets (GnPs) composite foam with graded microcellular structures via microcellular injection molding using $scCO_2$ as the blowing agent. The evolution of the cellular structure along the melt flow direction was investigated in detail. Cells ranging from shear-induced elongated morphology to isotropic morphology were obtained, which offered graded electrical and thermal properties to the HDPE/GnP composite foam. The electrical conductivity, permittivity, and thermal conductivity all improved with the foaming process, and the EMI shielding performance was raised up to 45% over the length of the foam.

Zhao and coworkers [44] fabricated poly(methyl methacrylate) (PMMA)/MWCNTs composite foams using a solution coating process and supercritical fluid-assisted foaming. The composite foam demonstrated excellent electrical conductivity under low filler content due to the concentration of MWCNTs nanoparticles within the cell walls. In addition, the gradient distribution of MWCNTs along the radial direction of PMMA microspheres, the composite foam showed a trimodal microcellular structure, which rendered the foam with a low density of 0.49 g cm^{-3}, a high electrical conductivity of 3.19 S m^{-1}, and a high EMI SE of 35.9 dB when the content of MWCNTs was 5.0 wt%. A specific EMI SE of 356.5 dB cm^2 g^{-1} was achieved due to the multireflection of EM waves within the gradient composite foam with trimodal cells (Figure 6.12).

In addition, Mi and coworkers [45] developed a facile approach to produce CPFs with gradient structure via multiple dip-coatings. As illustrated in Figure 6.13a, melamine foam (MF) is successively coated with MXene solutions with different concentrations to manipulate the loading content of MXene in different regions of the foam (Figure 6.13b). With this gradient design, the high porosity and low-conductivity surface layer facilitates the EM wave penetration due to the impedance matching, the gradual increased MXene content assisted the multiple reflection and attenuation of the EM waves, and the rGO film located on the back side of the foam effectively reflected the EM wave to the foam body (Figure 6.13c). Owing to the desired EMI shielding mechanism, an average SE of 78.8 dB and an R value of 0.022 were reached by the foam in the X band (Figure 6.13d).

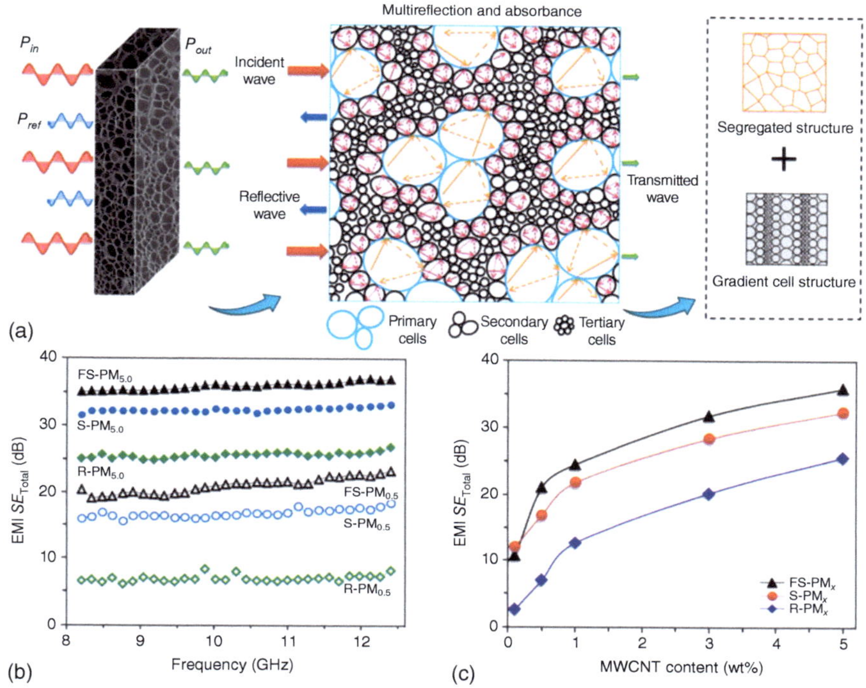

Figure 6.12 (a) Schematic of EM wave dissipation in the PMMA/MWCNTs nanocomposite foams, and (b) EMI SE_{Total} of PMMA/MWCNTs composite foams with different filler content, and the EMI SE_{Total} values of different composite foams with respect to the MWCNT content. Source: Li et al. [44]/with permission of eXPRESS Polymer Letters.

6.4.4.5 Layered Foam Structure

In conventional CPFs, conductive nanofillers are dispersed in the nonconductive polymer matrix, which inevitably restricts the formation of a continuous conductive network. In this regard, researchers developed a layered foam structure in which the conductive phase is produced into separate layers that are embedded in the polymer matrix. With the layered structure, EM waves can be reflected many times by several highly conductive planes, which could effectively enhance the EMI shielding performance. Zhao and coworkers [46] fabricated a composite layered foam composed of PVDF/SiCnw/MXene. A highly conductive PVDF/MWCNT/GnPs composite film was used for the reflection layer, and a SiCnw and MXene-containing foam layer was mainly used to absorb the EM waves. The impedance matching, the prolonged EM wave propagating path, and the EM wave reflection contributed to an EMI SE of 32.6 dB of an $R < 0.1$ at a thickness of 1.95 mm (Figure 6.14a). Song et al. [47] developed novel sandwich structures by employing electrical and magnetic loss spacers that possess tunable EMI shielding performance and can manipulate EM wave transport. The performance could be run by varying the layered structure and

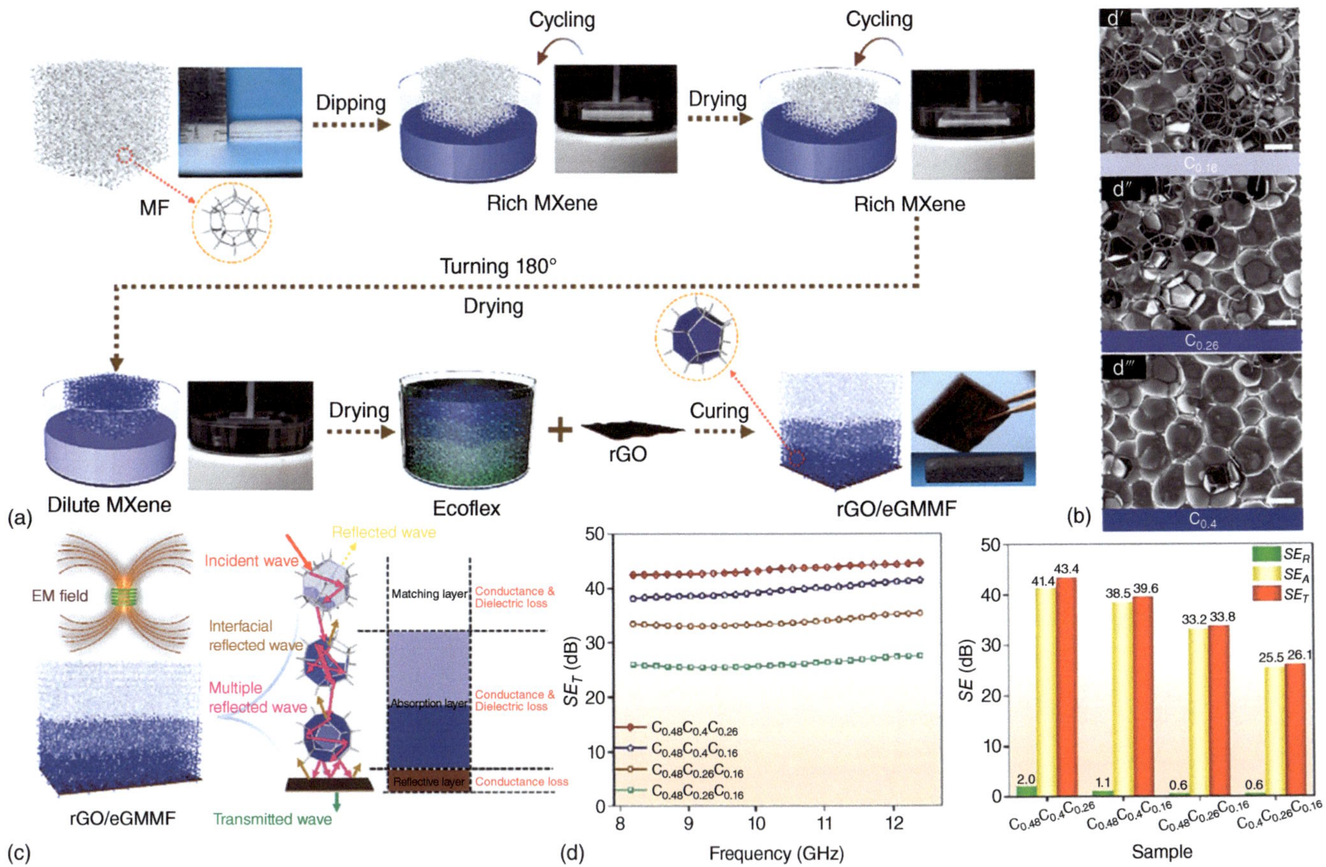

Figure 6.13 (a) The fabrication process of the MXene coated gradient composite foam, (b) SEM images in different locations of the composite foam, (c) EM wave absorption mechanism of the composite foam, and (d) EMI SE_T, SE_R, and SE_A results of different MXene coated gradient composite foams. Source: Du et al. [45]/with permission of Elsevier.

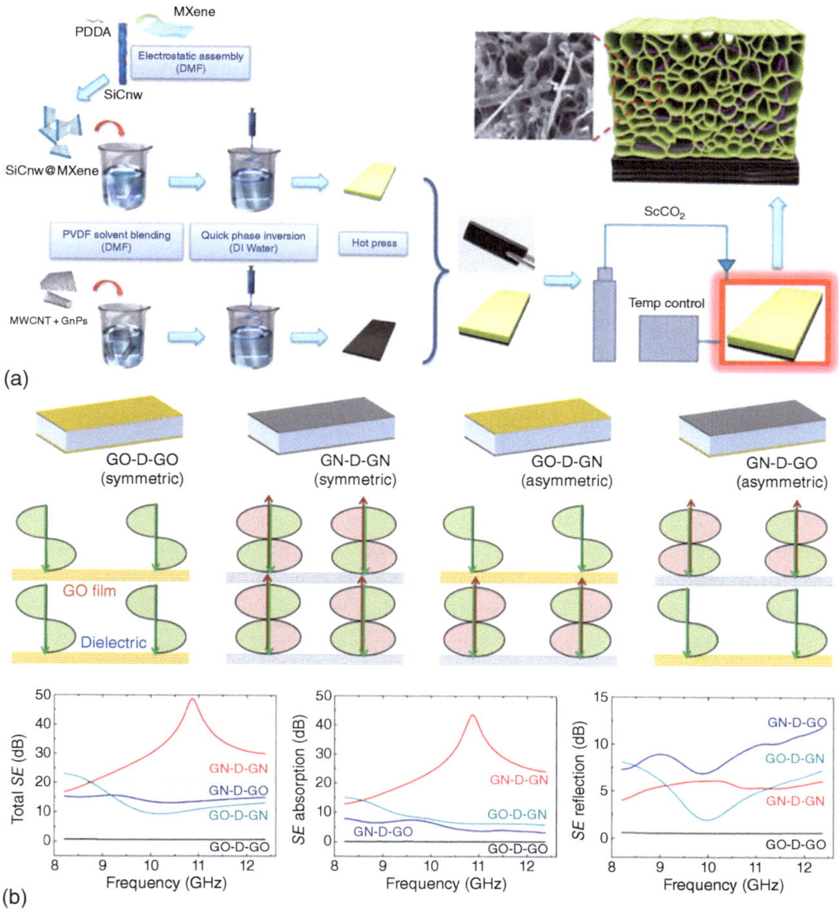

Figure 6.14 (a) Illustration of the fabrication process of the layered foam composed of PVDF/SiCnw/MXene, and (b) Two-layer sandwich structures with various graphene films and EMI shielding performance. The green and pink waves in the schemes indicate the transmittance and reflected waves. Source: (a) Ma et al. [46]/Springer Nature/CC BY 4.0. (b) Song et al. [47]/with permission of American Chemical Society.

composing materials, which provides a versatile strategy for designing and fabricating advanced EM attenuation materials (Figure 6.14b)

6.5 Advanced Research on EMI Shielding Porous Composites

Although the gas foaming technique has been recognized as one of the most promising methods for large-scale green production of EMI shielding foams, there are various methods and materials have been employed to develop EMI shielding and EM wave absorption porous materials with excellent properties. Some advanced progress will be briefly reviewed in this section.

6.5.1 EMI Shielding Aerogels

Aerogels as a class of porous material with high porosity and ultra-lightweight have been more and more used in various applications, including thermal insulation, sound absorption, structural materials, sensing applications, liquid absorption, and EMI shielding fields. Since aerogels are normally fabricated through a solution method, it is fairly easy to introduce conductive nanofillers into the aerogel matrix and achieve a uniform filler distribution.

For example, Xie and coworkers [48] fabricated an MXene/GR hybrid foam through a simple and feasible freeze-drying and reduction heat treatment strategy. The foam showed high electrical conductivity and highly efficient EM wave attenuation properties with a high EMI SE of 50.7 dB. Liu and coworkers [49] developed lightweight AgNW/MXene hybrid sponges featuring hierarchical structures by a combination of dip-coating and unidirectional freeze-drying methods. As shown in Figure 6.15a, a commercial MF foam is dip-coated by PDA and AgNW first and unidirectional freeze-dried in an MXene solution, which resulted in a complex porous structure with aligned MXene channels and AgNW skeleton networks (Figure 6.15b). The conductivity of the sponge was greatly enhanced with the increased dip-coating cycles, and the hierarchical porous structure of the hybrid sponge enabled multiscale EM wave attenuation through multiple reflections and dielectric loss (Figure 6.15c and d). The optimized hybrid sponge reached a high EMI SE of 52.6 dB with a low density of 49.5 mg cm^{-3} (Figure 6.15e).

6.5.2 EMI Shielding Fibrous Networks

Apart from cellular structured foams, fibrous structured membranes are also considered porous materials that are capable of being used in effective EMI shielding, especially in cases that require thin film shields with high flexibility. The fabrication of fibrous EMI shields mainly includes surface modification of pristine fabrics and directed generated fibrous networks through electrospinning or wetspinning. Zong et al. [50] fabricated a flexible wearable fabric with excellent EMI shielding and passive radiation heating functions by coating AgNW, Fe_3O_4 nanoparticles, and polydimethylsiloxane (PDMS) onto industrial fabrics. The highly conductive AgNW path and Fe_3O_4 nanoparticles rendered the fabric high EMI shielding performance (SE_T of 100.9 dB) at a thickness of only 450 μm. However, the shielding mechanism of the fabric is dominated by reflection due to the significant impedance mismatching on the EM wave incident surface.

To reduce the high reflection of common fabric-based EMI shields, Mi et al. designed a layered structure in which the conductive network is separated into several layers by the nonconductive electrospun fibrous membranes. By alternatively assembling the MXene conductive layer with the PLA fibrous layer, the composite-layered membrane achieved a high EMI SE of 55.4 dB owing to the multiple-reflection-absorption effect [51]. In addition, they investigated the effect of assembling configurations of the layered fibrous composite on the EMI shielding property. As shown in Figure 6.16a, the AgNW network is fabricated on the

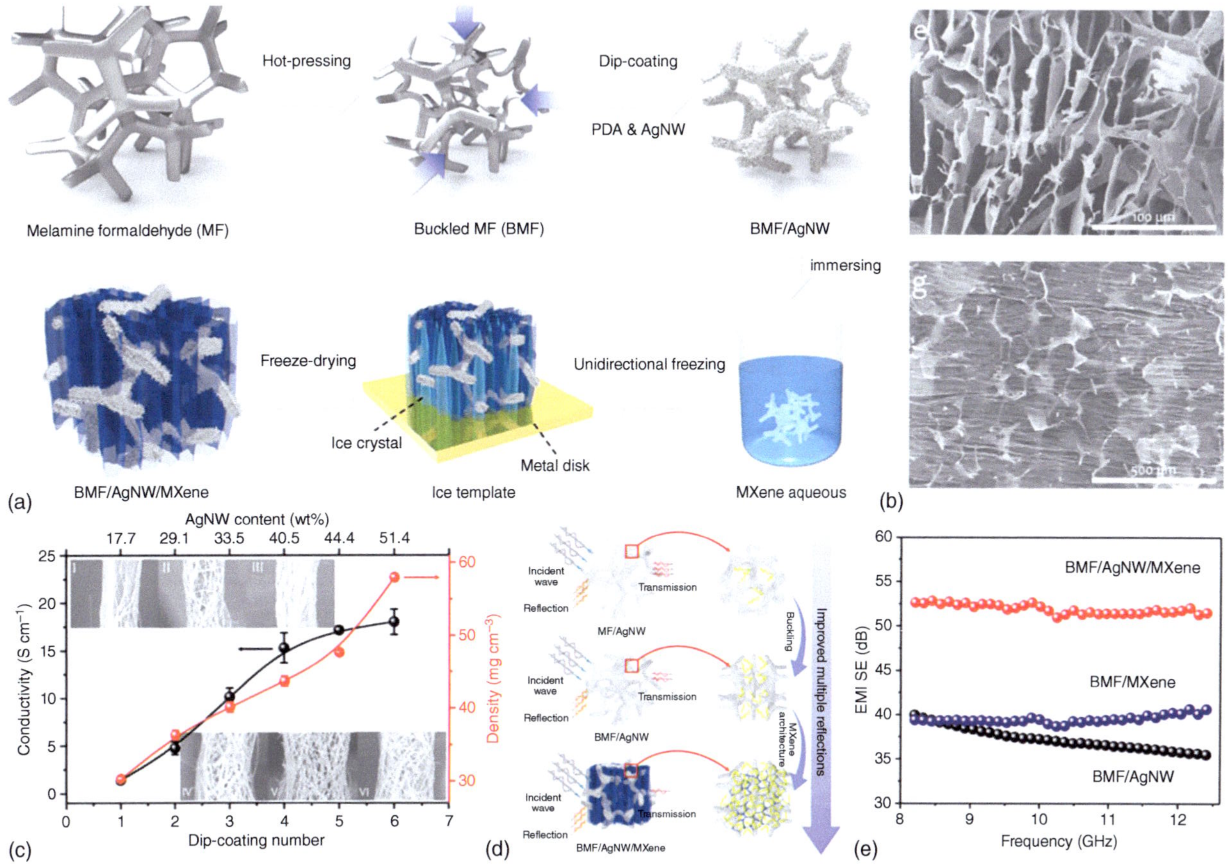

Figure 6.15 (a) Schematic illustrating the fabrication process of the BMF/AgNW/MXene hybrid sponge, (b) SEM images of the hybrid sponge, (c) Conductivity and density of the hybrid sponge as the increase in dip-coating cycles, (d) Schematic illustration of EMI shielding mechanisms, and (e) the comparison of EMI SE between different hybrid sponges. Source: Weng et al. [49]/with permission of Royal Society of Chemistry.

surface of the TPU electrospun membrane by vacuum-assisted filtration and bilayer membranes were assembled in different ways (Figure 6.16b). It was found that the continuous AgNW conductive layer and the sandwich structure facilitated electron transmission efficiency and EM wave dissipation through multiple reflections and resonance effect, which resulted in a high EMI SE of 97.3 dB mainly attributed to the improvement of SE_A (Figure 6.16c and d) [52].

6.5.3 Metal-Based EMI Shielding Porous Materials

Metal-based foams, such as nickel (Ni), aluminum (Al), and copper (Cu) foams, are another porous material used for EMI shielding due to their high conductivity and relatively low cost. To enhance the EMI shielding performance and EM wave absorption, metal foams are often combined with polymer materials and conductive/magnetic nanomaterials by wrapping a polymer layer on their skeleton. Although metal-based foams have high EMI shielding properties, they are still restricted by the disadvantages of poor flexibility, high reflection, and complex preparation procedures. For instance, Wang and coworkers [53] fabricated a highly porous and hierarchically structured Ni foam by solution combustion of Ni nitrate and ethylene glycol mixture. The porous Ni was then immersed in PDMS precursor followed by pressing and curing to develop the Ni-based composite foam. The foam showed excellent electrical conductivity, magnetic properties, and porous internal structure, which led to the conductive loss and magnetic loss of EM waves, as well as the reflection and scattering of EM waves in the porous interface. The Ni-based foam showed a high EMI SE of 56 dB with a large proportion of contribution from SE_A.

In addition to the direct use of a metallic skeleton, a metal layer is often plated on the material surface to enhance the conductivity or render the material with magnetic properties. Ma and coworkers [54] prepared the Cu- and Ni-plated polymethacrylimide (PMIF) composite foams via the electroless plating method. The surface electrical conductivity of the PMIF@Cu and PMIF@Ni composite foams reached $1.06 \times 104\,\mathrm{S\,m^{-1}}$ and $9.15 \times 103\,\mathrm{S\,m^{-1}}$ when the electroless plating time was 70 minutes, which corresponded to a high EMI SE of ~52 dB and ~43 dB, respectively. Moreover, the metalized PMI foams present remarkable radiation performances and good matching degrees with the feeder as vertically polarized monopole antennas, which can be used to replace traditional metal-based antennas.

6.5.4 Surface Coated/Modified Foams for EMI Shielding

Surface coating and modification is a widely used approach to fabricate EMI shielding porous foams. Through this method, almost all kinds of conductive and magnetic particles can be coated on the surface of the foam skeleton with or without the help of a resin binder. The content and proportion of different fillers can be easily controlled in the coating process, which allows the manipulation of conductivity and EMI shielding properties of the developed composite foams. Nevertheless, the porous structure of the foam is decided by the foam skeleton, and

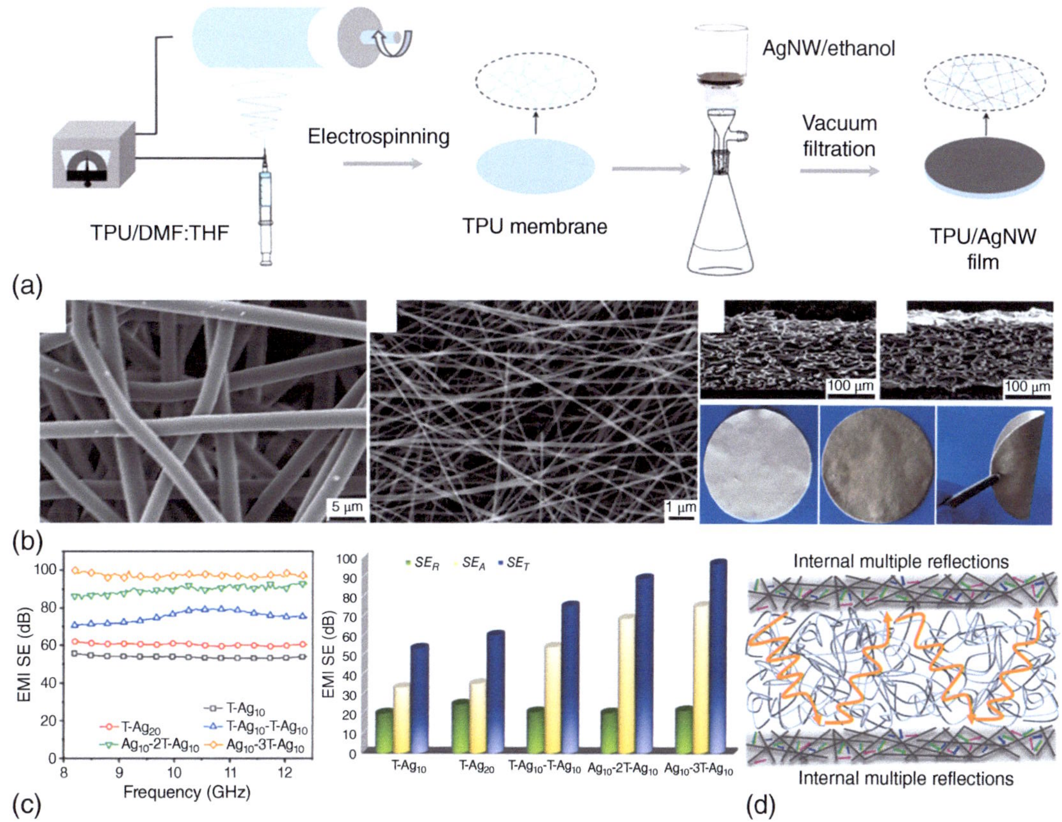

Figure 6.16 (a) Illustration of the fabrication process of the TPU-/AgNW-layered membranes, (b) SEM images of TPU electrospun fibers, AgNW networks, and the assembled composite films, (c) EMI SE results of TPU-/AgNW-layered membranes assembled in different ways, and (d) Schematic illustration of the multiple reflection and resonance effect of the TPU-/AgNW-layered membranes. Source: Zhang et al. [52]/with permission of Elsevier.

it is difficult to control in the postcoating process, and the stability of the coating layer on the skeleton surface is another defect that needs to be considered. For instance, Zhu et al. [55] coated PU foams with a D-gluconate-chitosan/single-wall CNT (SWCNT/GA-chitosan) layer and a crystalline paraffin layer derived from conductive biomass. The developed composite foam possesses EMI shielding properties and shape memory characteristics, which allows the adjustment of EMI SE from 18 to 30 dB by simply changing the overall thickness of the composite foam.

References

1 Sun, B., Sun, S., Guo, Y. et al. (2023). Asymmetric magnetic-electric dual-functional composite foams for ultra-efficient electromagnetic interference shielding with unprecedented low reflection. *Composites Part A: Applied Science and Manufacturing* 164: 107301.

2 Zhao, B. and Park, C.B. (2017). Tunable electromagnetic shielding properties of conductive poly (vinylidene fluoride)/Ni chain composite films with negative permittivity. *Journal of Materials Chemistry C* 5: 6954–6961.

3 Guo, Y., Pan, L., Yang, X. et al. (2019). Simultaneous improvement of thermal conductivities and electromagnetic interference shielding performances in polystyrene composites via constructing interconnection oriented networks based on electrospinning technology. *Composites Part A: Applied Science and Manufacturing* 124: 105484.

4 Pan, F., Liu, Z., Deng, B. et al. (2021). Lotus leaf-derived gradient hierarchical porous C/MoS_2 morphology genetic composites with wideband and tunable electromagnetic absorption performance. *Nano-Micro Letters* 13: 1–17.

5 Liu, T.-T., Cao, M.-Q., Fang, Y.-S. et al. (2022). Green building materials lit up by electromagnetic absorption function: a review. *Journal of Materials Science and Technology* 112: 329–344.

6 Wang, H., Li, S., Liu, M. et al. (2021). Review on shielding mechanism and structural design of electromagnetic interference shielding composites. *Macromolecular Materials and Engineering* 306: 2100032.

7 Pakdel, E., Wang, J., Kashi, S. et al. (2020). Advances in photocatalytic self-cleaning, superhydrophobic and electromagnetic interference shielding textile treatments. *Advances in Colloid and Interface Science* 277: 102116.

8 Zhao, B., Hamidinejad, M., Wang, S. et al. (2021). Advances in electromagnetic shielding properties of composite foams. *Journal of Materials Chemistry A* 9: 8896–8949.

9 Kumar, P., Shahzad, F., Yu, S. et al. (2015). Large-area reduced graphene oxide thin film with excellent thermal conductivity and electromagnetic interference shielding effectiveness. *Carbon* 94: 494–500.

10 Gupta, T.K., Singh, B.P., Dhakate, S.R. et al. (2013). Improved nanoindentation and microwave shielding properties of modified MWCNT reinforced polyurethane composites. *Journal of Materials Chemistry A* 1: 9138–9149.

11 Chung, D.D.L. (2000). Materials for electromagnetic interference shielding. *Journal of Materials Engineering and Performance* 9: 350–354.

12 Bera, R., Maitra, A., Paria, S. et al. (2018). An approach to widen the electromagnetic shielding efficiency in PDMS/ferrous ferric oxide decorated RGO–SWCNH composite through pressure induced tunability. *Chemical Engineering Journal* 335: 501–509.

13 Zhang, Y., Qiu, M., Yu, Y. et al. (2017). A novel polyaniline-coated bagasse fiber composite with core–shell heterostructure provides effective electromagnetic shielding performance. *ACS Applied Materials & Interfaces* 9: 809–818.

14 Ameli, A., Kazemi, Y., Wang, S. et al. (2017). Process-microstructure-electrical conductivity relationships in injection-molded polypropylene/carbon nanotube nanocomposite foams. *Composites Part A: Applied Science and Manufacturing* 96: 28–36.

15 Shen, B., Li, Y., Zhai, W., and Zheng, W. (2016). Compressible graphene-coated polymer foams with ultralow density for adjustable electromagnetic interference (EMI) shielding. *ACS Applied Materials & Interfaces* 8: 8050–8057.

16 Wang, L., Wang, D., Wu, Z. et al. (2020). Self-derived superhydrophobic and multifunctional polymer sponge composite with excellent joule heating and photothermal performance for strain/pressure sensors. *ACS Applied Materials & Interfaces* 12: 13316–13326.

17 Wu, X., Han, B., Zhang, H.-B. et al. (2020). Compressible, durable and conductive polydimethylsiloxane-coated MXene foams for high-performance electromagnetic interference shielding. *Chemical Engineering Journal* 381: 122622.

18 Zhang, S.J., Cheng, B., Gao, Z.G. et al. (2022). Two-dimensional nanomaterials for high-efficiency electromagnetic wave absorption: an overview of recent advances and prospects. *Journal of Alloys and Compounds* 893: 162343.

19 Lopez-Dominguez, V., Quesada, A., Guzmán-Mínguez, J.C. et al. (2018). A simple vibrating sample magnetometer for macroscopic samples. *Review of Scientific Instruments* 89: 034707.

20 Massa, A., Rocca, P., and Oliveri, G. (2015). Compressive sensing in electromagnetics-a review. *IEEE Antennas and Propagation Magazine* 57: 224–238.

21 Sheng, A., Ren, W., Yang, Y. et al. (2020). Multilayer WPU conductive composites with controllable electro-magnetic gradient for absorption-dominated electromagnetic interference shielding. *Composites Part A: Applied Science and Manufacturing* 129: 105692.

22 Liang, C., Gu, Z., Zhang, Y. et al. (2021). Structural design strategies of polymer matrix composites for electromagnetic interference shielding: a review. *Nano-Micro Letters* 13: 181.

23 Park, J.-B., Rho, H., Cha, A.-N. et al. (2020). Transparent carbon nanotube web structures with Ni-Pd nanoparticles for electromagnetic interference (EMI) shielding of advanced display devices. *Applied Surface Science* 516: 145745.

24 Ding, J.J., Wang, L., Zhao, Y.H. et al. (2019). Boosted interfacial polarization from multishell TiO_2@Fe_3O_4@PPy heterojunction for enhanced microwave absorption. *Small* 15: 1902885.

25 Mathews, S.A. and Babu, D.R. (2021). Analysis of the role of M-type hexaferrite-based materials in electromagnetic interference shielding. *Current Applied Physics* 29: 39–53.

26 Zhang, Y., Yang, Z.J., and Wen, B.Y. (2019). An ingenious strategy to construct helical structure with excellent electromagnetic shielding performance. *Advanced Materials Interfaces* 6: 1900375.

27 Du, Z.R., Zhang, G.Y., Chen, K. et al. (2022). MXene/polylactic acid fabric-based resonant cavity for realizing simultaneous high-performance electromagnetic interference (EMI) shielding and efficient energy harvesting. *ACS Applied Materials & Interfaces* 14: 14607–14617.

28 Ma, L., Hamidinejad, M., Wei, L.F. et al. (2023). Absorption-dominant EMI shielding polymer composite foams: microstructure and geometry optimization. *Materials Today Physics* 30: 100940.

29 Lee, S., Park, C.B., and Ramesh, N. (2006). *Polymeric Foams: Science and Technology*. Boca Raton, FL: CRC/Taylor & Francis.

30 Yang, Y., Gupta, M.C., Dudley, K.L., and Lawrence, R.W. (2005). Novel carbon nanotube–polystyrene foam composites for electromagnetic interference shielding. *Nano Letters* 5: 2131–2134.

31 Eswaraiah, V., Sankaranarayanan, V., and Ramaprabhu, S. (2011). Functionalized graphene–PVDF foam composites for EMI shielding. *Macromolecular Materials and Engineering* 296: 894–898.

32 Yang, J., Yang, Y., Duan, H. et al. (2017). Light-weight epoxy/nickel coated carbon fibers conductive foams for electromagnetic interference shielding. *Journal of Materials Science: Materials in Electronics* 28: 5925–5930.

33 Zhang, L., Liu, M., Roy, S. et al. (2016). Phthalonitrile-based carbon foam with high specific mechanical strength and superior electromagnetic interference shielding performance. *ACS Applied Materials & Interfaces* 8: 7422–7430.

34 Stankovich, S., Dikin, D.A., Dommett, G.H. et al. (2006). Graphene-based composite materials. *Nature* 442: 282–286.

35 Zhao, B., Wang, R.M., Li, Y. et al. (2020). Dependence of electromagnetic interference shielding ability of conductive polymer composite foams with hydrophobic properties on cellular structure. *Journal of Materials Chemistry C* 8: 7401–7410.

36 Gedler, G., Antunes, M., Velasco, J.I., and Ozisik, R. (2015). Electromagnetic shielding effectiveness of polycarbonate/graphene nanocomposite foams processed in 2-steps with supercritical carbon dioxide. *Materials Letters* 160: 41–44.

37 Cheng, H., Lu, Z., Gao, Q. et al. (2021). PVDF-Ni/PE-CNTs composite foams with co-continuous structure for electromagnetic interference shielding and photo-electro-thermal properties. *Engineered Science* 16: 331–340.

38 Kim, E., Zhang, H.M., Lee, J.H. et al. (2021). MXene/polyurethane auxetic composite foam for electromagnetic interference shielding and impact attenuation. *Composites Part A: Applied Science and Manufacturing* 147: 106430.

39 Fan, D.L., Li, N.X., Li, M.G. et al. (2022). Polyurethane/polydopamine/graphene auxetic composite foam with high-efficient and tunable electromagnetic interference shielding performance. *Chemical Engineering Journal* 427: 131635.

40 Zeng, Z., Wang, C., Wu, T. et al. (2020). Nanocellulose assisted preparation of ambient dried, large-scale and mechanically robust carbon nanotube foams for electromagnetic interference shielding. *Journal of Materials Chemistry A* 8: 17969–17979.

41 Jiang, Q.Y., Liao, X., Li, J.S. et al. (2019). Flexible thermoplastic polyurethane/reduced graphene oxide composite foams for electromagnetic interference shielding with high absorption characteristic. *Composites Part A: Applied Science and Manufacturing* 123: 310–319.

42 Ge, C., Wang, G., Zhao, G. et al. (2021). Lightweight and flexible poly (ether-block-amide)/multiwalled carbon nanotube composites with porous structure and segregated conductive networks for electromagnetic shielding applications. *Composites Part A: Applied Science and Manufacturing* 144: 106356.

43 Hamidinejad, M., Salari, M., Ma, L. et al. (2022). Electrically and thermally graded microcellular polymer/graphene nanoplatelet composite foams and their EMI shielding properties. *Carbon* 187: 153–164.

44 Li, T., Zhao, G., Zhang, L. et al. (2020). Ultralow-threshold and efficient EMI shielding PMMA/MWCNTs composite foams with segregated conductive network and gradient cells. *Express Polymer Letters* 14: 685–703.

45 Du, Z.R., Zhou, C., Mi, H.Y. et al. (2023). Recorded low-reflection in gradient MXene-decorated melamine microwave shielding foam integrated with piezoresistive sensing and energy absorption properties. *Composites Part A: Applied Science and Manufacturing* 173: 107694.

46 Ma, L., Hamidinejad, M., Zhao, B. et al. (2022). Layered foam/film polymer nanocomposites with highly efficient emi shielding properties and ultralow reflection. *Nano-Micro Letters* 14: 19.

47 Song, W.L., Gong, C.C., Li, H.M. et al. (2017). Graphene-based sandwich structures for frequency selectable electromagnetic shielding. *ACS Applied Materials & Interfaces* 9: 36119–36129.

48 Fan, Z.M., Wang, D.L., Yuan, Y. et al. (2020). A lightweight and conductive MXene/graphene hybrid foam for superior electromagnetic interference shielding. *Chemical Engineering Journal* 381: 122696.

49 Weng, C.X., Wang, G.R., Dai, Z.H. et al. (2019). Buckled AgNW/MXene hybrid hierarchical sponges for high-performance electromagnetic interference shielding. *Nanoscale* 11: 22804–22812.

50 Zong, J.-Y., Zhou, X.-J., Hu, Y.-F. et al. (2021). A wearable multifunctional fabric with excellent electromagnetic interference shielding and passive radiation heating performance. *Composites Part B: Engineering* 225: 109299.

51 Du, Z.R., Chen, K., Zhang, Y.X. et al. (2021). Engineering multilayered MXene/electrospun poly(lactic acid) membrane with increscent electromagnetic interference (EMI) shielding for integrated Joule heating and energy generating. *Composites Communications* 26: 100770.

52 Zhang, F.F., Hu, J.S., Zhao, P. et al. (2021). Multifunctional electromagnetic interference shielding films comprised of multilayered thermoplastic polyurethane membrane and silver nanowire. *Composites Part A: Applied Science and Manufacturing* 147: 106472.

53 Li, M.Y., Feng, Y.J., Zhong, Y. et al. (2023). Facile fabrication of novel high-performance electromagnetic interference shielding nickel foam. *Colloids and Surfaces A: Physicochemical and Engineering Aspects* 656: 130352.

54 Li, J.W., Wang, A.F., Qin, J.B. et al. (2021). Lightweight polymethacrylimide @copper/nickel composite foams for electromagnetic shielding and monopole antennas. *Composites Part A: Applied Science and Manufacturing* 140: 106144.

55 Zhu, S., Zhou, Q., Wang, M. et al. (2021). Modulating electromagnetic interference shielding performance of ultra-lightweight composite foams through shape memory function. *Composites Part B: Engineering* 204: 108497.

7

Polymer Foams for Tissue Engineering Scaffolds

7.1 Overview of Tissue Engineering

7.1.1 Basic Elements of Tissue Engineering

Biomedical polymer materials are usually used in the diagnosis and treatment of physiological system diseases, which can repair or replace the tissues or organs of living organisms and promote the function and functional recovery of human organs [1]. Some biomedical polymer materials somehow act on human tissue when implanted into the human body or after coating the body surface [2]. Modern bioengineering technology has brought biomedical polymer materials to a new space, and the significant biocompatibility and biodegradability have brought more attention to biomedical polymer materials.

In 1993, Robert and Joseph [3] proposed a concept called tissue engineering, in which cells were attached to a biodegradable artificial polymer in cell culture dishes, and then the polymer–cell scaffold was implanted into animals, becoming the prototype of a modern introduction to tissue engineering. It is generally recognized that tissue engineering is a subject that combines tissue cells and specific biological activities with biological materials, building tissues and organs *in vitro* or *in vivo* to maintain, repair, regenerate, or improve the function of damaged tissues and organs. A tissue engineering process usually refers to using a biodegradable polymer scaffold as a temporary carrier of cells to plant autologous homologous cells inducing the regeneration of new tissue under the appropriate growth factors *in vitro* or *in vivo*, and finally forming new healthy tissues and organs. This process includes three basic elements: seed cells, scaffold materials, and growth factors. The tissue engineering scaffold is the most important factor in the tissue engineering process since it provides residence for cells by mimicking the function of the extracellular matrix. The biological activities of the cell within the scaffold could gradually degrade the matrix biomaterial and maintain the long-term stability of the tissue structure, thus, making the regenerated tissue with specific physiological functions [4].

Biomaterial scaffolds acting as the three-dimensional (3D) morphological template of the constructed tissues and organs should provide mechanical support for tissue formation with a certain strength before their degradation [5]. The

Functional Polymer Foams: Green Fabrication Methods, Performance and Applications, First Edition. Haoyang Mi.

interaction between cells and biomaterials is the key to tissue formation and construction. To prevent the mechanical failure of tissue engineering scaffolds during *in vivo* implantation, cell inoculation must maintain a certain high density, and the degradation rate of biomaterials must match the cell growth and extracellular matrix synthesis rate.

For tissue engineering biomaterials, the following issues need to be considered. The biocompatibility of biodegradable materials not only limits the toxic and side effects of materials and their degradation products on the body, but the biochemistry and immune rejection generated between the material and the body should also be controlled, ensuring that the constructed tissues and organs can be accepted by the body and survive. The physicochemical properties of biodegradable materials, including geometric form plasticity, surface configuration, porosity, mechanical strength, tolerance to high-temperature and high-pressure treatment, and the degradation rate of biodegradable materials should match the tissue and organ formation rate [6].

7.1.2 Tissue Engineering Scaffold Materials

Artificial tissue repair materials can be divided into four categories according to the material components, namely, inorganic materials, natural polymer materials, synthetic polymer materials, and composite materials [7]. These materials are commonly used in the regeneration of human tissues, including bone, cartilage, nerve, vascular, organ, cornea, and skin.

Inorganic materials include metal and bioceramic materials. (i) Metal materials are the first used biological materials in plastic surgery and dentistry. They typically have high strength, poor compatibility, lack of biological activity, nondegradability, and mismatching mechanical properties. They are often used as permanent implants and may fall off after implantation in the body [8]. (ii) Bioceramic materials are divided into bioinert ceramics and bioactive ceramics according to their biological activity. Biological inert ceramics have thermodynamic stability, are chemically inert, and have good corrosion resistance, while they usually have poor mechanical compatibility and are easy to fracture due to their brittleness. Bioactive ceramic materials refer to ceramic materials that can form chemical bonds with living bone tissue and soft tissue. It can be gradually degraded and absorbed in the physiological environment and then replaced by new tissue. The scaffolds based on bioactive ceramics, such as hydroxyapatite (HA) and tricalcium phosphate, are used extensively in bone repair [9].

Natural polymers have the advantages of rich sources, good biocompatibility, nontoxic, hydrophilic, and cell affinity [10]. It has been found that natural polymers like gelatin, collagen, chitosan, cellulose, and alginate are suitable for the adhesion, proliferation, and differentiation of various cells. However, the degradation speed and mechanical strength of natural polymers are difficult to adjust, they usually degrade faster than the tissue generation rate. The processing ability of these materials is usually limited, which hinders their practical application in tissue engineering.

Synthetic polymers including polylactic acid (PLA), polyhydroxyacetic acid (PGA), and their copolymers such as copolymer (ethyl-propyl ester) (PLGA), polycaprolactone (PCL), polyurethane (PU), thermoplastic polyurethane (TPU), and polyhydroxybutyrate (PHB) are typically biodegradable polymers with high biocompatibility. They have the advantages of low cost, ease of process via various methods, large-scale production, adjustable mechanical properties, and degradation rates. However, the biocompatibility of synthetic polymers is usually lower than that of natural polymers. Thus, they are often combined to produce scaffolds with superior properties. In addition, synthetic polymers are also combined with bioactive ceramics to produce bone-repair composite scaffolds with tunable properties.

7.2 Fundamentals of Tissue Engineering

7.2.1 Typical Process of Tissue Engineering

In a typical tissue engineering process, there are five steps: (i) the acquisition of biological cells (Figure 7.1a), (ii) the culture and reproduction of cells (Figure 7.1b), (iii) the transplantation of cultured cells and bioactive factor to the scaffold (Figure 7.1c), (iv) further adhesion and growth of cells in the scaffold (Figure 7.1d), and (v) transplantation of the cell containing scaffold to the corresponding tissue and organ parts of the organism (Figure 7.1e). Therefore, the tissue engineering scaffold is the main part connecting external cells with the human body. Thus, the tissue engineering scaffolds not only need good mechanical strength to maintain the normal operation of body tissue but also must have specific shape and scaffold surface activity to ensure normal cell adhesion [12], migration, and growth [13].

The role of tissue engineering scaffold is to provide temporary support for cell growth and activity, which not only meet cell adhesion, differentiation, and reproductive growth but also achieve nutrient delivery and metabolism and induce the formation of specific shaped tissue organs. Therefore, the scaffold should have a special 3D structure, high porosity, suitable pore size and distribution, and high pore connectivity [14] so that the preparation of 3D porous scaffolds with a specific shape and suitable pore morphology structure becomes a key challenge in tissue engineering.

In addition, the surface of the scaffold is the main part of the interaction between the scaffold and cells or tissues, which significantly affects the cell adhesion and proliferation behavior. In tissue engineering, cells cannot achieve adhesion localization independently and must rely on the scaffold surface to form adhesion sites. A large specific surface area is essential to accommodate the cells required for the regenerated tissue. Moreover, the surface morphology [15] and the hydrophilic and hydrophobic [16] are also important. Various techniques have been used to selectively modify the scaffold surface to alter its chemical properties. For example,

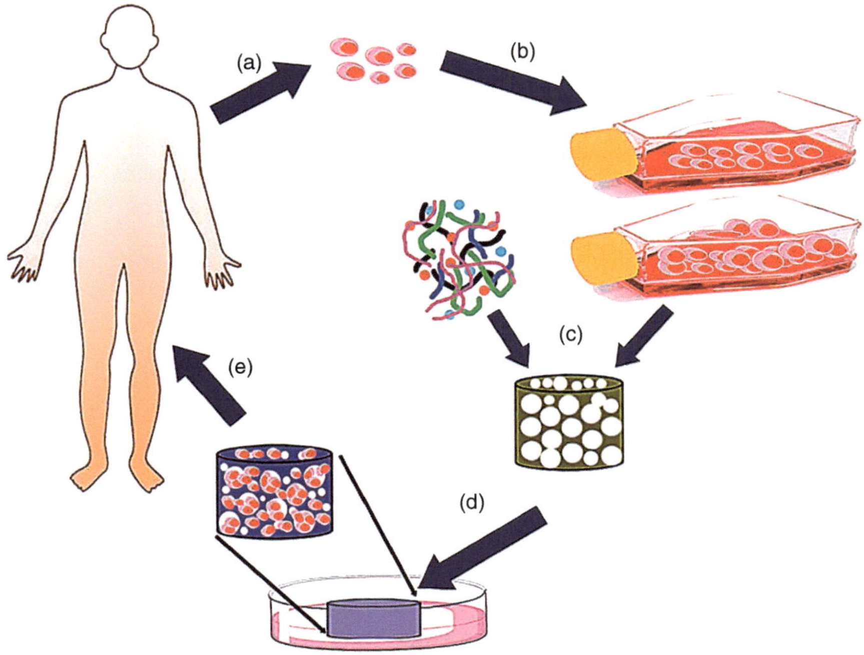

Figure 7.1 Steps in a typical tissue engineering process: (a) a small number of cells are isolated from the human body, (b) the cells are cultured to produce more cells, (c) these cells are seeded onto porous scaffolds together with growth factors, (d) the seeded scaffolds are further cultured to increase the number of cells, and (e) finally, the regenerated tissue is implanted into the site of the defect to integrate with the natural tissue. Source: Mi et al. [11]/with permission of Sage Publications, Inc.

surface grafting or surface coating of bioactive molecules could introduce certain functional groups or proteins on the scaffold surface to promote cell adhesion and diffusion. In addition, the roughness and hydrophilicity of the scaffold surface affect the cell adhesion as well. Most cells prefer rough surfaces with moderate hydrophilicity for better adhesion and proliferation.

The microstructure of tissue engineering scaffolds significantly affects cell growth and cell migration, especially the penetration of cells into the inner part of the scaffold. Typically, the pore size of a scaffold needs to be greater than a mature cell and the pores in the scaffold should be interconnected to allow the transfer of nutrient and metabolic waste [17, 18]. The mechanical property is another important factor for tissue engineering scaffolds, especially when the scaffolds are used for the regeneration of load-bearing tissues such as bone and cartilage. For the successful application of scaffolds, their elastic modulus, flexural modulus, and tensile strength are desired to be comparable to the target tissue. If the strength and stiffness are low, it is limited to the cartilage tissue application. In addition, the mechanical properties can be regulated by processing biomaterials into scaffolds with different pore sizes and pore orientations, and the degradation rate of scaffolds can be regulated by changing the proportion of copolymers and molecular weight.

7.2.2 Ideal Scaffold Conditions

The ideal tissue engineering scaffold [19] should have good biocompatibility with the surrounding tissue, their mechanical properties should be consistent with targeting tissues, they should be bioabsorbable with a degradation rate same as the new tissue formation rate, and they should have appropriate surface properties for cells to attach, grow, multiply, differentiate, and promote the formation of extracellular matrix. In addition, the scaffold should have optimal structural properties in aperture, porosity, pore connectivity, and permeability to effectively deliver nutrients and waste. As shown in Figure 7.2, some bioactive tissue engineering scaffolds are desired to release bioactive molecules and have biological motifs, which would stimulate the formation of new tissue. Physical stimulations, including anisotropic morphological stimulation, mechanical stimulation, electrical stimulation (ES), and magnetic stimulation have also been used in the development of advanced tissue engineering scaffolds.

Due to the different survival conditions of tissue and organ cells in different parts of the human body, the requirements for the pore structure of the porous scaffold are different. With a pore size of 300–500 μm, it is beneficial for the transport of nutrients and the exclusion of cell metabolites, which is conducive to the reproduction of human osteoblasts [20]. Whang et al. found that the best pore size was 5–15 μm for fibroblasts, 20 μm for hepatocytes, and 20–125 μm for adult mammalian skin cells [21]. Therefore, it is necessary to design proper scaffolds with pore sizes suitable for the targeting cells.

The mechanical properties of different human tissues, such as bone tissue, nerve tissue, blood vessels, and skin, could be significantly different. The mechanical

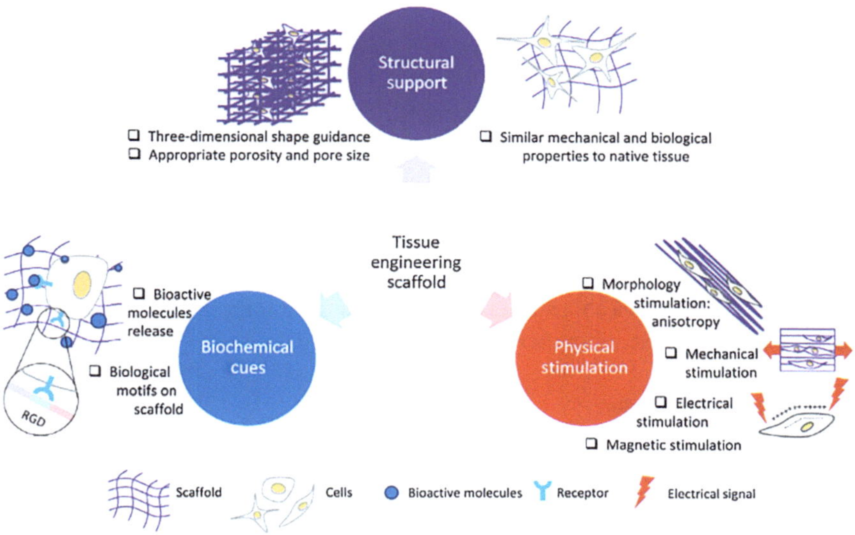

Figure 7.2 Factors to consider in the design of advanced tissue engineering scaffolds. Source: Zhang and King [13]/with permission of John Wiley & Sons.

property of the scaffold should match the tissue defect site to protect cells from destructive compression or stretching so that cells can develop into normal tissue during the healing process. Bone tissue supports the body to complete normal activities, and Young's modulus of human bone tissue is 0.008–33.7 GPa [22]. The tendon and ligament, as fibrous connective tissue, need to withstand the mechanical load between muscle and bone and between bone and bone. Thus, the desired scaffolds should have proper mechanical stiffness to fit the targeting tissues. In addition, the heart is an organ rich in mechanical signals, including electrical signals and contractile force generated by myocardial tissue, shear force generated by blood flow, and tension stress on vascular endothelial cells. The scaffolds for heart tissue should not only possess similar mechanical properties but also should support the regular contraction of myocardial cells. Tissue-engineered implants are wrapped by body fluids and tissues in the body. Moreover, the vascular scaffold is another example that requires strict mechanical and surface chemistry affinity before implantation. A mismatched vascular scaffold may cause the leak of blood, inflammation, neointimal hyperplasia, thrombosis, etc.

7.3 Characterization of Tissue Engineering Scaffolds

The properties normally characterized for tissue engineering scaffolds include scaffold microstructure, mechanical properties, surface properties, degradation properties, and biological properties. Sometimes, the drug release behavior, immune response behavior, and *in vivo* tissue generation behavior of the scaffolds are also investigated. In this section, some prevalent characterization methods for tissue engineering scaffolds are introduced.

7.3.1 Surface Property Testing

Water Contact Angle (WCA) Measurement. Different from the application of superhydrophobic foams, the WCA test is used to characterize the wettability of a scaffold surface to evaluate the ability of the surface to absorb protein and assist cell adhesion. The sessile drop method is frequently used in the WCA measurement for scaffolds. Basically, a droplet of deionized water is dropped on the surface of a flat sample, and the contact angle of the droplet is measured. In some cases, the water droplet would penetrate the scaffold in a short time, then the water penetration time is recorded to reflect the hydrophilicity of the scaffold.

7.3.1.1 Protein Absorption

Protein Absorption Test. The protein absorption test is used to evaluate the ability of the scaffold to absorb proteins on its surface since most scaffolds realize cell adhesion via the absorption of proteins that have specific receptors to anchor with the cell membrane. In a typical protein absorption test, a bicinchoninic acid (BCA) protein assay is used. The scaffolds are immersed in tissue culture plates containing minimum essential medium supplemented with 10% fetal bovine serum and 1%

antibiotic/antimycotic and incubated at 37 °C for different periods. The scaffolds are then rinsed with PBS solution and lysed with 0.1% Triton X-100 followed by adding the BCA reagent and further incubation. Then, the protein concentration can be obtained from the absorbance at 562 nm using a plate reader.

7.3.2 Mechanical Performance Testing

Tensile and Compression Test. Different from common materials, the mechanical property test of tissue engineering scaffolds is often required to be measured in a simulated body fluid solution or an aqueous solution at 37 °C to resemble the *in vivo* conditions. Therefore, environmental chambers are often equipped with mechanical property measurement instruments. Depending on the size and geometry of the scaffolds, different fixtures, and measurement methods can be selected.

Tensile and compression tests are the most frequently used mechanical property characterization means. The tests include one-way stretching or compression mode and cyclic stretching or compression mode. The one-way testing mode is used to obtain the tensile modulus, compressive modulus, ultimate strength, and strain-at-break values, while the cyclic mode is typically used to reflect the elasticity of the scaffold and the hysteresis loss during cyclic loading to resemble the repetitive activity of human being.

For some tissue engineering scaffolds, specific tests are required. For example, vascular tissue engineering scaffolds have strict requirements in terms of mechanical strength, suture retention strength, and burst pressure so as to meet the requirements of implantation surgery and withstand blood pressure.

7.3.3 Biological Performance Testing

Cell Viability Assay. Cell viability assay is the most frequently used method to evaluate the biocompatibility of a biomaterial. Fibroblast cells from mice or humans are typically used in the assay. A Live/Dead Viability/Cytotoxicity Kit is used in the assay. The scaffolds after culturing cells for a specific period (i.e. 3 days or 10 days) are stained with the assay following the protocol, and the living cells are stained with green fluorescent Calcein-AM, while the dead cells are stained with red fluorescent ethidium homodimer-1 (EthD-1). Then, the amount of live and dead cells can be counted using flow cytometry analysis or directly observed using a fluorescence microscope.

Cell Proliferation Assay. The amount of cells presented on the scaffolds needs to be counted to evaluate the cell proliferation rate on the scaffolds. This can be done by using a cell proliferation assay, such as the MTS assay. Prior to the MTS assay, a standard curve is established by performing the tests on cells seeded on the cell culture wells and confirmed by comparison to hemocytometer readings. In the MTS assay, cells were treated with an 83% media and 17% MTS solution and allowed to incubate for exactly one hour. After incubation, 100 μl of spent media were removed and added to a clear 96-well plate. The absorbance of this plate at the 450 nm wavelength was read with a GloMax-Multi + Multiplate Reader (Promega), and the subsequent number of cells was determined relative to the negative control.

Morphological characterization of cells on the scaffold. The cytoskeleton morphology of cells can be determined by phalloidintetramethylrhodamine B isothiocyanate staining. In the assay, cells are first fixed in 4% paraformaldehyde (EMS) diluted in PBS for 15 minutes at room temperature. Samples are then washed with PBS and permeabilized with 0.1% Triton-X in PBS for 5 minutes at room temperature. The cells are washed one more time and treated with 0.3 μM phalloidin–TMRho with DAPI for one hour at room temperature. At last, samples are washed with PBS and immediately imaged using a fluorescence microscope.

In addition to the cytoskeleton morphology characterization, the interaction between the cells and the scaffold can be investigated. To image the cell adhesion behavior, the cells cultured on the scaffold need to be fixed first and then dehydrated using a series of ethanol washes followed by sufficient drying and gold sputtering for the SEM imaging.

7.4 Preparation of Tissue Engineering Scaffolds by Gas Foaming

The artificial tissue or organ will be transplanted into the patient's body, and the macroscopic morphology of the scaffold should be similar to the damaged tissue or organ and have suitable mechanical properties [23]. Various approaches have been developed to fabricate tissue engineering scaffolds in the past decades, such as phase separation, electrospinning, 3D printing, decellularization, particulate leaching, and gas foaming. Among them, the gas foaming process has the highest productivity and the lowest cost and is believed to be a promising technique for the industrialization of tissue engineering scaffolds. Gas foaming is a complex thermodynamic process in which polymer melt forms a homogeneous system with a foaming gas. Under the sharp change of external pressure or temperature due to thermal instability, a large number of bubble cores were generated followed by the growth and fixing of the bubbles.

At present, the commonly used foaming techniques include the physical foaming method and chemical foaming method, which have been widely used to produce microporous foams of thermoplastic polymers. In 1996, Mooney et al. [24] first prepared three porous scaffolds, PGA, PLLA, and PLGA, using gas as the pore-causing agent. Scaffolds with up to 93% porosity and pore size up to 100 μm were obtained. In 1999, Nam et al. [25] used the chemical foaming method and ammonium bicarbonate particles as pore-forming agents to prepare scaffolds with porosity up to 90% and 200–500 μm pore size. However, the porous scaffolds prepared by single gas foams are usually closed pore structures, and the connection rate of bubble pores is low. Therefore, gas foams are often combined with other technologies to prepare porous scaffolds.

7.4.1 Preparation of Tissue Engineering Scaffolds by Chemical Foaming

The chemical foaming method usually uses chemical foaming agents (CFAs) as gas sources to prepare foaming materials. Some reactions that release volatile small molecules or gases are also used as gas sources to prepare foaming materials. Commonly used CFAs are carbonates, ammonia, azo compounds, etc. CFAs are usually directly mixed with the polymer matrix and generate gas in the material when the foaming agents decompose at high temperatures. In addition, the CFA is self-nucleating, which allows a more extensive processing window.

However, due to the limitation of biocompatible CFAs and the low-pore connectivity of the foams, chemical foaming is rarely used to prepare tissue engineering porous scaffolds. One of the main advantages of using CFAs is that there is no need to modify the existing equipment, and the resulting gas can be evenly distributed in the polymer by premixing.

Shen et al. [26] used sodium bicarbonate/sodium carbonate ($NaHCO_3/Na_2CO_3$) particles as porogen and CFA to react with hydrochloric acid (HCl) to prepare porous polymer scaffolds, and the reaction residual can be easily removed by distilled water. It was found that the porosity of the scaffold increased with the increase of the mass ratio of CFA to polymer, and the pore size increased with the increase of the size of particles. The porosity of the scaffold was up to 94.8%, the size of the macropores ranged from 100 to 400 μm, and the scaffold had a uniform interconnected open pore structure, which was conducive to the growth of cells and tissues. The scaffolds fabricated by this method showed more pores, more uniform distribution, and more interconnected and open pores than those prepared by the traditional salt leaching method.

Hu and coworkers [27] developed a water-assisted foaming process to fabricate PU foams containing HA as bioactive fillers for bone tissue repair. In the synthesis, HA nanoparticles were firstly grafted with isocyanate functional groups using hexamethylene diisocyanate (HDI) . Then, the modified HA nanoparticles were reacted with PCL diols in the presence of castor oil and catalyst. After the addition of water to the system, CO_2 gas was emitted, and the material was foamed. The scaffolds were obtained after curing for 24 hours, which presented a highly porous structure that can be tuned by changing the HA content. The composite foam exhibited particular shape memory behavior, which makes it possible to implant and recover *in vivo* at body temperature. Kazimierczak et al. [28] fabricated highly porous bone scaffolds of chitosan–agarose matrix reinforced with nHA. The pores of the scaffolds were obtained and controlled by the combination of gas foaming and freeze-drying (Figure 7.3a). The bioactivity, cytotoxicity, and cell growth results indicated that the best porosity is obtained when solvent (CH_3COOH) and foaming agent ($NaHCO_3$) are controlled at a ratio of 1 : 1, and the optimum scaffold was nontoxic to osteoblasts and showed favorable cell attachment (Figure 7.3b and c).

In addition, Nublat et al. [29] used ammonium bicarbonate as CBA to synthesize PLA membranes with open porous structures. Kucharska et al. [30] synthesized

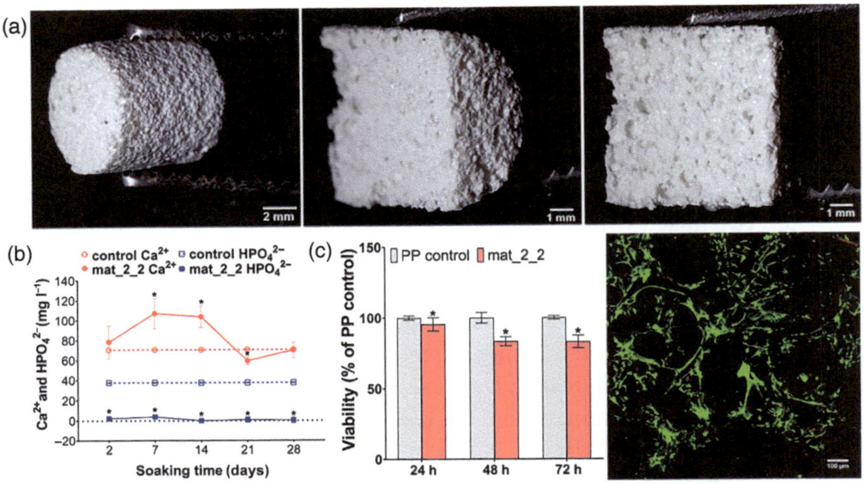

Figure 7.3 (a) Images of different nHA containing chitosan–agarose scaffolds, (b) Ca^{2+} and HPO_4^{2-} concentrations change of different scaffolds when soaked in SBF, and (c) cytotoxicity evaluation of the scaffold against MC3T3-E1 cells. Source: Kazimierczak et al. [28]/MDPI/CC BY 4.0.

chitosan/β-tricalcium phosphate (TCP) scaffolds by adding sodium bicarbonate to an acetic acid solution mixed with chitosan and β-TCP powder, followed by freeze-drying. Salgado et al. [31] prepared a starch/cellulose acetate porous framework by chemical extrusion foaming method using 2% CFA. The prepared porous scaffolds had a porosity of 70%, a pore diameter of 200–500 μm, and spatial connectivity.

7.4.2 Preparation of Tissue Engineering Scaffolds by Physical Foaming

Physical foaming technology has achieved great success in the preparation of high-performance thermoplastic foam materials with microporous or nanoporous structures. Various foaming systems and parameters have been optimized to maximize foam performance, and various foaming methods have been developed to obtain the desired foam shape. The physical foaming agent (PFA) is also known as the volatile foaming agent. The mechanism for physical foaming is the phase change of PFAs from a high-density state (e.g. solid, liquid, and supercritical states) to a low-density gas state by the elevating of temperature or the reducing of pressure. In the process, pores nucleate in the polymer matrix and gradually grow as the gases accumulate in the pores until the pressure is balanced or the material is solidified.

In the preparation of tissue engineering scaffolds, supercritical CO_2 ($scCO_2$) is the most commonly used PFA due to its nontoxicity, nonresidual, high solubility in the polymer matrix, mild temperature and pressure conditions to maintain supercritical state, and wide availability. However, the precise control of the porous structure, the manipulation of the mechanical property, and the increase of interpore connectivity are the main challenges for the $scCO_2$-based physical foaming process.

Zhao and coworkers [32] prepared PCL/PLA blends at various ratios using a co-rotating twin-screw extruder and further fabricated PCL/PLA foams via batch foaming using $scCO_2$ as the foaming agent. A highly open-cellular structure was obtained through foaming due to the high tensile stress during cell expansion. Using this method, they fabricated a small-diameter tubular scaffold with a tensile toe region of approximately 40%. The tubular scaffold supported the growth of human umbilical vein endothelial cells (HUVECs), which emphasized the potential applicability of the scaffold for vascular tissue engineering. Similarly, Mi et al. [33] developed TPU/PLA blend tissue engineering scaffold materials with different proportions and adjustable properties by melt blending and microcellular injection molding. The PLA phase was dispersed inside the TPU matrix, which contributed to various porous structures (Figure 7.4a and c) with the increase of TPU to PLA ratio. The rough surface constructed on the pore surface supported the proliferation and migration of fibroblast cells well (Figure 7.4b and d). In addition, the mechanical property of the developed TPU/PLA scaffold also varies in a large range as the

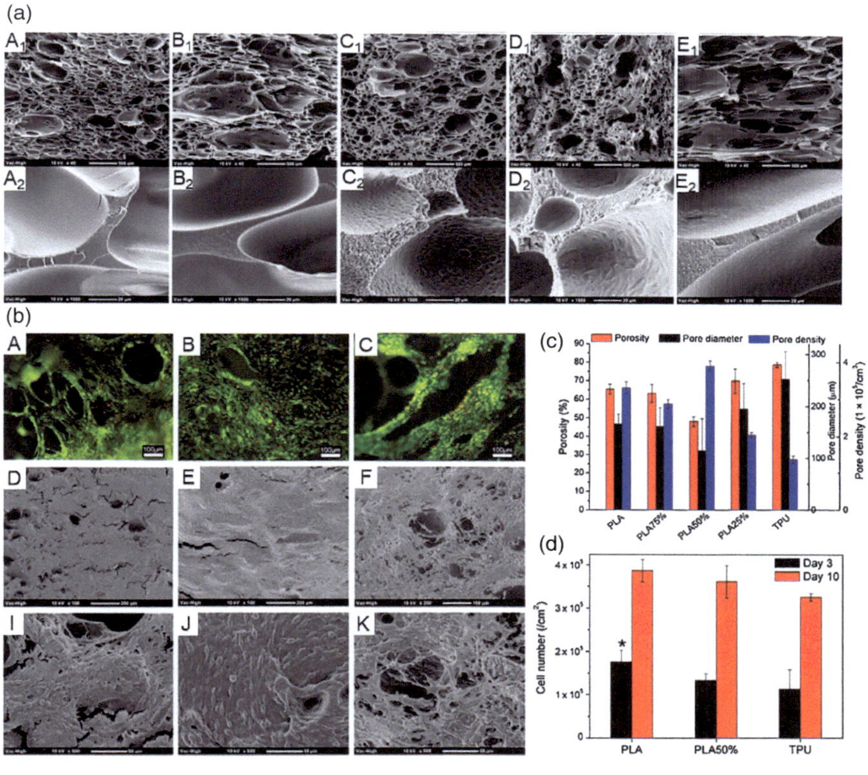

Figure 7.4 (a) SEM images of TPU/PLA scaffolds with different TPU to PLA ratios, (b) fluorescent and SEM images of fibroblast cells cultured on different TPU/PLA scaffolds for 10 days, (c) porosity, pore diameter, and pore density statistical results of different TPU/PLA scaffolds, and (d) MTS assay cell count statistical results of PLA, PLA50%, and TPU scaffolds at day 10. Source: Mi et al. [33]/with permission of Elsevier.

change of TPU to PLA ratio, which demonstrates the ability of the scaffolds to be used as artificial scaffolds in a variety of tissue engineering applications.

The porous scaffolds prepared by gas foaming typically suffer from low pore connectivity. Many researchers have combined the gas foaming method with the particle or sacrificial phase leaching method to enhance the pore connectivity of the scaffolds. The particles or sacrificial phases used are typically water-soluble materials that can be easily removed by water rinsing. NaCl is a commonly used porogen particle, and pore size and distribution can be adjusted by the size and content of the porogen. Polyethylene glycol (PEG), polyethylene oxide (PEO), and polyvinyl alcohol (PVOH) are commonly used water-soluble sacrificial phases.

Chen et al. [34] prepared a PLA/PEG blend by melt blending in a triple-screw dynamic extruder. The blends were further foamed using $scCO_2$ via batch foaming at different temperatures, and the effect of foaming conditions on the microstructure of the scaffold was investigated. They found that the content of PEG had a significant influence on the cell opening rate. In a later study, they combined NaCl particles in the PLA/PEG blend system as another porogen phase [35]. Without the foaming process, the NaCl particles were difficult to remove completely (Figure 7.5a). With the increasing foaming temperature, the foams showed higher porosity and controllable porous morphology (Figure 7.5b). After leaching with circulating water, both pore size and connectivity were enhanced (Figure 7.5c), which resulted in an interconnected bimodal porous scaffold with a maximum porosity of 89% and also facilitated the increase in WCA (Figure 7.5d).

Teng et al. [36] prepared a series of PLA-based porous scaffolds via $scCO_2$ foaming and salt leaching methods. The morphology and porosity of the scaffolds were adjusted by changing the foaming pressure, temperature, and saturation time, and the connectivity between pores was regulated by controlling the size and content of salt particles. Similarly, Salerno et al. [37] prepared PCL–HA composite scaffolds using the mixed gas of CO_2/N_2 as the co-blowing agent and NaCl as the porogen. They produced porous scaffolds with both macropores and micropores by adjusting the content of NaCl particles and the components of gas. Annabi et al. [38] prepared a macroporous network based on PCL and elastin by simultaneously using NaCl particles as the porogen and CO_2 as the foaming agent, resulting in a uniformly distributed macroporous structure.

To further enhance the pore connectivity, Mi et al. [39] combined microcellular injection molding with a particle leaching method to produce a porous TPU tissue engineering scaffold using water-soluble PVOH and NaCl as the sacrificial phase and porogen. As shown in Figure 7.6a, the TPU/PVOH/NaCl composite was foamed using $scCO_2$ gas, and the foamed composite was leached via circulating water, which resulted in a porous structure with open pores (Figure 7.6b). The pore size of the scaffolds can be controlled in the range of 50–120 μm by varying the foaming conditions and the material composition, and the scaffolds were found to be suitable for 3T3 fibroblast growth (Figure 7.6c). In addition, the effect of PVOH and NaCl components on the thermal property, and hydrophilicity were investigated. They found that with the increase of preconnectivity and hydrophilicity, the cell proliferation rate was enhanced due to the improvement in cell attachment (Figure 7.6d).

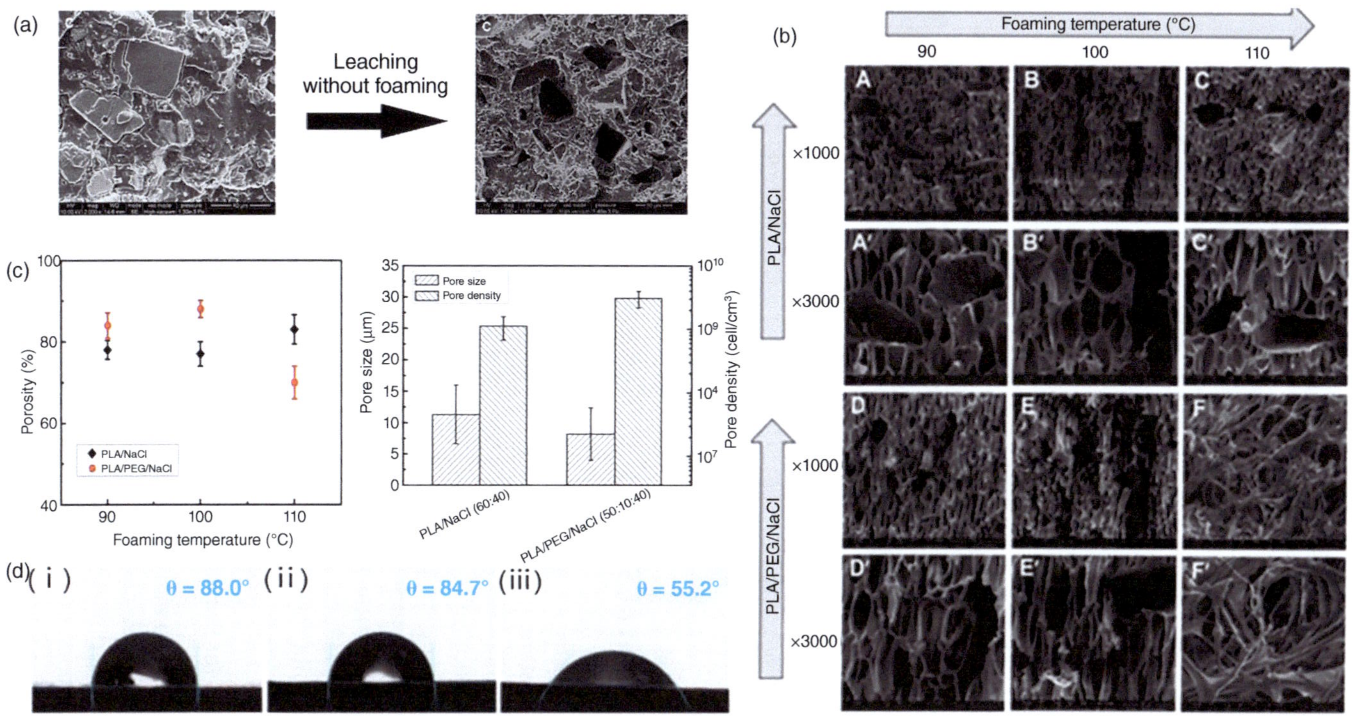

Figure 7.5 (a) SEM images of the PLA/PEG/NaCl blends before and after leaching, (b) foam morphology of PLA/NaCl and PLA/PEG/NaCl foams after leaching as a function of foaming temperatures, (c) average porosity, cell size, and cell density of PLA/NaCl and PLA/PEG/NaCl foams before and after leaching, and (d) water contact angles of neat PLA and PLA blend scaffold after leaching. Source: Chen et al. [35]/with permission of John Wiley & Sons.

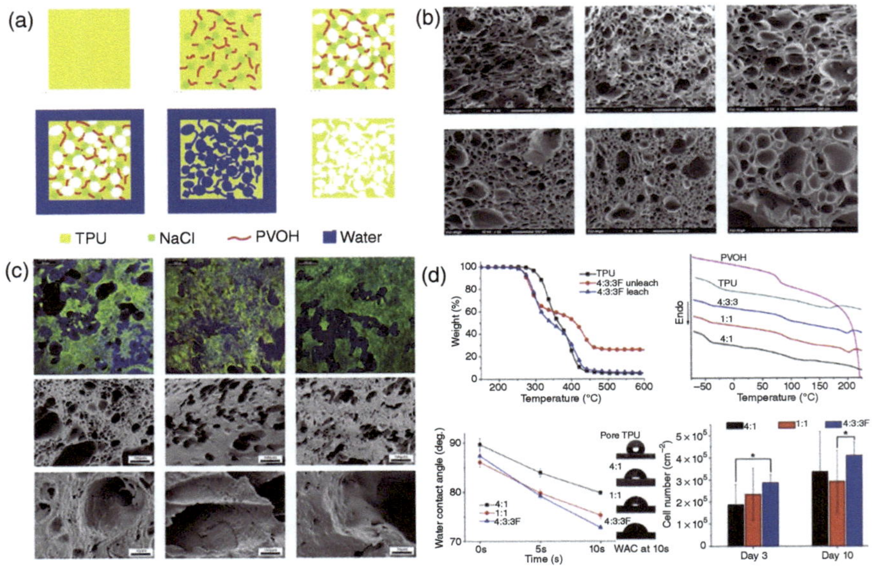

Figure 7.6 (a) Illustration of the fabrication process of the interconnected TPU scaffolds, (b) SEM images of prepared TPU scaffolds in cross-section and melt flow direction, (c) day 10 fibroblast cell culture results on different TPU scaffolds, and (d) TGA, DSC, water contact angle, and cell proliferation (days 3 and 10) results of different TPU scaffolds. Source: Mi et al. [39]/with permission of Springer Nature.

7.5 Application of Scaffolds in Tissues

Tissue engineering is an interdisciplinary field that aims to develop biological substitutes to restore, maintain, and improve tissue function, with potential applications in cardiovascular, skin, bone, nerve, and other tissues. It has received much attention because it provides a means of developing artificial tissue as an alternative to autologous and allogeneic transplantation. In tissue engineering, scaffolds serve as cell carriers to generate an environment conducive to cell growth and reproduction. The use of tissue engineering scaffolds can alleviate the shortage of organs, such as kidneys and hearts, reduce the use of laboratory animals, and improve the reliability of the experimental results.

7.5.1 Scaffolds for Bone Tissue Engineering

With the increase in human life expectancy and the aggravation of population aging, some traumatic injuries and bone pathological diseases have gradually increased in recent years. These diseases and injuries can lead to serious consequences, such as fractures and bone defects. Traditional treatment methods include autologous bone grafting, allogeneic bone grafting, and artificial bone material replacement [40]. Autologous bone is considered the gold standard of bone transplantation, but it is accompanied by some possible complications and limited donor sites, which restricts its clinical application. The emergence of bone tissue engineering provides a feasible solution for the surgical field.

Bone is similar to other connective tissues and comprises three components: cells, fibers, and matrix. The most significant feature of bone is that the cellular matrix has a large amount of calcium salt deposition, which becomes a very hard tissue and constitutes the skeletal system of the body. In the fabrication scaffolds for bone tissues, calcium salts such as HA and TCP are often used. The introduction of the bioactive components could not only improve the mechanical properties of the scaffolds but also enhance the bioactivity, facilitate the differentiation of stem cells, and stimulate the formation of new bone tissues. Mi et al. [41] investigated the effect of two types of HA with micro or nano sizes on the property of TPU/HA composite scaffolds that are prepared via thermally induced phase separation using dioxane or dioxane/water solvent systems. As shown in Figure 7.7a, micro- or nano-sized HA particles were exposed on the pore surface of the scaffolds, and the scaffolds prepared using dioxane/water as the solvent showed higher porosity, high mechanical properties, and superior cellular structure compared to the scaffolds using only dioxane as solvent (Figure 7.8b and c). With the addition of HA particles, the TPU/HA composite scaffolds could effectively be mineralized in a simulated body fluid solution. The formed

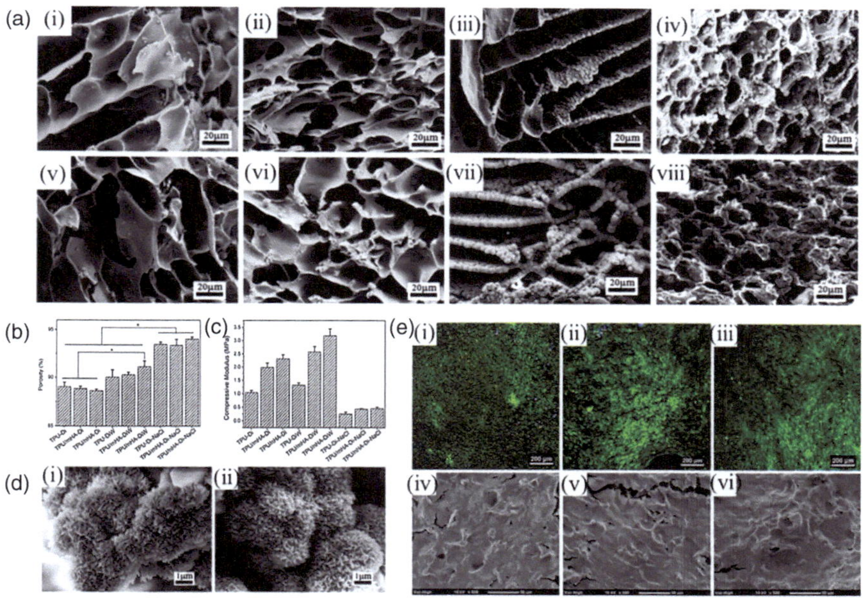

Figure 7.7 (a) SEM images of TPU/HA composite scaffolds containing micro HA or nano HA that are fabricated using dioxane or dioxane/water as the solvents after incubation in simulated body fluid for 1 week (i–iv) and 3 weeks (v–viii), (b) The porosity, (c) the compressive modulus of the scaffolds fabricated with different methods and material formulae, (d) the morphology of the newly formed apatite cluster on the TPU/HA scaffold containing nano HA after mineralization test for 4 weeks, and (e) cell viability and cell morphology of fibroblast cells cultured on different TPU/HA composite scaffolds for 10 days. Source: Mi et al. [41]/with permission of Springer Nature.

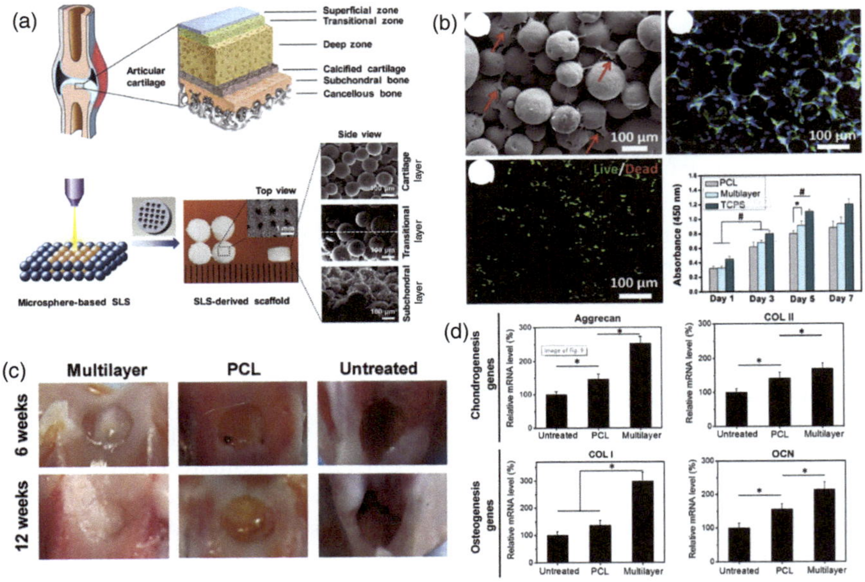

Figure 7.8 (a) The architecture of the natural osteochondral unit and the HA/PCL composite microspheres-based multilayer scaffolds fabricated via SLS technique, (b) SEM image showing cells attached to the multilayer scaffold, confocal fluorescence image showing the cell morphology on the scaffold, and the live/dead assay results of cells cultured on the multilayer scaffold for 7 days, (c) *in vivo* evaluation of the HA/PCL SLS-derived scaffolds at 6 and 12 weeks postimplantation, and (d) gene analysis for chondrogenesis and osteogenesis in newly formed tissue at week 12.

mineral crystal clusters showed flower-like morphologies (Figure 7.7d). In addition, the TPU/HA composite scaffolds demonstrated excellent biocompatibility to fibroblast cells (Figure 7.7e). In another work, they fabricated fibrous TPU/HA composite scaffolds via electrospinning [42]. The effect of TPU hardness and HA particle size were investigated. It was found that micro HA is mostly embedded in the fibers to reduce the diameter of the electrospun fibers, while some nano HA could present both in the fiber and on the fiber surface, which increased the fiber diameter for both soft and hard TPU materials. In addition, soft scaffolds with nHA added further enhanced the osteogenic ability of human stem cells compared with scaffolds without nHA added. With the stimulation effect of HA particles, the stem cells tend to differentiate into osteoblast cells.

Similarly, Lin et al. [43] tested different alginate/HA ratios to obtain bone-like structures. When the alginate–HA ratio was 75/25 and 50/50, the attachment and proliferation abilities of rat osteosarcoma cells were better than those of pure alginate. Although the fabricated scaffolds all exhibited a good interconnected porous structure with an average pore size of 150 μm and porosity of more than 82%, the 50/50 alginate/HA scaffold showed a high-compressive modulus of ~18 MPa and compressive strength of ~150 MPa, making it a promising scaffold for bone applications.

7.5.2 Scaffolds for Soft Tissue Repair

The majority of the human body is composed of soft tissues, such as skin, muscle, tendon, and cartilage. Articular cartilage is composed of chondrocytes, type II collagen, and proteoglycans and is hyaline cartilage covering the bony end of the synovial joint of the joint cavity. However, the self-repair ability of cartilage is very limited due to its avascular nature, less migration of chondrocytes surrounded by extracellular matrix, and limited proliferation of mature chondrocytes [44]. Therefore, developing artificial scaffolds for cartilage repair is highly desired.

Although skin and muscles can repair themselves in most cases, the regeneration ability would be greatly reduced in large damages. With the assistance of artificial scaffolds, which have bioactivity and can support the growth of new tissue, the recovery of the damaged tissue would be greatly accelerated. Since the properties of different soft tissues differ significantly, it requires tissue engineering scaffolds to possess a wide range of complex properties including mechanical properties, porosity, degradation properties, etc. Therefore, the fabrication of biocompatible scaffolds with controllable microstructures and properties is a critical challenge for soft tissue engineering scaffolds.

The addition of HA can not only stimulate the osteogenesis of osteoblast but also facilitate osteochondral repair. Du et al. [45] fabricated HA/PCL scaffolds using PCL microspheres-based selective laser sintering (SLS) method. As shown in Figure 7.8a, the fabricated scaffold has a gradient structure that mimics the structure of natural osteochondral. The scaffold was prepared by layering PCL microspheres on HA/PCL microspheres to promote an osteogenic phenotype. The scaffold had a controllable structure and mechanical comparable to native cartilage and showed excellent cell affinity with high cell viability and desirable cell morphology (Figure 7.8b). The multilayered HA/PCL scaffold demonstrated superior properties than the PCL scaffold in the *in vivo* test, as shown in the macroscopic observation at 6 and 12 weeks postimplantation (Figure 7.8c). In week 12, the multilayer group displayed cartilage-like tissue with a similar appearance to the surrounding native cartilage. In addition, histological analysis revealed that the gradient scaffold significantly increased bone and cartilage repair compared with PCL-only and untreated controls. The overall filling of the defect, scaffold degradation, and bone and cartilage morphology were improved considerably with the use of gradient scaffolds (Figure 7.8d).

To alter the mechanical property and degradation rate of soft tissue engineering scaffolds, Mi et al. [46] synthesized three types of biodegradable TPU using different chain extenders. All synthesized TPUs have sufficient molecular weights for the electrospinning process. Fibrous scaffolds with random structures and aligned structures were prepared using different collection setups (Figure 7.9a). The aligned fiber membranes showed superior mechanical strength and tensile modulus, and the TPU synthesized using a degradable diol (bis(2-hydroxyethyl) terephthalate, BET) showed high elasticity and lower strength (Figure 7.9b). All scaffolds were biocompatible with fibroblast cells, and the cells showed aligned cytoskeleton morphology and better proliferation on the aligned scaffolds (Figure 7.9c and d). The *in vitro*

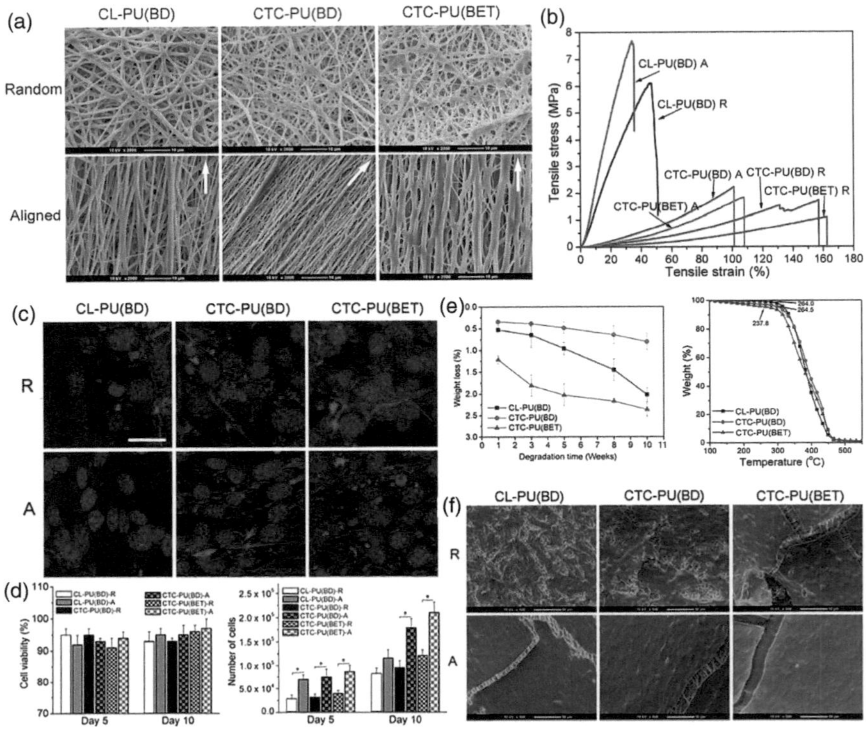

Figure 7.9 (a) SEM images of the random fibrous scaffolds and aligned fibrous scaffolds fabricated using TPUs synthesized with different chain extenders, (b) the tensile test results of different TPU scaffolds with random and aligned fibrous structures, (c) the cytoskeleton morphology of fibroblast cells cultured on different TPU scaffolds with random and aligned fibrous structures, (d) corresponding cell viability, and cell proliferation statistical results, (e) *in vitro* biodegradation and thermal decomposition results of scaffolds fabricated using different TPUs, and (f) SEM images of cells cultured on random and aligned scaffolds for 5 and 10 days. Source: Mi et al. [46]/with permission of Royal Society of Chemistry.

degradation test proved that the BET chain extender dramatically accelerated the initial degradation rate of the material, although they have similar thermal decomposition behavior (Figure 7.9e). The cells cultured on the three types of scaffolds had tight adhesion and formed a dense cell membrane after only 10 days of culturing. Their work verified the feasibility of altering the mechanical property, degradation behavior, and cellular response of scaffolds by changing the chemical structure of matrix polymer and the microstructure of the scaffolds.

7.5.3 Scaffolds for Vascular Tissue Engineering

Blood vessels are weight-bearing tissues designed to withstand normal blood pressure and maintain structural integrity [47]. Replacement vascular grafts for vascular regeneration must accomplish this task without inducing thrombosis, which is the formation of blood clots, or hemolysis, which is the destruction of red blood cells, and vascular graft engineering must minimize both conditions [48]. Vascular grafts made of nondegradable polymers, including polyethylene terephthalate (PET) and expanded polytetrafluoroethylene (ePTFE), are used for small-diameter vascular grafts but have low long-term patency rates.

Vascular tissue engineering scaffolds refer to tubular grafts that can mimic the properties of native blood vessels, accommodate the growth of endothelial cells, and withstand blood pressure while preventing thrombosis. It is known that native blood vessels are composed of three major layers: tunica intima, tunica media, and tunica adventitia [49]. It has a special nonlinear mechanical behavior where it exhibits a "toe-region" in strain–stress curves when stretched in the circumferential direction [50]. Therefore, closely mimicking the structural, mechanical, and biological properties of native blood vessels is the key target for the fabrication of biodegradable vascular grafts.

Coaxial electrospinning is a technique that allows the production of nanofibers with a core–shell structure, which has been used in the fabrication of vascular grafts. Ferreira and coworkers [51] fabricated a core–shell fibrous meshes, composed of a PCL core and a functionalized gelatin shell, by coaxial electrospinning and further photocrosslinking. The fiber uniformity was enhanced, and the fiber diameter was reduced as the increase of gelatin content and a mean diameter value of 600 nm was achieved. This special feature is crucial for the high biological properties when in contact with blood and with normal human dermal fibroblast cells.

To mimic the wavy inner structure of native blood vessels, Turng and coworkers [52] invented a novel method to fabricate wavy structured tubular scaffolds. They assembled a bundle of copper rods as the collector in the electrospinning of TPU/PCL blend fibers. The expansion of the satellite rods when the collection bundle was rotated at high speed resulted in the formation of a tubular structure with a large inner diameter. As electrospun fiber was deposited, the satellite rods were wrapped tightly, which resulted in a wavy inner-structured tubular scaffold. The scaffold maintained a highly porous fibrous structure (Figure 7.10a). This special wavy structure contributed to the special nonlinear mechanical property of the scaffold, and the mechanical performance was retained within the range of the internal mammary artery (IMA) (Figure 7.10b).

Inspired by the configuration and function of collagen and elastin, as well as the triple-layered structure in native blood vessels, Mi et al. [53] developed two types of triple-layered vascular grafts (TLVGs) by combining the electrospinning and phase separation methods. The electrospinning TPU inner layer facilitated the growth of HUVECS, and the outer porous layer of TPU and PPC layers could provide mechanical strength and fibroblast cell accommodation. To further enhance the suture retention and burst pressure of the vascular scaffolds, Mi et al. [54] invented a new TLVG by combining braided silk as the inner layer, polyacrylamide (PAM) hydrogel as the middle layer, and electrospun TPU as the outer layer (Figure 7.11a). Figure 7.11b shows the appearance of the fabricated TLVG. After freeze drying, the TLVG demonstrated a fibrous inner and outer layer with a porous aerogel middle layer (Figure 7.11c). The woven-structured silk fibers could mimic the properties of the loosely distributed collagen fibers, while the highly elastic PAM hydrogel

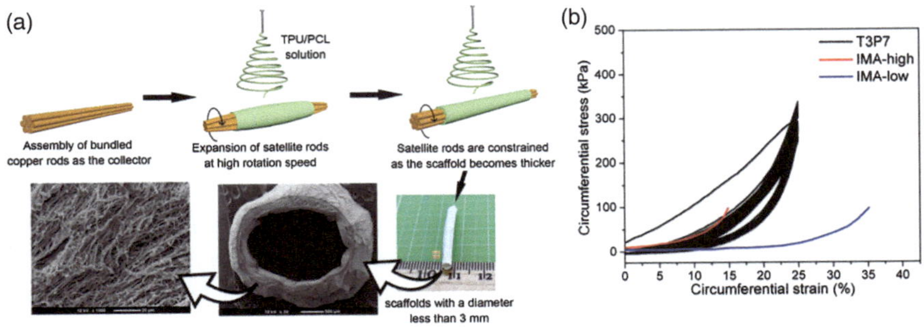

Figure 7.10 (a) Schematic of preparation of vascular grafts using assembled copper rods as a rotating collector and the appearance and cross-sectional structure of a TPU/PCL tubular scaffold, and (b) circumferential cyclical tensile properties of the TPU/PCL tubular scaffold. Source: Mi et al. [52]/with permission of Elsevier.

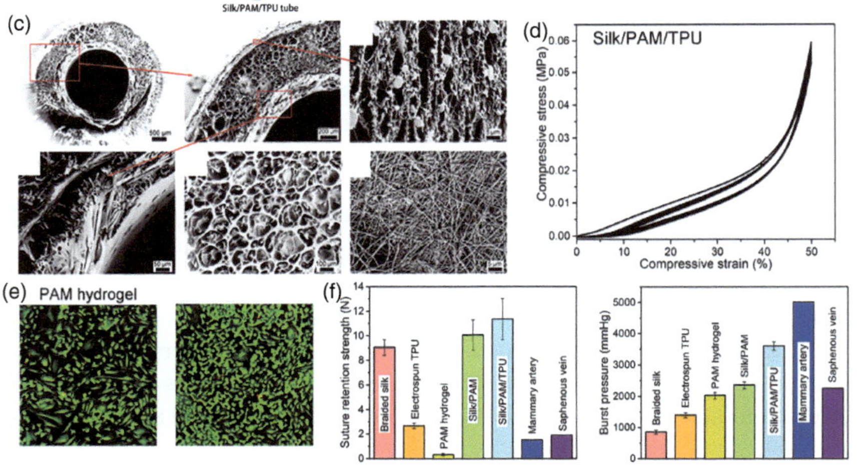

Figure 7.11 (a) The fabrication process of triple-layered vascular grafts (TLVGs), (b) photo of different TLVGs, (c) the microstructure of the TLVGs showing the morphology of different layers, (d) cyclical compressive properties of TLVGs; (e) HUVECs cultured on the PAM hydrogel layer, and (f) the suture retention strength and burst pressure results of TLVGs. Source: Mi et al. [54]/with permission of Elsevier.

and TPU nanofibers mimicked the elasticity of elastin in the blood vessel. With this special structure, the scaffold exhibited excellent mechanical properties with a toe region similar to the native blood vessel (Figure 7.11d). In addition, the TLVG demonstrated excellent biocompatibility with HUVECs and showed high suture retention strength and burst pressures comparable or superior to IMA (Figure 7.11e and f).

7.5.4 Scaffolds for Nervous Tissue Engineering

The peripheral nervous system has an inherent capacity to regenerate after injury, but in severe nerve injury, axons, and endonedulites are destroyed, and axons lose their established guidance pathways for self-regeneration, requiring surgical intervention. In such cases, direct engagement of the proximal and distal nerve stump is the preferred treatment for small discontinuities (<5 mm). For large discontinuities (>10 mm), it is not possible to suture directly without causing tension, which negatively affects nerve regeneration.

In nervous tissue engineering, the nerve conduit should be biocompatible, have sufficient strength and flexibility without exerting pressure on the nerve, be low immunogenic, have good cell adhesion in the desired direction, and have an appropriate degradation time with nontoxic degradation products [55]. The degradation rate, permeability, surface structure, and diameter of nerve conduits have important impacts on their clinical applications. The ideal nerve conduit would degrade according to nerve regeneration at a controlled rate to maintain its shape and strength throughout the regeneration period and should not induce an inflammatory response. Ideally, the nerve conduit should have a high porosity (>80%) to provide sufficient permeability to allow the infiltration of nutrients from the outside to the inside of the conduit while inhibiting the infiltration of unnecessary tissues [56]. Moreover, the precise control of the microstructure, mechanical properties, and surface chemistry of the scaffolds is important to affect the cellular–substrate interaction and the growth behavior of the cells.

TPU, as an elastic biopolymer, has been frequently used in the fabrication of scaffolds for soft tissue repair. To investigate the effect of TPU harness and surface chemistry on cell attachment and growth behavior, Mi et al. [57] prepared TPU electrospun nanofibrous scaffolds with different mechanical and surface properties by combining a soft hydrophilic TPU with a relatively hard hydrophobic TPU (Figure 7.12a). They found that the hard segment content in the scaffold increased with the hard TPU ratio, which resulted in improved hydrophobicity and mechanical properties. All materials could form uniform fibrous structures via electrospinning, but the fiber diameter gradually increased with the increase of the hard TPU ratio. Noticeably, the fibroblast cells cultured on those scaffolds showed different attachment and spreading behavior. After culture for three days, the cells formed were round in shape and aggregated into clusters on soft TPU scaffolds (Figure 7.12b). However, on hard TPU scaffolds, the cells exhibited a spindle shape and spread out on the scaffolds, indicating preferred cell–substrate interaction. After 10 days of culture, cells located in the center of the cell clusters died on the soft

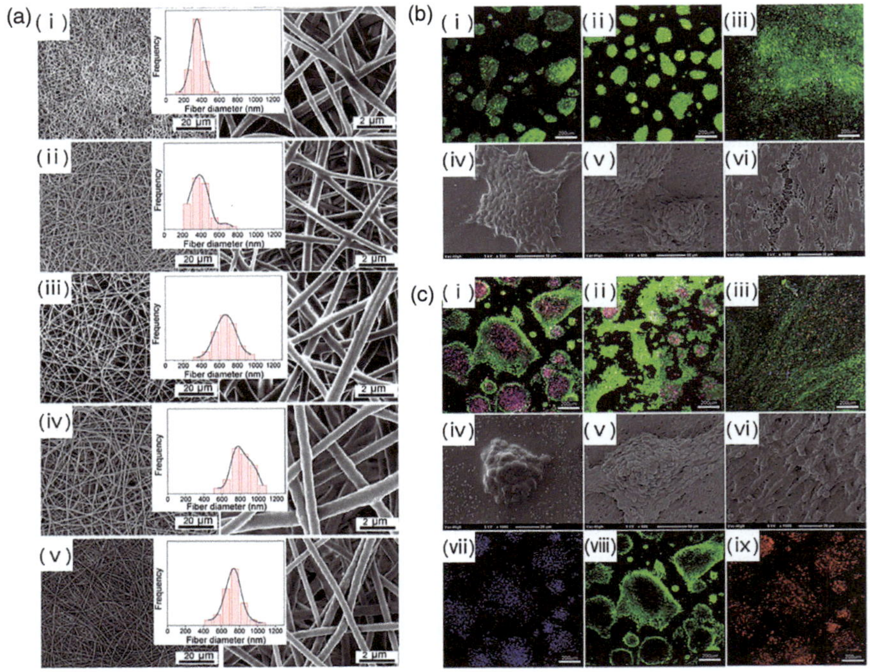

Figure 7.12 (a) SEM images of electrospun scaffolds: from i to v are the scaffolds composed of soft TPU, 25% of soft TPU, 50% of soft TPU, 25% of hard TPU, and hard TPU; fluorescence images and SEM images of cells cultured on the surface of soft TPU, 50% soft TPU, and hard TPU scaffolds for (b) 3 days and (c) 10 days. Source: Mi et al. [57]/with permission of John Wiley & Sons.

TPU scaffolds, while the cells maintained high viability on the hard TPU scaffolds, which means that the properties of the scaffolds have a significant influence on the cell response (Figure 7.12c).

Repairing peripheral nerve injury, especially long-range defects of thick nerves, is a great challenge in the clinic due to their limited regeneration capability. Fan and coworkers [58] developed a self-forming multichannel nerve guidance conduit with topographical cues for peripheral nerve injury repair. Poly(lactide-*co*-trimethylene carbonate) (PLATMC), which is a kind of shape memory polymer based on lactic acid (LA) and trimethylene carbonate (TMC), was used to produce the tubular scaffold. The stimulation temperature of the PLATMC was controlled to be around the body temperature by varying the LA to TMC ratio. As illustrated in Figure 7.13a, bilayer PLATMC nanofiber mats with random and aligned fibrous structures were prepared by electrospinning. The permanent tubular shape was programed at 80 °C, and the temporary flat shape was fixed at 25 °C, which facilitates the cell loading to realize the uniform distribution of cells. The scaffolds could automatically restore their permanent tubular shape at 37 °C, which allows the construction of a special multichannel conduit structure (Figure 7.13b and c). The multichannel conduit exhibits better performance in terms of cell growth and the repair of rat sciatic nerve

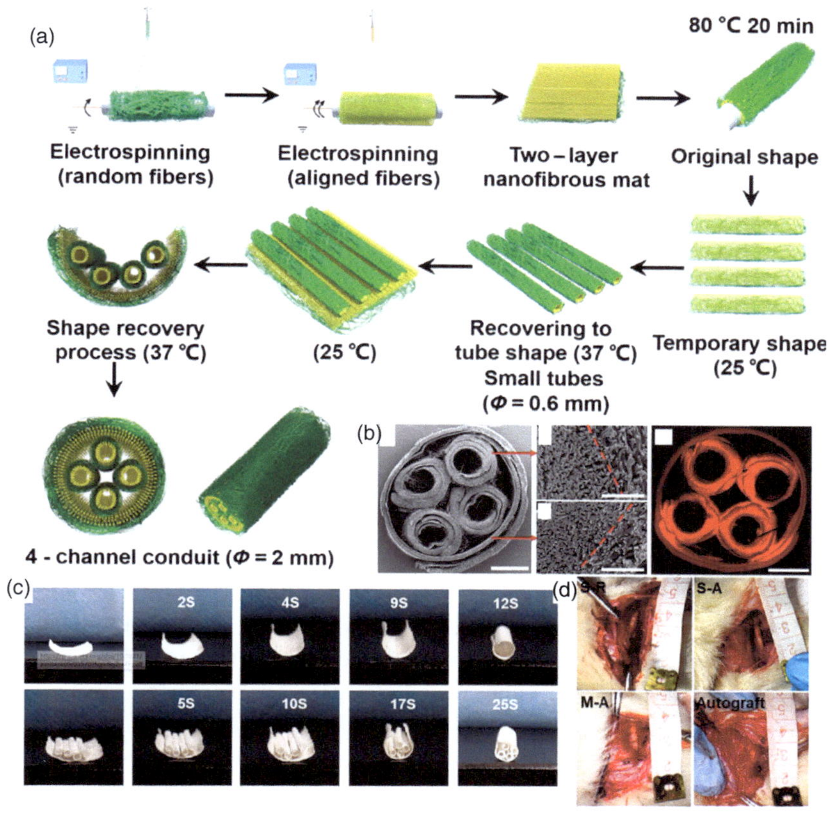

Figure 7.13 (a) Schematic diagram of the fabrication of the four-channel conduit based on shape memory nanofibrous mat, (b) SEM images of the cross-section of the fabricated multichannel nerve conduit in different magnifications, and the fluorescence micrograph of the cross-section of the conduit after staining with Nile red dye, (c) shape recovery process of PLATMC tubes from their deformed temporary planar shapes to permanent tubular shapes, and (d) the site of nerve conduit transplantation in different groups at 12 weeks postsurgery. Source: Wang et al. [58]/with permission of American Chemical Society.

defects. As shown in Figure 7.13d, the nerve conduit transplantation at 12 weeks postsurgery indicated that the self-forming nerve conduits have great potential in peripheral nerve regeneration.

In general, semipermeable nerve conduits will have more effective regenerative properties. The diameter of the nerve conduit should match the diameter of the distal and proximal stump. Diameters that are too small or too large may lead to chronic compression and loosening, thus failing to support nerve regeneration. A flexible nerve conduit with a thin wall is required to withstand the surrounding muscular forces. This problem can be solved by minimizing the diameter of the internal nerve conduit. If the inner diameter is too large, fibrous tissue will grow into the nerve conduit, leading to the arrest of nerve regeneration and maturation. In addition, conductive polymers, such as polypyrrole (PPy), have been frequently used as a surface modification polymer in nerve tissue engineering to facilitate the transmission of electrical signals between the nerve cells.

For example, Huang and coworkers [59] fabricated carboxylic GO (C-GO)/PPy/PLA films fabricated by electrochemical deposition of C-GO-composited PPy on PLA fibrous films. The film showed a high conductivity of ~4.6 S/cm and tensile strength of 26.4 MPa even after immersion in PBS for four weeks. The addition of PPy and GO layer enhanced the surface area and pore volume. Then, the C-GO/PPy/PLLA conducting conduit with a 2 mm inner diameter was used to bridge the 10 mm sciatic nerve defect of rats. ES of 1 V and 20 Hz through the conducting conduit exerted on the defect site successfully enhanced the nerve repair and improved re-innervated gastrocnemius muscle and nerve conduction by the electrophysiological and histological evaluation. Similarly, Yu et al. [60] fabricated a poly(L-lactic acid-*co*-ε-caprolactone)/silk fibroin (PLCL/SF) composite scaffold by electrospinning followed by PPy coating for nerve guidance conduit (NGC). The PPy-coated NGC was used to repair a 10 mm sciatic nerve gap *in vivo*. It was found that the scaffold could promote stem cell proliferation in early postsurgery and enhance myelin formation in later postsurgery, which verified the favorable effect of PPy on the regeneration of nerve tissues.

Moreover, biomolecular functionalized, conductive, biodegradable conduits with electrically active can be filled with biodegradable gels in the internal lumen and nanoporous sheath, which allows nutrients to enter but limits muscle cell penetration and may facilitate faster regeneration of large gaps. Developing effective bioactive nerve tissue engineering scaffolds is still a challenge so far.

References

1 Englert, C., Brendel, J.C., Majdanski, T.C. et al. (2018). Pharmapolymers in the 21st century: synthetic polymers in drug delivery applications. *Progress in Polymer Science* 87: 107–164.

2 Kupikowska-Stobba, B. and Lewińska, D. (2020). Polymer microcapsules and microbeads as cell carriers for in vivo biomedical applications. *Biomaterials Science* 8: 1536–1574.

3 Langer, R. and Vacanti, J.P. (1993). Tissue engineering. *Science* 260: 920–926.

4 Chandra, P.K., Soker, S., and Atala, A. (2020). Chapter 1 – Tissue engineering: current status and future perspectives. In: *Principles of Tissue Engineering*, 5e (ed. R. Lanza, R. Langer, J.P. Vacanti, and A. Atala), 1–35. Academic Press.

5 Brion, A. and Rammal, H. (2020). Chapter 7 – Cell–biomaterials interactions: the role of growth factors. In: *Handbook of Biomaterials Biocompatibility* (ed. M. Mozafari), 111–137. Woodhead Publishing.

6 Abbasi, S. and Ammar Haeri, S. (2021). Biodegradable materials and their applications in sample preparation techniques—a review. *Microchemical Journal* 171: 106831.

7 Kirillova, A., Yeazel, T.R., Asheghali, D. et al. (2021). Fabrication of biomedical scaffolds using biodegradable polymers. *Chemical Reviews* 121: 11238–11304.

8 Tharani Kumar, S., Prasanna Devi, S., Krithika, C., and Raghavan, R.N. (2020). Review of metallic biomaterials in dental applications. *Journal of Pharmacy & Bioallied Sciences* 12.

9 Koons, G.L., Diba, M., and Mikos, A.G. (2020). Materials design for bone-tissue engineering. *Nature Reviews Materials* 5: 584–603.

10 Halake, K., Kim, H.J., Birajdar, M. et al. (2016). Recently developed applications for natural hydrophilic polymers. *Journal of Industrial and Engineering Chemistry* 40: 16–22.

11 Mi, H.-Y., Jing, X., and Turng, L.-S. (2014). Fabrication of porous synthetic polymer scaffolds for tissue engineering. *Journal of Cellular Plastics* 51: 165–196.

12 Zheng, Y., Hong, X., Wang, J. et al. (2021). 2D nanomaterials for tissue engineering and regenerative nanomedicines: recent advances and future challenges. *Advanced Healthcare Materials* 10: 2001743.

13 Zhang, F. and King, M.W. (2020). Biodegradable polymers as the pivotal player in the design of tissue engineering scaffolds. *Advanced Healthcare Materials* 9: 1901358.

14 Björninen, M., Gilmore, K., Pelto, J. et al. (2017). Electrically stimulated adipose stem cells on polypyrrole-coated scaffolds for smooth muscle tissue engineering. *Annals of Biomedical Engineering* 45: 1015–1026.

15 Ivanovic, J., Rezwan, K., and Kroll, S. (2018). Supercritical CO_2 deposition and foaming process for fabrication of biopolyester–ZnO bone scaffolds. *Journal of Applied Polymer Science* 135: 45824.

16 Li, Y., Chu, Z., Li, X. et al. (2017). The effect of mechanical loads on the degradation of aliphatic biodegradable polyesters. *Regenerative Biomaterials* 4: 179–190.

17 O'Brien, F.J. (2011). Biomaterials & scaffolds for tissue engineering. *Materials Today* 14: 88–95.

18 Feng, P., Wu, P., Gao, C. et al. (2018). A multimaterial scaffold with tunable properties: toward bone tissue repair. *Advanced Science* 5: 1700817.

19 Sultana, N. (2018). Chapter 1 – Mechanical and biological properties of scaffold materials. In: *Functional 3D Tissue Engineering Scaffolds* (ed. Y. Deng and J. Kuiper), 1–21. Woodhead Publishing.

20 Yin, N., Stilwell, M.D., Santos, T.M.A. et al. (2015). Agarose particle-templated porous bacterial cellulose and its application in cartilage growth in vitro. *Acta Biomaterialia* 12: 129–138.

21 Whang, K., Healy, K.E., Elenz, D.R. et al. (1999). Engineering bone regeneration with bioabsorbable scaffolds with novel microarchitecture. *Tissue Engineering* 5: 35–51.

22 Kovács, K., Váncsa, S., Agócs, G. et al. (2023). Anisotropy, anatomical region, and additional variables influence Young's modulus of bone: a systematic review and meta-analysis. *JBMR Plus* 7: e10835.

23 Nawroth, J.C. and Parker, K.K. (2013). Design standards for engineered tissues. *Biotechnology Advances* 31: 632–637.

24 Mooney, D.J., Baldwin, D.F., Suh, N.P. et al. (1996). Novel approach to fabricate porous sponges of poly(D,L-lactic-*co*-glycolic acid) without the use of organic solvents. *Biomaterials* 17: 1417–1422.

25 Nam, Y.S., Yoon, J.J., and Park, T.G. (2000). A novel fabrication method of macroporous biodegradable polymer scaffolds using gas foaming salt as a porogen additive. *Journal of Biomedical Materials Research* 53: 1–7.

26 Shen, X., Ruan, J., Zhou, Z. et al. (2007). Fabrication of porous scaffolds using $NaHCO_3$ particulates as the porogen material. *Journal of Wuhan University of Technology Materials Science Edition* 22: 279–283.

27 Xie, R.Q., Hu, J.L., Ng, F. et al. (2017). High performance shape memory foams with isocyanate-modified hydroxyapatite nanoparticles for minimally invasive bone regeneration. *Ceramics International* 43: 4794–4802.

28 Kazimierczak, P., Palka, K., and Przekora, A. (2019). Development and optimization of the novel fabrication method of highly macroporous chitosan/agarose/nanohydroxyapatite bone scaffold for potential regenerative medicine applications. *Biomolecules* 9: 434.

29 Nublat, C., Braud, C., Garreau, H., and Vert, M. (2006). Ammonium bicarbonate as porogen to make tetracycline-loaded porous bioresorbable membranes for dental guided tissue regeneration: failure due to tetracycline instability. *Journal of Biomaterials Science. Polymer Edition* 17: 1333–1346.

30 Kucharska, M., Butruk, B., Walenko, K. et al. (2012). Fabrication of in-situ foamed chitosan/β-TCP scaffolds for bone tissue engineering application. *Materials Letters* 85: 124–127.

31 Salgado, A.J., Coutinho, O.P., Reis, R.L., and Davies, J.E. (2007). In vivo response to starch-based scaffolds designed for bone tissue engineering applications. *Journal of Biomedical Materials Research. Part A* 80A: 983–989.

32 Lv, Z., Zhao, N., Wu, Z. et al. (2018). Fabrication of novel open-cell foams of poly(ε-caprolactone)/poly(lactic acid) blends for tissue-engineering scaffolds. *Industrial and Engineering Chemistry Research* 57: 12951–12958.

33 Mi, H.Y., Salick, M.R., Jing, X. et al. (2013). Characterization of thermoplastic polyurethane/polylactic acid (TPU/PLA) tissue engineering scaffolds fabricated by microcellular injection molding. *Materials Science & Engineering, C: Materials for Biological Applications* 33: 4767–4776.

34 Chen, B.-Y., Jing, X., Mi, H.-Y. et al. (2015). Fabrication of polylactic acid/polyethylene glycol (PLA/PEG) porous scaffold by supercritical CO2 foaming and particle leaching. *Polymer Engineering and Science* 55: 1339–1348.

35 Chen, B.-Y., Wang, Y.-S., Mi, H.-Y. et al. (2014). Effect of poly(ethylene glycol) on the properties and foaming behavior of macroporous poly(lactic acid)/sodium chloride scaffold. *Journal of Applied Polymer Science* 131: https://doi.org/10.1002/app.41181.

36 Teng, X., Ren, J., and Gu, S. (2007). Preparation and characterization of porous PDLLA/HA composite foams by supercritical carbon dioxide technology. *Journal of Biomedical Materials Research Part B: Applied Biomaterials* 81B: 185–193.

37 Salerno, A., Zeppetelli, S., Di Maio, E. et al. (2011). Processing/structure/property relationship of multi-scaled PCL and PCL–HA composite scaffolds prepared via gas foaming and NaCl reverse templating. *Biotechnology and Bioengineering* 108: 963–976.

38 Annabi, N., Fathi, A., Mithieux, S.M. et al. (2011). Fabrication of porous PCL/elastin composite scaffolds for tissue engineering applications. *The Journal of Supercritical Fluids* 59: 157–167.

39 Mi, H.-Y., Jing, X., Salick, M.R. et al. (2014). Fabrication of thermoplastic polyurethane tissue engineering scaffold by combining microcellular injection molding and particle leaching. *Journal of Materials Research* 29: 911–922.

40 Jiang, S., Wang, M., and He, J. (2021). A review of biomimetic scaffolds for bone regeneration: toward a cell-free strategy. *Bioengineering & Translational Medicine* 6: e10206.

41 Mi, H.-Y., Jing, X., Salick, M.R. et al. (2014). Morphology, mechanical properties, and mineralization of rigid thermoplastic polyurethane/hydroxyapatite scaffolds for bone tissue applications: effects of fabrication approaches and hydroxyapatite size. *Journal of Materials Science* 49: 2324–2337.

42 Mi, H.-Y., Palumbo, S., Jing, X. et al. (2014). Thermoplastic polyurethane/hydroxyapatite electrospun scaffolds for bone tissue engineering: effects of polymer properties and particle size. *Journal of Biomedical Materials Research Part B: Applied Biomaterials* 102: 1434–1444.

43 Lin, H.-R. and Yeh, Y.-J. (2004). Porous alginate/hydroxyapatite composite scaffolds for bone tissue engineering: preparation, characterization, and in vitro studies. *Journal of Biomedical Materials Research Part B: Applied Biomaterials* 71B: 52–65.

44 Demoor, M., Ollitrault, D., Gomez-Leduc, T. et al. (2014). Cartilage tissue engineering: molecular control of chondrocyte differentiation for proper cartilage matrix reconstruction. *Biochimica et Biophysica Acta (BBA) - General Subjects* 1840: 2414–2440.

45 Du, Y., Liu, H., Yang, Q. et al. (2017). Selective laser sintering scaffold with hierarchical architecture and gradient composition for osteochondral repair in rabbits. *Biomaterials* 137: 37–48.

46 Mi, H.-Y., Jing, X., Napiwocki, B.N. et al. (2017). Biocompatible, degradable thermoplastic polyurethane based on polycaprolactone-block-polytetrahydrofuran-block-polycaprolactone copolymers for soft tissue engineering. *Journal of Materials Chemistry B* 5: 4137–4151.

47 Mokhtari-Jafari, F., Amoabediny, G., and Dehghan, M.M. (2020). Role of biomechanics in vascularization of tissue-engineered bones. *Journal of Biomechanics* 110: 109920.

48 Xue, L. and Greisler, H.P. (2003). Biomaterials in the development and future of vascular grafts. *Journal of Vascular Surgery* 37: 472–480.

49 Wu, J.X., Hu, C.M., Tang, Z.C. et al. (2018). Tissue-engineered vascular grafts: balance of the four major requirements. *Colloid and Interface Science Communications* 23: 34–44.

50 Mi, H.Y., Jing, X., Yu, E. et al. (2015). Fabrication of triple-layered vascular scaffolds by combining electrospinning, braiding, and thermally induced phase separation. *Materials Letters* 161: 305–308.

51 Coimbra, P., Santos, P., Alves, P. et al. (2017). Coaxial electrospun PCL/gelatin-MA fibers as scaffolds for vascular tissue engineering. *Colloids and Surfaces B: Biointerfaces* 159: 7–15.

52 Mi, H.Y., Jing, X., Yu, E. et al. (2018). Manipulating the structure and mechanical properties of thermoplastic polyurethane/polycaprolactone hybrid small diameter vascular scaffolds fabricated via electrospinning using an assembled rotating collector. *Journal of the Mechanical Behavior of Biomedical Materials* 78: 433–441.

53 Mi, H.Y., Jing, X., McNulty, J. et al. (2016). Approaches to fabricating multiple-layered vascular scaffolds using hybrid electrospinning and thermally induced phase separation methods. *Industrial and Engineering Chemistry Research* 55: 882–892.

54 Mi, H.-Y., Jiang, Y., Jing, X. et al. (2019). Fabrication of triple-layered vascular grafts composed of silk fibers, polyacrylamide hydrogel, and polyurethane nanofibers with biomimetic mechanical properties. *Materials Science and Engineering: C* 98: 241–249.

55 Chrząszcz, P., Derbisz, K., Suszyński, K. et al. (2018). Application of peripheral nerve conduits in clinical practice: a literature review. *Neurologia i Neurochirurgia Polska* 52: 427–435.

56 Ijkema-Paassen, J., Jansen, K., Gramsbergen, A., and Meek, M.F. (2004). Transection of peripheral nerves, bridging strategies and effect evaluation. *Biomaterials* 25: 1583–1592.

57 Mi, H.-Y., Jing, X., Salick, M.R. et al. (2015). Properties and fibroblast cellular response of soft and hard thermoplastic polyurethane electrospun nanofibrous scaffolds. *Journal of Biomedical Materials Research Part B: Applied Biomaterials* 103: 960–970.

58 Wang, J., Xiong, H., Zhu, T. et al. (2020). Bioinspired multichannel nerve guidance conduit based on shape memory nanofibers for potential application in peripheral nerve repair. *ACS Nano* 14: 12579–12595.

59 Chen, X.C., Liu, C.Y., Huang, Z.B. et al. (2019). Preparation of carboxylic graphene oxide-composited polypyrrole conduits and their effect on sciatic nerve repair under electrical stimulation. *Journal of Biomedical Materials Research Part A* 107: 2784–2795.

60 Sun, B.B., Zhou, Z.F., Li, D.W. et al. (2019). Polypyrrole-coated poly(L-lactic acid-*co*-ε-caprolactone)/silk fibroin nanofibrous nerve guidance conduit induced nerve regeneration in rat. *Materials Science and Engineering C* 94: 190–199.

8

Flexible Sensors Based on Porous Polymer and Polymer Foams

8.1 Overview of Flexible Strain Sensors

In recent decades, with the improvement of people's life quality, the development of wearable devices has been extended to various fields such as solar cells, mobile phones, smart watches, displays, batteries, sensors, and radio frequency identification devices (RFID) as shown in Figure 8.1 [1]. Among them, sensors are a class of important devices that convert various types of stimuli signals from the external environment, such as chemical, physical, or biological signals, into collectible electrical signals. Flexible pressure sensors are capable of detecting the change of subtle pressure in real time and can be mounted on the human body easily. They have been extended to applications such as electronic skin, artificial intelligence, soft robotics, health monitoring, and other fields [2].

According to the working mechanism, flexible pressure sensors can be divided into resistive, capacitive, piezoelectric, and other types [3]. Flexible capacitive pressure sensors convert mechanical signals into capacitance signals by sensing changes in the capacitance of the sensing device, and they are typically sensitive to the action of small forces and capable of detecting small static forces. Flexible piezoelectric pressure sensors have the characteristics of high sensitivity and fast response. Thus, they are widely used in real-time monitoring of dynamic mechanical changes, but cannot monitor constant pressure. Flexible pressure sensors based on organic field-effect transistors generally have poor stability, and measurement signal drift may occur during pressure monitoring. Compared with the aforementioned types of flexible pressure sensors, flexible resistive pressure sensors are considered to be the ideal choice for the next generation of flexible pressure sensors due to their simple device structure, high sensitivity, fast response, low manufacturing cost, and good stability [4].

Flexible resistive pressure sensors require both flexibility and good sensing performance, allowing them to adapt to different application scenarios. Although significant progress has been made in the research of flexible resistive pressure sensors, it remains a huge challenge to mass produce low cost, highly sensitive devices with stable long-term performance through the selection of active layer

Functional Polymer Foams: Green Fabrication Methods, Performance and Applications, First Edition. Haoyang Mi.

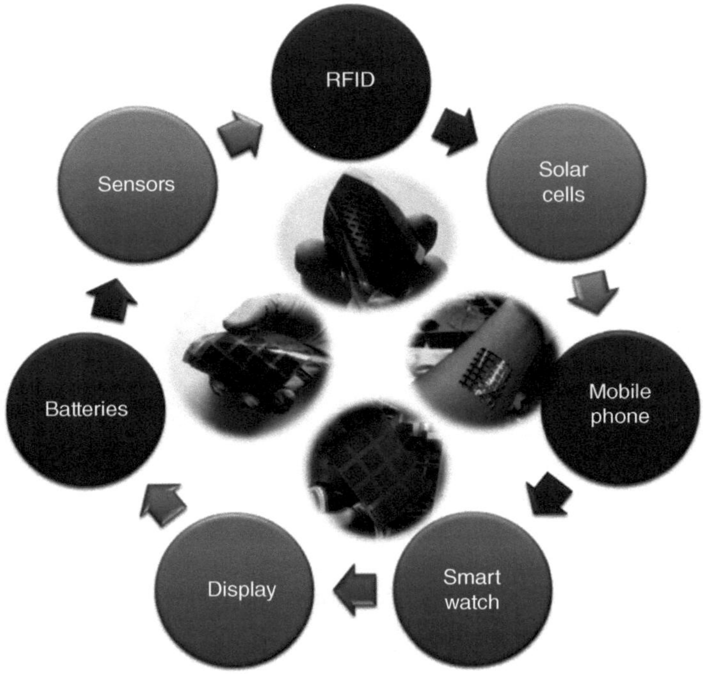

Figure 8.1 The fields where flexible electronic products are currently applied. Source: Han et al. [1]/with permission of John Wiley & Sons.

materials and the design of flexible sensing structures. Traditional resistive sensors are often based on materials such as single-crystal silicon, metals, and metal oxide films. Due to the inherent rigidity and brittleness of inorganic materials, traditional resistive sensors are very sensitive but have limited flexibility and strain range, which restricts their applications in the field of flexible sensing scenarios.

In recent years, porous polymer foams have been more and more used in advanced fields as functional materials such as electromagnetic interference shielding, biomedical scaffolds, superabsorbents, etc., due to the advantages of their lightweight, high specific surface area, and high porosity. CPFs have also demonstrated excellent ability to be used as flexible strain or pressure sensors attributing to their great resistance variation upon deformation due to their highly porous internal structure.

8.1.1 Substrate Materials for CPF Sensors

The materials used for CPF-based piezoresistive sensors usually include two fractions, namely, the substrate material and the sensitive material. The substrate material is usually selected first based on the requirements for the sensing applications. The mechanical property, stretchability, hardness, flexibility, recoverability, and tactile impression are often considered in the selection of substrate material. Then, the sensitive material is selected based on the properties of the substrate material and the application requirements. If the extensibility of the sensitive material does not match the substrate and device requirements, special structural designs or the addition of a third proportion of ingredients of the sensitive material are sometimes needed.

Among various substrate materials, melamine foam and polyurethane (PU) foams are the most frequently used substrate skeleton for piezoresistive pressure sensors. Melamine foam is a kind of thermosetting elastic organic rubber foam made of melamine polyformaldehyde resin by microwave foaming, which has a fine three-dimensional mesh structure with high porosity and open-cell structure [5]. PU foams have a wide density range and high rebound rate and can be manufactured on a large scale through simple foaming techniques. They are ideal candidate materials for large deformation and high-sensitivity piezoresistive sensors due to their high recovery rate and stability [6].

Polyimide (PI) with aromatic structures in its main chain is considered an attractive high-performance engineering polymer because it combines extraordinary thermal stability, flame retardancy, excellent mechanical properties, and good environmental adaptability. When PI foam is synthesized with elastic polyether chains, it can obtain highly elastic properties, which makes them excellent candidates as substrate materials for piezoresistive pressure sensors [7].

Other thermosetting elastic materials such as rubbers have also been used as substrate materials for piezoresistive sensors. Among them, silicone rubber is the most frequently used. Polydimethylsiloxane (PDMS) has been used in numerous reports to develop piezoresistive sensors due to their high elasticity, abrasion resistance, and easy processability [8].

For thermoplastic polymers, thermoplastic polyurethane (TPU) is the most frequently used substrate material for piezoresistive sensors. TPU is synthesized by the polycondensation reaction of isocyanate and polyol and can be plasticized and dissolved in many solvents owing to their linear molecular structure. TPU has excellent properties, such as a wide range of hardness, wear resistance, transparency, oil resistance, and good elasticity. Its hardness is mainly controlled by changing the ratio of various reactive components of TPU. As the hardness increases, the resulting products can still maintain excellent elasticity and wear resistance. Therefore, many researchers have considered preparing flexible pressure sensors based on TPU to achieve better sensing performance [9].

8.1.2 Active Conductive Materials

In piezoresistive flexible strain sensors, the active material is the core component of the flexible sensor. The electrical property, micromorphology, and distribution of the active conductive materials have a significant influence on the sensitivity, response time, and strain coefficient of the resulting pressure sensors. Active conductive materials that are frequently used will be discussed in this section.

8.1.2.1 Carbon-Based Materials

Carbon-based materials possess excellent mechanical and electrical properties, making them widely applicable in the fabrication of pressure-sensitive flexible sensors. Common carbon-based materials include carbon nanotubes (CNTs), graphene, carbon black (CB), and Mxene.

CNTs exhibit good conductivity, large specific surface area, controllable pore size through synthesis, and 100% utilization of specific surface area, making them excellent candidates for flexible electrodes and electronic devices. Wang et al. [10] placed a single-walled carbon nanotube (SWCNT) film between two layers of flexible PDMS film to prepare a highly sensitive and stable sandwich-structured flexible sensor. The prepared sensor has the advantages of low-pressure detection, fast response, high stability, and sensitive detection of human physiological activity signals.

Graphene is a planar film composed of hexagonal lattices formed by carbon atoms with sp^2 hybrid orbitals. Graphene nanosheets have excellent physical properties including large specific surface area, high carrier mobility, high mechanical strength, and high thermal stability, which makes graphene a promising active material for developing high-performance pressure sensors when combined with various substrate materials. There are various types of graphene derivatives with different chemical structures including graphene oxide (GO), reduced graphene oxide (rGO), and functionalized graphene. GO exploited from graphite has good dispersion ability in various solvents and excellent compatibility with most substrate materials due to the presence of oxygen-containing groups on the surface of GO sheets. However, GO has poor conductivity due to the discontinuous π–π electron cloud. GO typically needs to be reduced to rGO when combined with the substrate polymer since deoxygenation and reduction of oxygen-containing groups on the sp^2 structure of graphene are usually essential for π electrons to move freely on the graphene structure to obtain excellent conductivity and piezoresistivity [11].

Other carbon-based materials such as CB and carbon fiber, have also been used to prepare high-sensitivity piezoresistive strain sensors. CB is an amorphous form of carbon, which is a black powder with low density and a large surface area. It has a much lower cost compared with CNT and graphene, thus, CB is believed as an important active conductive material in the practical production of piezoresistive sensors.

Moreover, pyrolysis is another approach for conductive carbon sources. Polymers rich in carbon and nitrogen atoms and lacking oxygen atoms can be carbonized at high temperatures with the presence of inter-gas protection. These carbon sources are typically similar to the CB, but they could usually maintain their pristine porous or fibrous structures [12]. Therefore, the carbonized active carbon materials are called architected conductive networks, which can be directly combined with substrate materials to result in piezoresistive sensors.

MXene is a 2D transition metal carbide, nitride, or carbonitride, represented by the formula $M_{n+1}X_nT_x$. As a member of the 2D material family, MXene has the advantages of high mechanical strength, metallic conductivity, and large surface area. Although both graphene and MXene are conductive 2D nanomaterials, the surface of MXene is rich in functional groups and has a hydrophilic nature, which makes MXene bind more easily with most substrate materials than graphene. The metal conductivity and interlayer spacing of MXene give it a huge adjustable resistance range, making it highly sensitive to pressure, thus becoming a popular material for preparing flexible resistive pressure sensor active layers [13]. Nonetheless, the relatively small planar size and high rigidity of MXene restricted their stability when coated on the surface of a substrate material like PU foam.

8.1.2.2 Metallic Nanomaterials

Metallic nanomaterials refer to a class of metal particles or wires with sizes at the nanometer scale. Metallic nanomaterials usually possess high specific surface areas, good mechanical strength, and excellent electrical conductivity. Depending on the dimension of the particles, they can be classified into 0D particles or spheres, 1D rods or wires, and 2D flakes or sheets. Sometimes, nanoparticles can be synthesized as microaggregates, which form special three-dimensional (3D) clusters or 3D flowers [14].

In the application of flexible piezoresistive sensors, metallic nanowires are one of the most popular metallic nanomaterials due to their high aspect ratio, which makes the contraction and separation of the conductive path more sensitive. Among all metallic nanomaterials, silver nanowires (AgNWs) exhibit superior electrical conductivity, higher transparency, as well as good flexibility and corrosion resistance [15]. Thus, AgNW has been widely used in the field of flexible electronics, such as flexible circuit boards, touch screens, smart gloves, and human health monitoring.

8.1.2.3 Conducting Polymers

In recent years, new sensors based on conductive polymers as active conductive materials, such as polythiophene, polyaniline (PANI), and polypyrrole (PPy), have received much attention due to their high performance, lightweight, good flexibility, biocompatibility, and ease of surface modification. These sensors have shown great potential in areas such as human motion monitoring and healthcare.

PPy is synthesized from pyrrole monomers using either chemical oxidation or electrochemical methods. Pure pyrrole monomers are five-membered carbon and nitrogen heterocyclic molecules that are slightly soluble in water and nontoxic. Oxidized PPy exhibits excellent conductivity, and biodegradability, and is easier to deposit. It has been widely used in ion detection, antistatic materials, electromagnetic shielding materials, electrocatalysis, conductive polymer composite (CPC) materials, and other fields [16].

Compared to PPy, poly(3,4-ethylenedioxythiophene) (PEDOT) has better conductivity and thermal stability, making it suitable for developing highly conductive flexible sensors for various applications. When PEDOT is doped with a polyanion, such as polystyrene sulfonate (PSS), it forms PEDOT:PSS. Different ratios of PEDOT and PSS can result in solutions with varying conductivities.

In recent years, PEDOT:PSS has been used to produce highly conductive and stretchable sensors for various applications such as human motion monitoring, intelligent robot control, and solar cells [17].

PANI is a typical conductive polymer that has been widely used in the electronic and optical fields. It is obtained by chemically or electrochemically oxidizing aniline monomers in an acidic aqueous solution, with ammonium persulfate (APS) often used as the oxidant. The resulting conductive PANI is a dark green emulsion that decomposes when the temperature exceeds 120 °C. Due to its high conductivity, stability, and ease of preparation, PANI has also been used in the development of piezoresistive sensors [18].

8.1.2.4 Hybrid Active Conductive Materials

As mentioned earlier, different active materials have their unique properties and advantages. Therefore, different active materials are often combined in the development of piezoresistive sensors to utilize their advantages to achieve superior performance compared to using a single type of active material. For example, nanofillers with different dimensions are sometimes combined together to elevate the repetitive segregation and connection of the conductive path, thus, enhancing the sensing sensitivity. Tewari et al. [19] synthesized a rGO ink with a hydrophilic nature and mixed with multiwalled carbon nanotubes (MWNTs) to fabricate a pressure sensor by coating the hybrid ink on to a PU foam, which could monitor small-scale and large-scale motions. Wang et al. [20] fabricated a foam-based strain sensor composed of TPU, MWCNTs, and $Ti_3C_2T_x$ MXene by combining salt-templating and dip-coating methods, which obtained a wide strain sensing range and a high sensitivity (GF = 363). Chen et al. [21] used the sacrificial template method and chemical crosslinking method to prepare ethylene vinyl acetate copolymer (EVA) foams with shape memory capability and used PPy and AgNPs as the conductive layer to decorate the foam frames for the preparation of smart wearable sensors. Therefore, the rational combination of different active materials in a proper proportion is a feasible approach to boost the performance of piezoresistive sensors.

8.2 Fundamentals of Piezoresistive Sensors

8.2.1 Sensing Mechanism of Flexible Piezoresistive Sensors

The resistance-type pressure sensor consists of an active layer made of elastic conductive or semiconductor material, which is in contact with two electrodes. The sensor is capable of converting the applied pressure stimulus into a change in resistance signal, thus, the pressure level on the sensor can be evaluated by analyzing the resistance signal obtained.

The total resistance of the sensor consists of two parts: electrode resistance and active layer resistance. For a specific resistance-type pressure sensor, the electrode resistance is generally a fixed value. The change in the resistance signal of the sensor mainly depends on the change in the resistance value of the active layer.

The calculation formula for the resistance (R) of a resistance-type pressure sensor is expressed as follows,

$$R = \frac{\rho L}{A} \tag{8.1}$$

In this equation, ρ represents electrical resistivity, L represents length, and A represents cross-sectional area [22]. Generally speaking, the resistance change of resistive pressure sensors is caused by macro-geometric parameter changes triggered by external stimuli, while the resistance changes of flexible resistive pressure sensors based on piezoresistive effects mainly depend on changes in resistivity. By selecting types of nanocomposites, controlling the amount of active materials used, and designing the microstructure of the substrate and the active layer, flexible piezoresistive pressure sensors with different performance emphases can be prepared.

8.2.1.1 Percolation Theory Based on Conductive Polymer Composites

Most pressure sensors consist of a structure composed of an elastic material with low Young's modulus and a surface conductive layer, but this two-layer structure model imposes higher requirements on the long-term stability of high-performance pressure sensors since the attached conductive layer on the substrate surface may crack or even detach from the substrate under long-term cyclic stress due to the Young's modulus mismatch between the deformation layer and the surface conductive layer [23]. To ensure the cyclic stability of flexible resistive pressure sensors, the percolation theory based on CPC materials has been proposed [24]. A permeable conductive composite nanomaterial with high-electrical conductivity and low Young's modulus as the sensing layer of flexible resistive pressure sensors is developed to replace the traditional two-layer structure model.

CPCs consist of a polymer matrix and conductive fillers, and the magnitude of the electrical resistivity depends on the volume fraction of the conductive fillers. When the volume fraction of conductive fillers is extremely low, the average distance between conductive particles in the polymer matrix is large, and no conductive path is formed, leading to limited conductivity, resulting in high electrical resistivity. As the volume fraction of conductive fillers increases, the average distance between conductive particles decreases, forming partial conductive paths, and the electrical resistivity decreases. When the volume fraction of conductive fillers is sufficiently large, the conductive filler particles come close to each other in the polymer matrix, forming a complete conductive network that passes through the entire material. Electrically CPCs will transform from initially insulating polymer matrices to conductive matrices as the construction of the conductive network in the matrix. According to percolation theory, the critical volume fraction of conductive fillers is known as the percolation threshold. Therefore, when designing flexible resistive pressure sensors based on percolation theory, it is necessary to ensure the sensitivity and pressure range of the sensor. As shown in Figure 8.2, to render a CPC with pressure-sensing properties, the volume fraction of conductive fillers must be controlled to form a partial conductive path in the matrix so that the change in

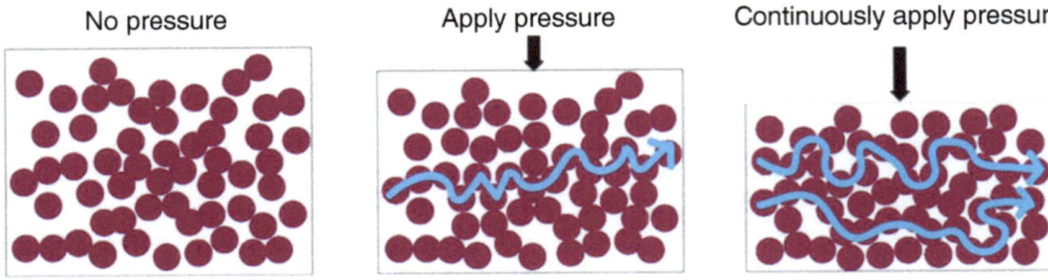

Figure 8.2 Schematic diagram of reducing resistivity by pressing through a conductive polymer composite (CPC). Source: Wang et al. [25]/with permission of John Wiley & Sons.

the average distance between conductive particles upon pressure can significantly influence the overall resistivity of the CPC material [25].

8.2.1.2 Tunneling Effect Based on Conductive Polymer Composites

The tunneling effect based on CPCs refers to the phenomenon that electrons between conductive fillers pass through the otherwise impassable "energy barrier" [26]. Under this effect, electrons have a certain probability of transitioning from a low-energy state to a high-energy state when the distance between conductive particles is within a certain range, thus forming a tunneling current. Amjadi et al. [27] reported a composite sandwich structural material based on the tunneling effect (i.e. silver nanowire film embedded between two layers of PDMS elastoids). As shown in Figure 8.3, when the distance (d) between the center lines of two adjacent nanowires (NWs) was less than or equal to the diameter of the NWs (D) ($d \leqslant D$), they were completely connected without contact resistance, and the total resistance was the sum of the resistance R_1 and R_2 of the two NWs. When the d was greater than a truncation (C) ($d > C$), two NWs completely disconnected, total resistance value was infinite; However, when d was larger than the diameter of the NWs and smaller than the cut-off distance ($D < d \leqslant C$), the electrons could pass through the polymer matrix and form the quantum tunneling effect.

Yang et al. [28] established a representative volumetric element numerical model (RVE model) based on graphene rubber composites to characterize the influence of the tunneling effect more accurately. Using this model, they analyzed the calculation

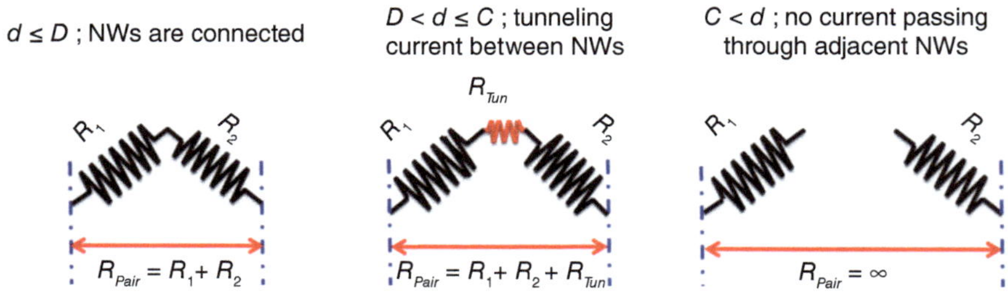

Figure 8.3 Schematic diagram of tunneling effect in CPCs. Source: Amjadi et al. [27]/with permission of American Chemical Society.

results of the current density of composites with 2.3% graphene content (mass fraction) when the tunneling effect was ignored and considered, respectively. It was found that the electron migration ability in graphene-rubber composites is closely related to the distance between graphene sheets, and some graphene sheets with small distances may have electron migration as well. When the tunneling effect is not taken into account, the current density of the rubber matrix is the lowest among the composites, and the graphene rubber composite is insulated in this state. On the other hand, when the tunneling effect is considered, the tunneling current density of composite materials and rubber matrix increases significantly, reaching about 106 times the case without considering the tunneling effect, which indicates that the tunneling effect is more conducive to the formation of conductive paths in composites.

8.2.1.3 Variation Based on Interface Contact Resistance

The interface contact resistance is an important mechanism affecting the sensing property of piezoresistive sensors. Since all solid surfaces are rough at the microscopic scale, the interfacial contact between two materials occurs at the point of contact generated by mechanical contact between the slightly convex bodies of two surfaces, and the actual contact area is only a fraction of the nominal contact area. The current is transmitted through these independently separated contact points (Figure 8.4a) [32]. Therefore, the change in the number and size of contact points

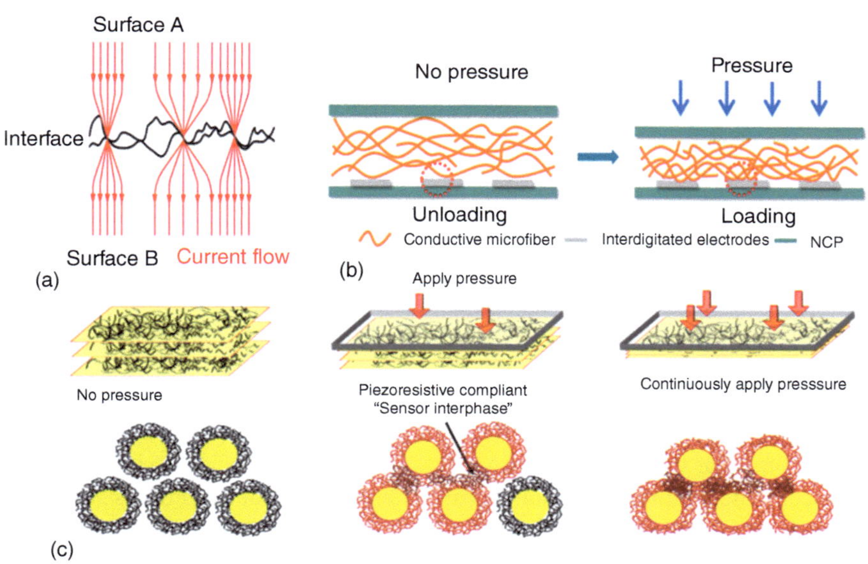

Figure 8.4 Change of contact resistance based on interface contact resistance change mechanism: (a) Schematic diagram of the contact interface of two solid materials, (b) working mechanism of electrode–active layer contact type pressure sensor, and (c) sensing mechanism of nonconductive fibers coated with PEI–CNT under applied pressure. Source: (a) Hamedi and Atashparva [29]/with permission of Springer Nature. (b) Gao et al. [30]/with permission of American Chemical Society. (c) Doshi and Thostenson [31]/with permission of American Chemical Society.

directly determines the resistance changing rate so that the performance of the piezoresistive sensors can be optimized by adjusting the size and number of contact micro-convex bodies, increasing the amount of contact junctions, and unifying the distribution of the micro-contact points of conductive fillers [29].

Flexible piezoresistive pressure sensors based on the interface contact resistance change mechanism can be divided into two categories. The first design utilizes the interface contact resistance change between the electrode and the active layer material, as shown in Figure 8.4b, and the second design employs the interface contact resistance change among the active materials located within different layers of the material as illustrated in Figure 8.4c. By designing the microstructure of the active material layer and the electrode, the interface contact resistance can be controlled to meet the requirements for various sensing applications.

8.2.2 Sensing Mechanism of Conductive Foams

Polymer foams are presently used materials as substrates for piezoresistive pressure sensors. Although the neat foam skeletons are nonconductive, they can be modified or coated with an active conducting layer to render them with excellent conductivity. In this mean, the loading content of the active conductive material to establish a conductive network can be reduced since they are solely located on the surface of the foam skeleton. Therefore, the modified polymer foams typically have much better conductivity and sensitivity compared to traditional foams developed using CPCs.

Figure 8.5 illustrates the graphene foam (GF) variation upon compression and the working mechanism of the CPF-based pressure sensors. Depending on the strain range and GF values, the sensing range of the sensors can be divided into three regions, which are closely related to the change in the porous structure. In region I ($0 < \varepsilon < 0.6\%$), the resistance increases sharply over a very small strain range due to the rupture of the conductive coating layer on the skeleton of polymer foams. During the external compression process, the foam structure is deformed, and the bending of the porous skeleton leads to the opening and propagation of cracks, resulting in the fracture of the local conductive path and a sharp increase in the resistance. This "crack effect" resulted in an initial high GF.

In region II ($0.6 < \varepsilon < 70\%$), as the conductive foam is further compressed, the "contact effect" occurs at the same time as the "crack effect," which means that the adjacent struts of the foam substrate start to contact with each other which leads to the construction of a better conductive path and the reduction of resistance. In this region, the foam skeleton gradually transfers from a "point contact" mode to an "areal contact" mode due to the increase of the contact struts in the foam. In the "area contact" mode, the electrical network generally has reached a stable state, and the resistance will not change significantly with further compression. Thus, the resistance change signal showed a rapid decrease in the relatively small strain, and the decrease rate became slow when the strain exceeded 30%.

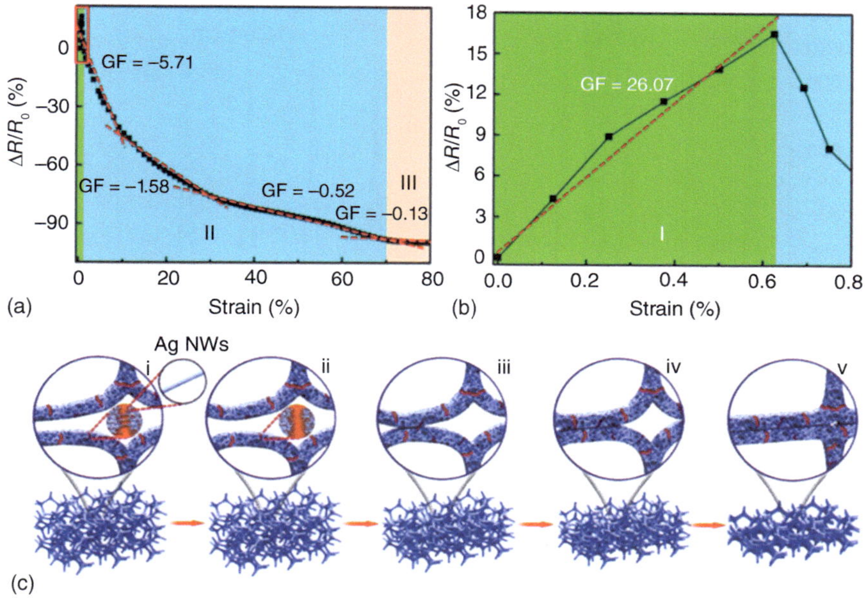

Figure 8.5 (a) $\Delta R/R_0$ results of conductive polymer foam with increased compressive strain, (b) Enlarged image of the rectangle in panel (a), and (c) schematic diagram of the evolution of porous structure with increasing compressive strain. Source: Zhang et al. [33]/with permission of American Chemical Society.

In region III ($70\% < \varepsilon < 80\%$), the "area contact" mode is dominant due to the densification of the porous sponge at high strains. At this point, the cells in the foam are mostly collapsed and almost all struts of the skeleton touch each other, thus, the tunnel distance between the conductive coating layers decreases with external compression, resulting in a decrease in the resistance change rate in large strain range [33].

8.2.3 The Influencing Factors of Foam-Based Piezoresistive Sensors

The microstructure of pressure sensors based on CPFs directly affects the resistance change, sensitivity, and stability of piezoresistive sensors. These factors generally include the homogeneity and interconnectivity of the foam structure, the cell size and thickness of the cell wall, and the surface microstructure of the active conductive materials.

8.2.3.1 Cell Size and Cell Wall Thickness of CPFs

The cell size and cell wall thickness are critical factors influencing the performance of CPFs. At low pressure, the sensitivity decreases first and then increases with the

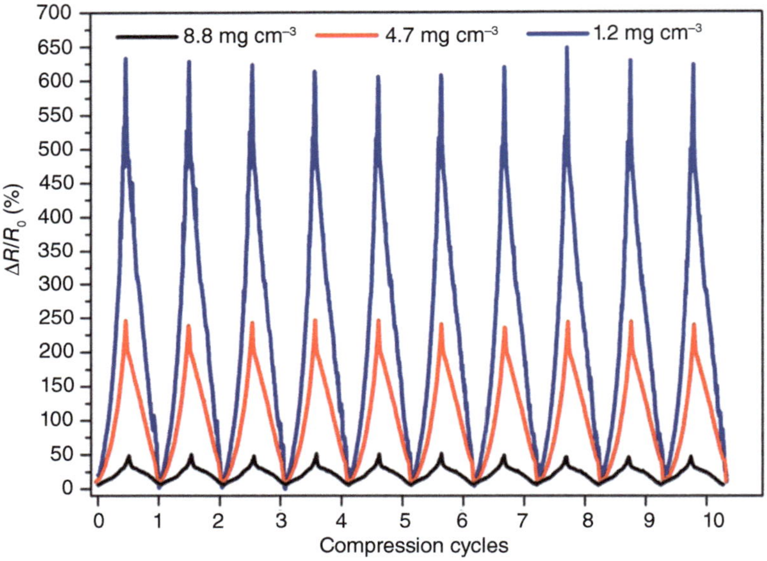

Figure 8.6 Compression resistance changes of three graphene foams with different densities. Source: Samad et al. [34]/with permission of John Wiley & Sons.

increase of cell size. At higher pressure, the sensitivity is usually unchanged with the increase of cell size.

The foam size and wall thickness have a great influence on the performance of the foam sensor. When the aperture is too small, the foam material may not deform sufficiently to accurately detect environmental changes, resulting in decreased sensitivity. On the contrary, if the aperture is too large, the foam may be too fragile and easily burst, affecting the stability and durability of the sensor. The cell wall thickness also affects the sensitivity and stability of the sensor. The thicker the cell wall is, the higher the mechanical strength and the lower the sensitivity of the sensor. The thinner the cell wall, the higher the sensitivity of the sensor and the lower the mechanical strength and stability. For example, Samad et al. [34] prepared an independent, mechanically stable, and highly conductive graphene foam through a simple two-step method, which can be used as a high-low strain/pressure strain/pressure sensor and has the advantage of adjustable density. This study shows that the sensitivity of the composite decreases with the increase in graphene foam density (Figure 8.6). This is because high-density Graphene foam has thicker cell walls than low-density Graphene foam and requires greater compression load to produce the same resistance change.

8.2.3.2 Surface Microstructure of Materials

The surface microstructure of CPFs is important for the performance of a piezoresistive sensor since the surface topography greatly influences the contact state and contact area of the foam in the cyclic deformation and recovery process. Therefore, the CPFs with a hierarchical and complex surface structure tend to possess superior sensitivity than the foam that has fewer surface microstructures.

Enhancing the surface roughness of sensing material is a prevalent approach to boost the sensing performance. Micrometer-sized patterns such as micro pyramids have been designed on the surface of sensing material. As shown in Figure 8.7, upon compression, the pyramid deforms to a more rectangular shape, which increases its contact area with the electrode [35]. Due to the high stress concentration at the tip of the pyramid structure, a large shape change would occur under a certain external pressure, which is responsible for a more significant resistance change compared with an unstructured surface [36]. Other microstructures such as microdomes and micron-semi-cylinders have also shown effective enhancement for the sensing performance. Compared with pyramid structures, microdomes show greater variation in contact area for a given applied pressure. The micron-semi-cylinders reported by Peng et al. [37] showed even greater pressure changes in contact areas than pyramid and microdome structures. By analyzing the Von Mises pressure distribution stress fields of three characteristic types, it was found that the stress levels experienced by pyramidal structures were almost an order of magnitude higher than those experienced by semicylindrical lines.

The simulation results revealed that the greater the contact area increased by structural deformation, the higher the sensitivity the sensor would have. Because the sensitivity of piezoresistive sensors is determined by the change in contact area rather than the change in distance between electrodes, the more the contact area changes under a given deformation strain or a given pressure, the greater the sensitivity of the sensor will be. Therefore, constructing hierarchical structures on the cell walls and struts of CPF-based sensors is a highly effective approach to enhance sensitivity. Approaches including etching, coating (dip coating, spray coating), deposition, *in situ* synthesis, and self-assembly have been used to generate a structured surface on the CPF-based sensors [38]. However, the long-term stability of these structures becomes a critical issue for the practical applications of these sensors since the physically coated or chemically grafted active conducting layers may be destroyed over time and detached from the substrate material.

8.2.3.3 Interconnections of CPFs

Foam interconnections can significantly affect the sensing performance of foam-based sensors, especially in applications such as pressure and strain sensing. The interconnections of a polymer foam refer to the percentage of open cells in the foam matrix.

By applying external pressure, the wall of the cell is stretched or bent, resulting in a highly conductive path within the abundant conductive region. Under a constant cell size and porosity, the foam with more interconnections usually possesses a higher surface area, which means that the skeleton of the foam would be more likely to contact each other under the same pressure when compared with a foam with fewer interconnections. Therefore, increasing the percentage of open cells is a feasible way to enhance the sensitivity of CPF-based sensors.

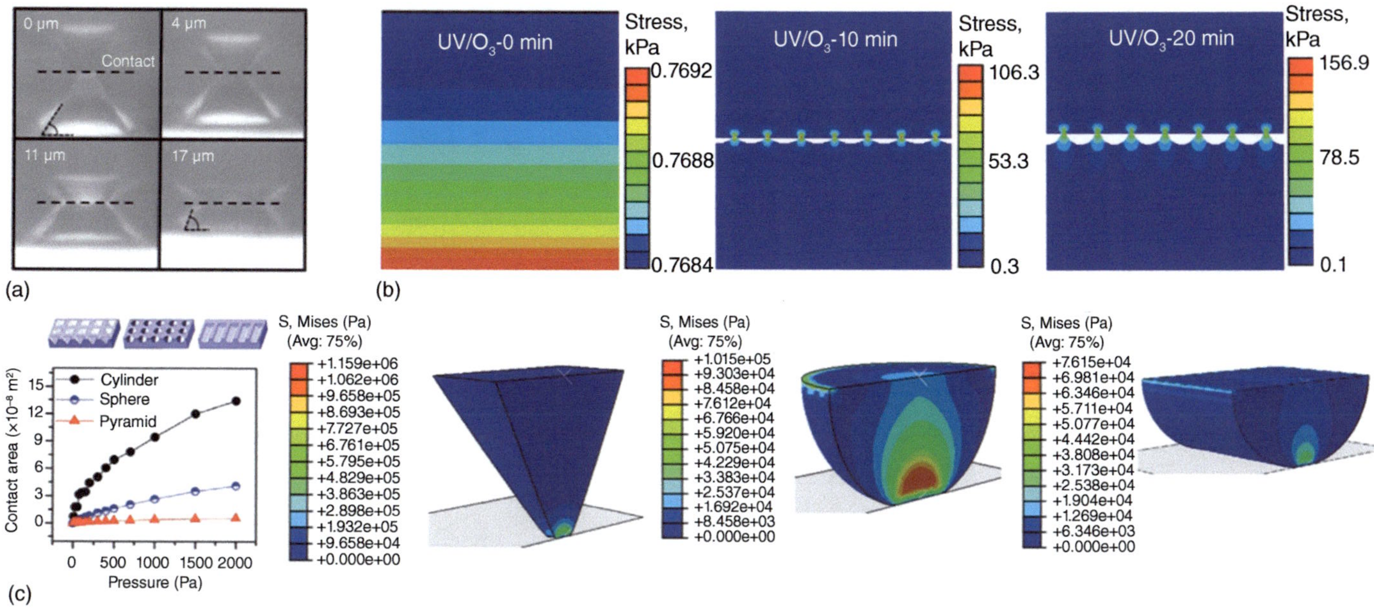

Figure 8.7 (a) Optical side-view images during compression of the microstructures, (b) finite-element simulations that show the local stress distribution of samples with and without microstructured CNT/PDMS arrays under a pressure of 10 kPa, (c) finite element modeling of micropatterned structures of pyramids, domes, and semicylinders show that cylinders have the greatest change in contact area. Source: (a) Tee et al. [35]/with permission of John Wiley & Sons. (b) Yu et al. [36]/with permission of IOP Publishing. (c) Peng et al. [37]/with permission of John Wiley & Sons.

8.3 Performance and Characterization of Piezoresistive Sensors

8.3.1 Sensitivity

Sensitivity (S) refers to the slope of the relative change in resistance with the applied pressure and is a key performance parameter describing the ability of a pressure sensor to convert a pressure signal into a resistance signal. The formula for the sensitivity of a flexible piezoresistive pressure sensor is:

$$S = (\Delta R/R_0)/\Delta P \tag{8.2}$$

where ΔR is the change in resistance, R_0 is the initial resistance value without applied pressure, and ΔP is the applied pressure. Therefore, to increase the sensitivity, the $\Delta R/R_0$ ratio must be increased at a certain ΔP.

In addition, the sensitivity of a flexible sensor can be evaluated by the GF, which is expressed as the ratio between the relative change in resistance and the applied strain and is one of the most important evaluation criteria for sensors. GF is usually tested by a combination of universal material testing machine and electrochemical testing equipment (electric meter, electrochemical workstation). The sensitive factors of flexible piezoresistive sensors are commonly calculated using the following formula:

$$GF = \frac{\Delta R}{\varepsilon R_0} \varepsilon = \Delta L/L_0 \tag{8.3}$$

where R_0 is the initial resistance; R is the immediate resistance changing under strain; ε is the strain, which represents the ratio of ΔL to the original length (L_0). It reflects the response-ability of the sensor to the strain.

In some cases, when the relative change of resistance and strain do not show a regular linear relationship, the following formula is introduced to calculate GF under the real strain,

$$G_{FT} = \frac{\ln\left(\frac{R}{R_0}\right)}{\ln\left(\frac{L}{L_0}\right)} = (1 + 2v) + \frac{1}{\varepsilon_T} \ln\left(\frac{\rho}{\rho_0}\right) \tag{8.4}$$

where G_{FT} is the true sensitivity; v is the Poisson ratio of the sensor, ρ_0 is the initial resistivity of the sensor; ρ is the resistivity of the sensor at the true strain $\varepsilon_T = \ln(L/L_0)$. The sensitivity of sensors made of different materials often depends on the main factors such as the layered structure of the design material, and the structure and dimension of the conductive network (Figure 8.8).

8.3.2 Response Time and Recovery Time

Response time is often expressed as the time difference between strain and resistance change which refers to the sensing lag behavior of the sensor. Response time

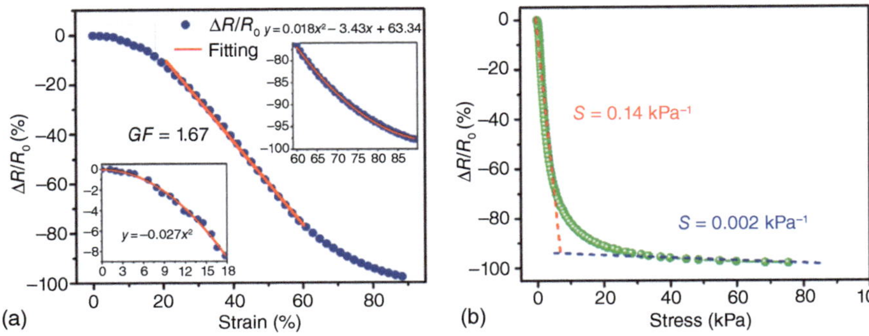

Figure 8.8 The Schematic diagram of (a) GF and (b) sensitivity with the increase in strain and pressure. Source: Liu et al. [39]/with permission of John Wiley & Sons.

is measured by the time required when the resistance signal of the flexible piezoresistive pressure sensor reaches 90% of the stable output value. Recovery time refers to the time required for the resistance signal of the flexible piezoresistive pressure sensor to change from the stable output value to the initial value after removing the applied pressure (Figure 8.9) [40]. Response time and recovery time are important parameters to evaluate the real-time dynamic pressure sensing performance of the sensor. The faster the signal response and recovery, the more accurate the detection would be. For polymer-based flexible sensors, the response and recovery time are related to the viscoelasticity of the substrate material and the active material, as well as the contact and separation rate between the active materials and between the active material and the electrode. In practice, the response and recovery time are closely related to the resolution of the device reading the resistance value of the sensor [41]. The sensor with a long response and recovery time cannot be used in cases where deformation occurs at a high frequency.

8.3.3 Hysteresis

During the cyclic sensing process, the output signal might have slightly different values under the same strain level in the loading and unloading cycles, and this difference is called hysteresis [42]. The electrical hysteresis (EH) of a strain sensor can be evaluated by the area under the stress–strain curves during its tension/release, as follows [43]:

$$EH(\%) = \frac{A_R - A_S}{A_S} \times 100\% \tag{8.5}$$

where A_S and A_R denote the area of the stress–strain curves at stretch and release, respectively.

Temperature and mechanical deformation are typically the main causes of hysteresis. Temperature lag is the output deviation of an input point before and after the

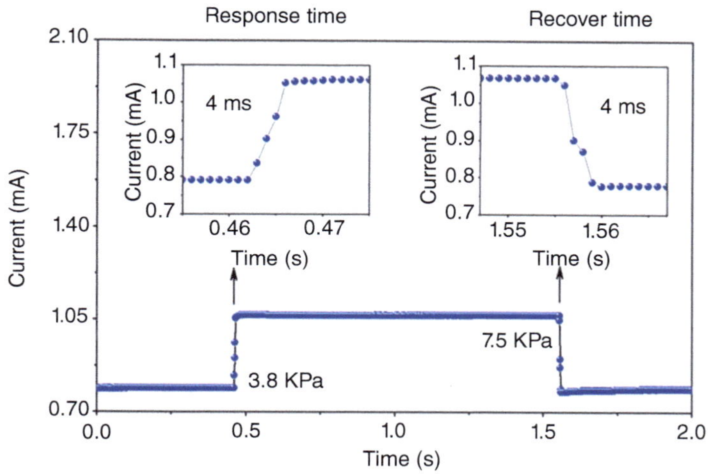

Figure 8.9 Diagram of response time and recovery time for a piezoresistive strain sensor. Source: Luo et al. [40]/with permission of John Wiley & Sons.

temperature change. Mechanical hysteresis may be due to the viscoelasticity of the substrate material, the unrecordable deformation of the active conducting materials, and the interfacial gap between the sensing material and the substrate.

The hysteresis is commonly observed for CPC-based piezoresistive sensors due to the hysteresis behavior of polymer materials in the process of stretching and releasing. Therefore, developing substrate materials with no or low hysteresis is an important task in the flexible strain sensing field. In practice, the flexible piezoresistive sensors are usually calibrated twice in the stretching and releasing periods to achieve more accurate sensing results (Figure 8.10).

Figure 8.10 The diagram shows the hysteresis for a piezoresistive strain sensor. Source: Xia et al. [43]/with permission of Elsevier.

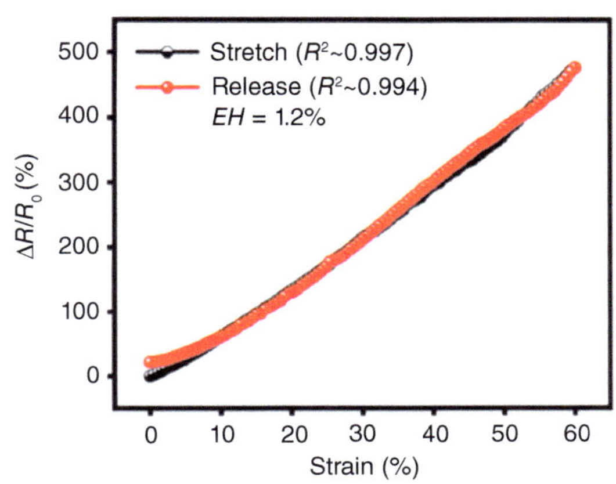

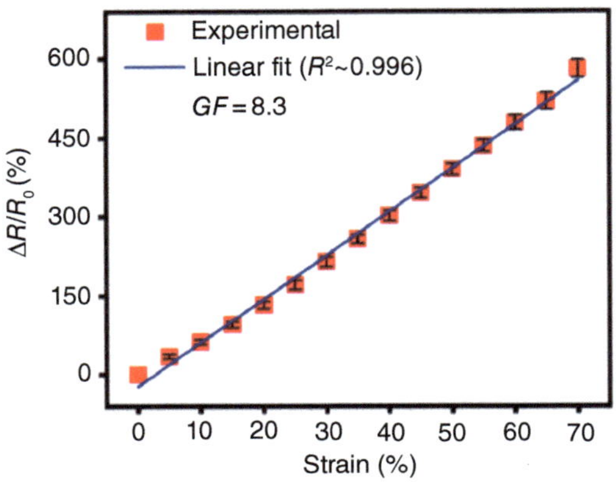

Figure 8.11 The diagram of linearity for a piezoresistive strain sensor. Source: Xia et al. [43]/with permission of Elsevier.

8.3.4 Linearity

Linearity refers to the ratio of the maximum deviation (ΔY_{max}) between the sensor calibration curve and the fitted line to the full-scale output (Y) under specified conditions. The smaller the value, the better the linearity [44]. Metal-based strain sensors usually have good linearity ($R^2 \approx 1$) since their measurement factors only depend on Poisson's ratio [45]. For semiconductor-based sensors, the resistivity is not proportional to the temperature, and the linearity is lower than metal [46]. Flexible CPC-based sensors usually have multiple linear regions, which is due to the rearrangement of the conductive network and the recombination of the sensor with the substrate material during the stretching or compression process. Therefore, the sensing curve of CPC-based sensors is typically divided into several regions with different linearity since the sensing mechanism might be different when the sensor is stretched or compressed to different strains. Generally, the sensing signal in the initial strain has high linearity because the internal structure of the sensor is the most stable within this range, the structure of the sensor is almost not damaged, and the sensor stability is the best (Figure 8.11).

8.3.5 Strain Response Range

Strain response range is an important index for sensor performance evaluation since it decides the application range of the sensor. Most sensors have excellent response within the normal strain range, but the sensor would not respond at a strain lower than its detection limit. When the strain is too high, the sensor may undergo permeate deformation and even failure, which causes damage to the sensor [47].

To be used as a piezoresistive sensor in various applications, ultra-low detection limit, and wide induction range are two basic requirements. The response range of a piezoresistive sensor is mainly decided by the physical properties of the substrate material and the active conducting material. The failure of the sensor comes from the failure of the substrate material and the active conducting network. To enhance the strain response range of a sensor, it is necessary to improve the elasticity of the substrate material and enhance the stretchability and adhesion of the active material with the substrate material.

8.3.6 Reliability and Stability

Reliability and stability are important indicators of the quality of the sensor, which is usually tested by cyclic application and release of pressure to the sensors. Reliability means that the sensor can provide correct measurement results without damage in multiple tests. Stability indicates that the sensor can maintain a stable sensing performance for a long period. These are the critical factors limiting the wide application of CPC-based flexible piezoresistive sensors. Generally, to test the stability of the sensor, it is necessary to carry out continuous loading–unloading test cycles for more than 10,000 cycles. The higher the number of stable loading–unloading cycles, the better the stability of the pressure sensor [48].

In long-term repetitive tests, the performance of flexible sensors may be influenced by environmental influence, nonmonotonic behavior, creeping of substrate materials, irreversibility of interface bonding between the active material and the substrate material, and permanent damage to the active layers [49]. Studies of stretchable strain sensors for polymer/nanomaterial composites have shown that instability appears to be common in CPC-based flexible piezoresistive sensors. A common phenomenon for flexible piezoresistive sensors is the drifting of the baseline in long-term measurement, which is associated with the creeping of the substrate material or the sliding of the active material relative to the substrate (Figure 8.12) [50].

8.4 Preparation of Flexible Sensors Based on Porous Foams

8.4.1 Physical Foaming

The most frequently used physical foaming approach in the fabrication of flexible CPF-based sensors is supercritical CO_2 ($scCO_2$) foaming. By dissolving $scCO_2$ in the polymer matrix under high pressure, the polymer matrix can be foamed via condition change (e.g. temperature rise or pressure drop) to trigger the phase change of $scCO_2$ from the supercritical state to the gas state. The $scCO_2$ foaming has many advantages, such as being environment friendly, no residue in polymer, nontoxic, noncombustible, low cost, high solubility in polymer, microcellular structure, and stable performance of the prepared foam products, but it also has some problems, such as harsh foaming environment (high temperature and high pressure), cell size variation in the cross-section and the solid skin layer.

Li et al. [51] constructed segregated composite foams using poly(ether block amide) (PEBA) and carbon nanostructure (CNS) by combining coating, heat compression molding, and supercritical CO_2 ($scCO_2$) foaming, as shown in Figure 8.13a. The segregated structure endowed the composite foam with better

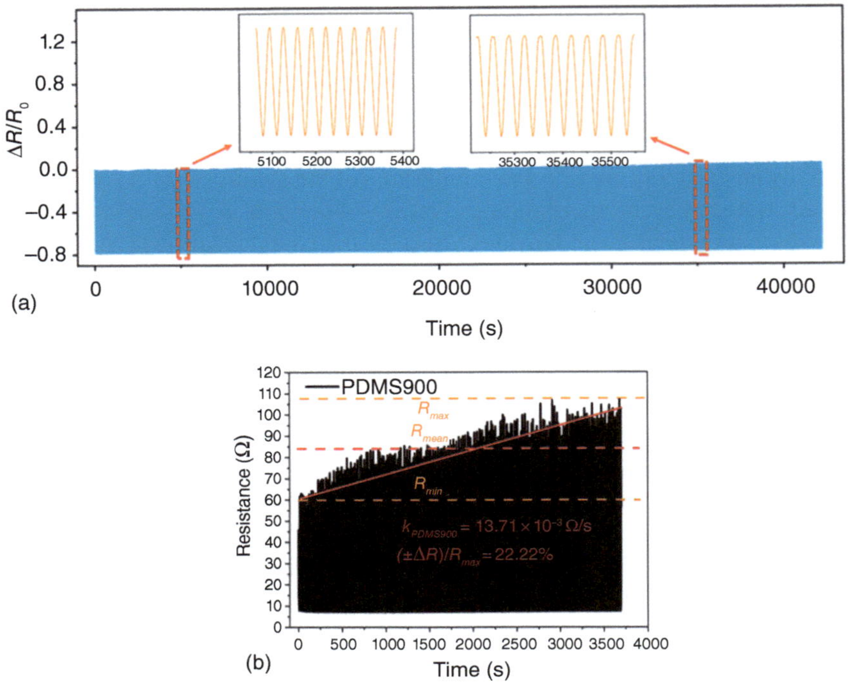

Figure 8.12 The diagrams show the: (a) operation stability [48], and (b) the baseline shift behavior of a piezoresistive strain sensor. Source: Wang et al. [50]/MDPI/CC BY 4.0.

conductivity and sensing capability compared with the randomly distributed structure. After $scCO_2$ foaming, the percolation threshold was further reduced, and the foam gained excellent compressibility. The segregated PEBA/CNS foam with optimum conductivity exhibited a high sensitivity of GF = 3.5, a large work range of 80% strain, high compression strength of ~0.5 MPa at 50% strain, and a stable, reproducible, and durable sensing response (Figure 8.13b). The composite foam demonstrated the ability to be used as a sensing function solely for monitoring human motions. In addition, Yang et al. [52] constructed a CPF with segregated conductive network by coating the regular-shaped poly(butylene adipate-co-terephthal-ate) (PBAT) beads with CNT nanoparticles through ball miling followed by $scCO_2$ foaming (Figure 8.13c). The obtained lightweight and flexible sensor foam showed uniform cell structures and a high conductivity of 6.5 S m^{-1} with a 3 wt% CNT loading, which rendered the foam high sensitivity, good stability, and long-term durability.

8.4.2 Chemical Foaming

Chemical foaming is a method using chemical blowing agents to generate microcells in the polymer matrix. The chemical blowing agent could be an alien reagent that decomposes at a certain temperature and releases gas. The foaming gas can also be

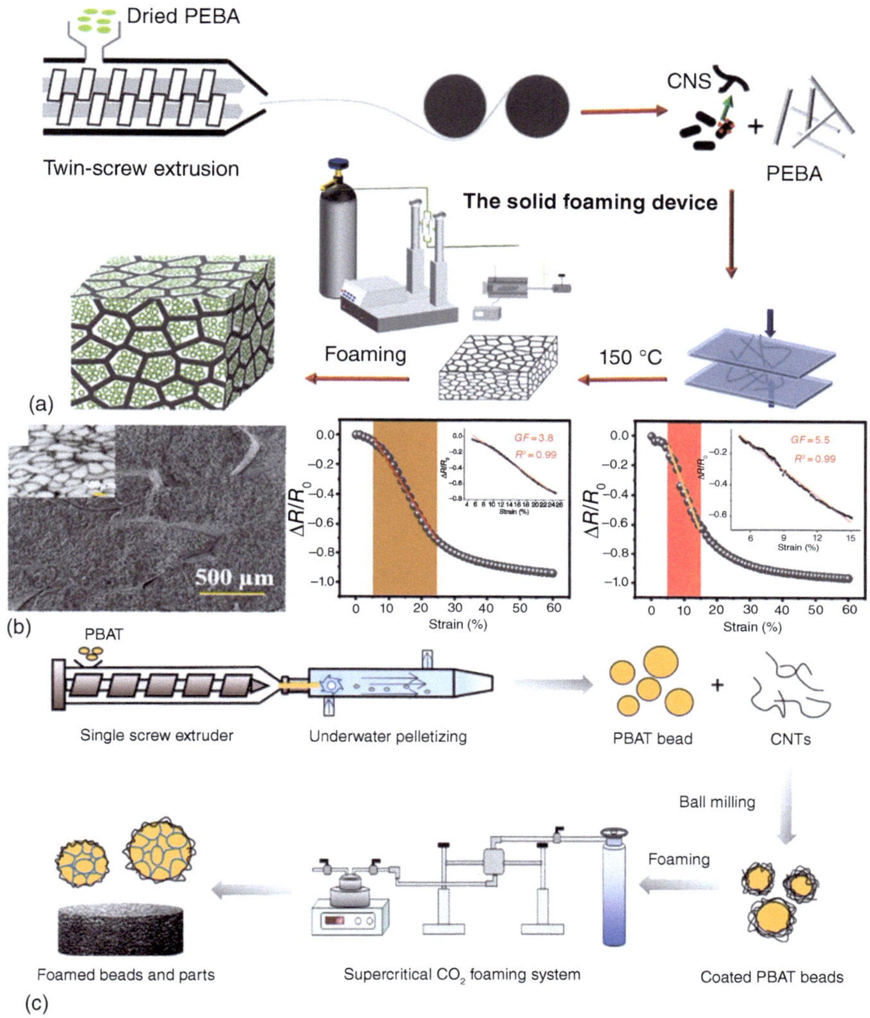

Figure 8.13 (a) Schematic diagram illustrating the preparation process for segregated PEBA/CNS composite foam, (b) $\Delta R/R_0$ as a function of compression strain for different composite foams, and (c) illustration of preparation process of PBAT/CNT composite foam.

generated by the reaction of the polymer matrix, such as the process of preparing PU foams. The chemical foaming method is simple and convenient, and the technology is mature, but it has some disadvantages, such as environmental pollution, toxic gas release, and low recycling ability.

For instance, Wang et al. [53] prepared porous graphene nanoparticles (GNP)/silicone rubber composites using a foaming method based on the decomposition of the chemical foaming agent, which allows the GNPs to be uniformly distributed within the silicone rubber matrix (Figure 8.14a). By using a patterned mold, the chemical-foamed sample obtained a truncated pyramid-shaped structure,

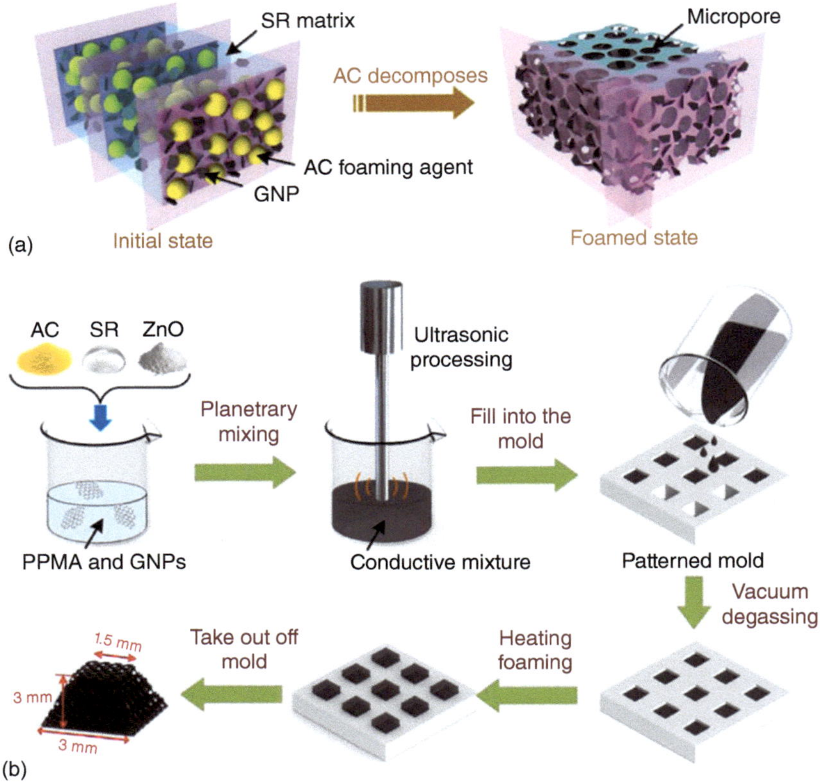

Figure 8.14 (a) Foaming principle of porous GNP/SR composite and (b) diagram of the fabrication process to prepare the truncated pyramid-shaped porous GNP/SR composites. Source: Wang et al. [53]/with permission of Elsevier.

which rendered the foam sensor a high sensitivity of 195.02 kPa^{-1} at 55–80 kPa for pressure sensing and can endure large strain over 60% (Figure 8.14b). The proposed tactile sensor also demonstrated the potential for human motion and health monitoring.

8.4.3 Freeze-Drying Method

Freeze drying is the process that combines the effect of thermally induced phase separation and the sublimation of solvent. In the process, the solvent phase in a polymer solution (i.e. water) is frozen at a low temperature, and then the solvent is removed under vacuum by the sublimation phase transition, which leaves the porous polymer phase. The obtained highly porous polymer foam is often called aerogel since the solution system usually goes through a hydrogel or solvogel state before being frozen, and the liquid phase in the gel is ultimately replaced by air in the final foam. Although the freeze-drying method can easily introduce various conductive components in the polymer matrix, preserve the chemical structure of

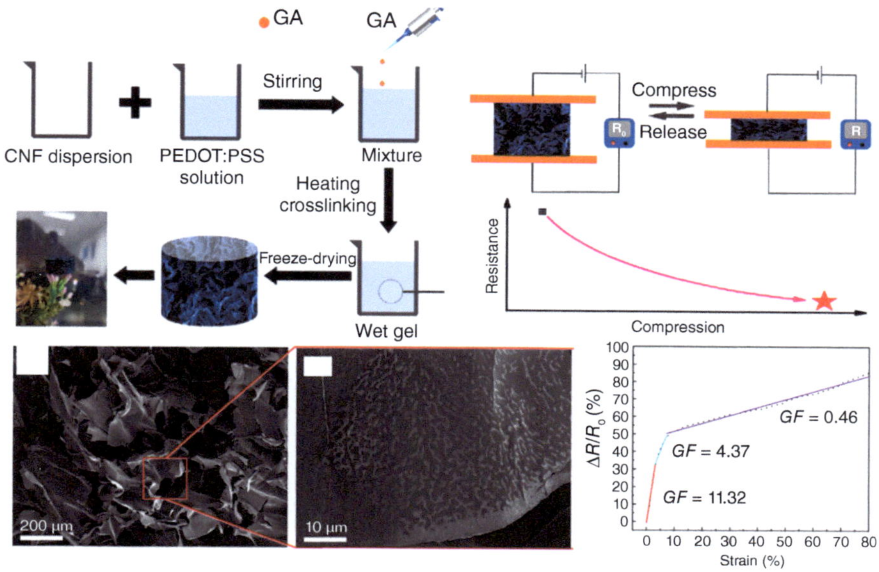

Figure 8.15 Schematic of the preparation of a CNF/PEDOT:PSS composite foam, the diagram of piezoresistive sensing mechanism, the microstructure and the GF results against compressive strain results of the composite foam. Source: Wu et al. [54]/with permisson of American Chemical Society.

the polymer, and maintain a high porosity, it is also associated with high costs and limitations in polymer species.

Wu et al. [54] fabricated cellulose-nanofiber (CNF)-based conductive composite foams by combing with PEDOT:PSS as the conductive filler and glutaraldehyde (GA) as the cross-linking agent using freeze drying. The developed foams showed excellent flexibility and piezoresistive properties with a high GF of 11.32 (Figure 8.15). The incorporation of GA enhanced the sensitivity from 10.65 to 15.18 kPa^{-1}, which makes the foam capable to detect various human motion behaviors. Wang et al. [55] reported the fabrication method of an anisotropic micro-honeycomb aerogel composed of cellulose nanofibers and CNTs. The micro-honeycomb aerogel contains unidirectionally aligned penetrating microchannels, which are prepared using a unidirectional freeze-drying method. Due to the obvious anisotropic microporous structure, the conductivity and Young's modulus of the sensor along the direction of the penetration microchannel are about twice that of the orthogonal direction. The sensor displayed varying sensitivities when subjected to different directions of strain, which allows the design of special "directional" strain sensors.

8.4.4 Coating of Commercial Foams

Dip coating of commercially available foams is a method of impregnating pristine foam in a solvent containing conductive fillers, either with or without the presence of a binder resin. The even coating of conductive filler on the surface and inside of

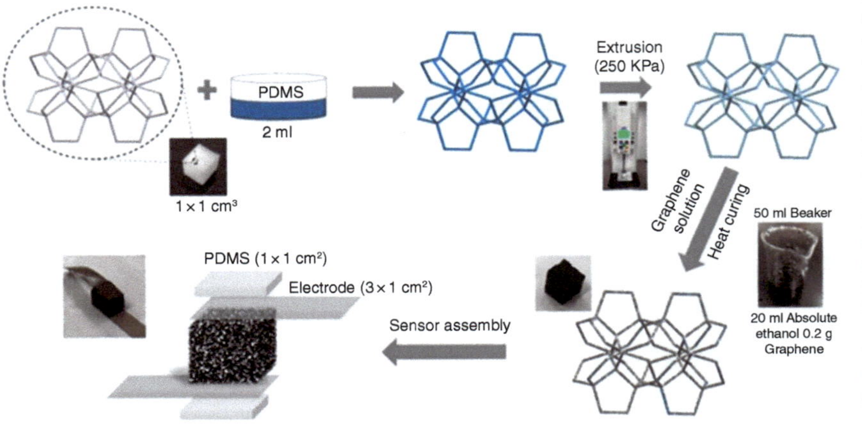

Figure 8.16 The dip coating fabrication process of piezoresistive sensors that are based on graphene-PDMS coated sponge. Source: Jing et al. [56]/with permission of IOP Publishing Ltd.

the foam skeleton renders the foam excellent conductivity and sensitivity when used in the application of pressure sensors due to the repetitive contact and separation of the coated conductive network upon pressing and releasing of the foam. Generally, the coating method is a simple and efficient approach to fabricating piezoresistive sensors with diverse structures and material compositions, while it usually suffers from the disadvantages of use of organic solvent and low stability of the coated nanoparticles under cyclical mechanical loading.

Jing et al. [56] prepared a piezoresistive sensor by coating a graphene-PDMS hybrid layer on a commercial sponge skeleton (Figure 8.16). Owing to the high elasticity of the sponge skeleton and PDMS coating and the conductivity of the graphene coating, the coated foam showed a highly elastic with a strain capacity of up to 85%, an outstanding sensitivity of 0.075 kPa^{-1}, a wide response range of 0–50 kPa, and a stable sensing performance in 2000 cycles. The piezoresistive sensor can be used to detect blood pressure, heart rate, and human movements, showing great potential applications in medical testing and electronic skin. Similarly, Xu et al. [57] improved the interaction between cellulose nanofibrils (CNFs)/CB inks and PU foam by introducing oxygen-containing groups on the foam's surface through low-pressure oxygen plasma treatment prior to the coating and drying process. The enhanced interfacial bonding rendered the composite foam a high sensitivity of 0.35 kPa^{-1}, a conductivity of 0.047 $S\,m^{-1}$, and good stability and durability in more than 1000 cycles. The conductive foam sensor could monitor several human activities such as breathing and swallowing in real time.

8.4.5 Skeleton Etching Method

The skeleton etching method is a method of using porous material as the skeleton that is coated with a conductive material or nanoparticles followed by removing the skeleton by a specific etching method to leave out a highly porous conductive frame.

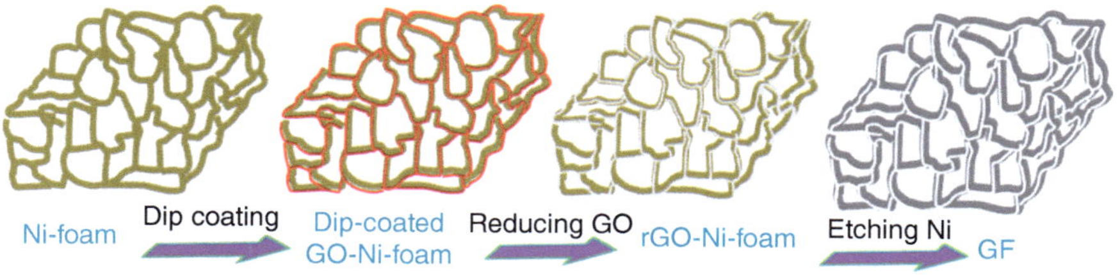

Figure 8.17 Schematic diagram showing the fabrication process of graphene foam using the skeleton etching method. Source: Samad et al. [58]/with permission of American Chemical Society.

The structure of the porous foam prepared by skeleton etching is highly controllable, and the foam usually has a low density and high surface area due to the hierarchical structure generated in the etching process. However, the approach is seldom used due to its high cost and limitation in the skeleton selection.

Samad et al. [58] imitated the nickel foam (Ni-F) structure into an independent 3D graphene foam by combining the dipping, *in situ* reduction, and skeleton etching methods (Figure 8.17). Then, a PDMS layer is coated on the graphene foam surface to obtain a novel graphene foam composite material with adjustable sensitivity. The obtained graphene foam-PDMS composites showed a high sensitivity of 60 kPa^{-1} and durability. When attached to human skin, the sensor can be used to measure human blood pressure.

8.4.6 Template Sacrificial Method

The template sacrificial method in the preparation of CPFs refers to the use of water-soluble or solvent-dissolvable porogen as a template to generate porous structures in a conductive polymer matrix. In the process, polymer melt or polymer solutions were mixed or infiltrated into the template aggregates to form a continuous phase. Numerous pores can be generated after removing the sacrificial porogens and the solvent phase via drying and leaching. The pore size, porosity, and pore morphology can be easily controlled by regulating the porogen property, and this approach is widely applicable to various polymers.

Wu et al. [59] used sacrificial template and coating methods to prepare conductive PDMS foams as piezoresistive sensors. As shown in Figure 8.18, NaCl particles were added to the PDMS precursor as porogen. After curing and leaching, a PDMS foam is obtained. By coating the porous PDMS with a conductive graphene layer, the foam gained piezoresistive sensing ability with a sensitivity of 17.9 kPa^{-1} in 0–30 kPa pressure load and 79 kPa^{-1} in 30–100 kPa pressure load. In addition, the conductive foam showed a fast response time of 20 ms and long-term work stability of more than 10,000 cycles, which makes it suitable for accurate and real-time monitoring of tiny physiological signals, vocalization, and facial emotions, as well as violent joint and limb movements. Similarly, Wan et al. [60] prepared porous conductive PU sponges by combining a sugar-templated sacrificial method. The PU solution is infiltrated into sugar cubes and then obtained PU sponges after leaching out the

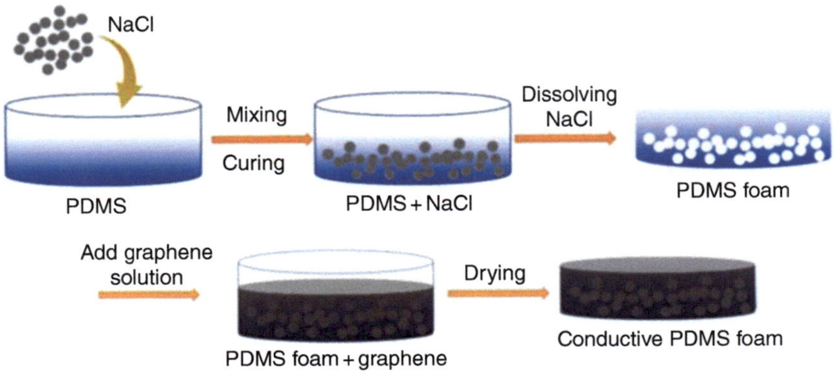

Figure 8.18 The template sacrificial fabrication process of piezoresistive sensors that are based on conductive PDMS foam using NaCl as the porogen. Source: Kaiqiang and Xingyang [59]/IOP Publishing, CC BY 4.0.

sugar particles. Then, the PU sponges are coated with a pyrrole layer via *in situ* polymerization, resulting in a uniform porous structure similar to the sugar cube templates and a unique microwrinkled structure of PPy. The as-prepared PPy–PU sponge sensors exhibited excellent sensitivity and stability over a force range from 0.1 to 200 N, which could accurately monitor both huge and tiny human motion in real time.

8.5 Applications of Flexible Piezoresistive Sensors

Flexible sensors are favored by people due to their mechanical flexibility, lightweight, and wearable advantages. Flexible piezoresistive pressure sensors are designed to improve sensitivity by considering the fusion between substrate and conductive materials, as well as the various microstructures that can be integrated into their simple structure. They are widely used in areas such as life, industry, and medicine due to their low cost and ease of preparation. The flexibility of these sensors is a key feature, allowing them to bend without losing their functionality. The flexible piezoresistive pressure sensor can also be applied to robots to mimic real skin and give them tactile perception. With the rapid development of IoTs technology and artificial intelligence, there is an increasing demand for flexible piezoresistive pressure sensors in the industrial field. This allows for the monitoring and precise control of equipment through mechanical signal sensing, electrical signal processing, and mechanical signal feedback.

8.5.1 Human Motion Monitoring

Human skin is soft, elastic, and flexible. Wearable devices that monitor health, such as breathing, pulse, blood pressure, pronunciation, and limb movement, can be attached to the skin surface due to their compatibility with human skin. To detect pressure, the sensor must have high sensitivity as the vibration signal of the epidermis is extremely weak.

For example, Yang and coworkers [61] proposed a pressure sensor made of graphene-modified PU foam. They designed a fractal electrode configuration, which enlarged the resistance change upon compression. Assembled with serrated electrodes, the sensitivity of the foam reached 134.75 kPa^{-1} in a wide linear range of 0–35 kPa. The response and recovery times of the foam sensor were 83 and 166 ms, respectively. Owing to its high sensitivity and flexibility, this sensor is capable of detecting subtle human movements, such as cheek puckering, head nodding, finger movement, and pulse detection. Sengupta et al. [62] prepared conductive porous PDMS foams by coating PDMS foams with six layers of graphene via solution coating and drying. The graphene–PDMS foam is used as a piezoresistive sensor and showed a GF of 8.77 over a strain range of 0–50%. The sensor is capable of monitoring the complete human gait motion. By integrating these sensors into the sole and conducting real-time gait experiments, the sensor can distinguish between various types of human gait, including walking and running and identify potential fall conditions (Figure 8.19).

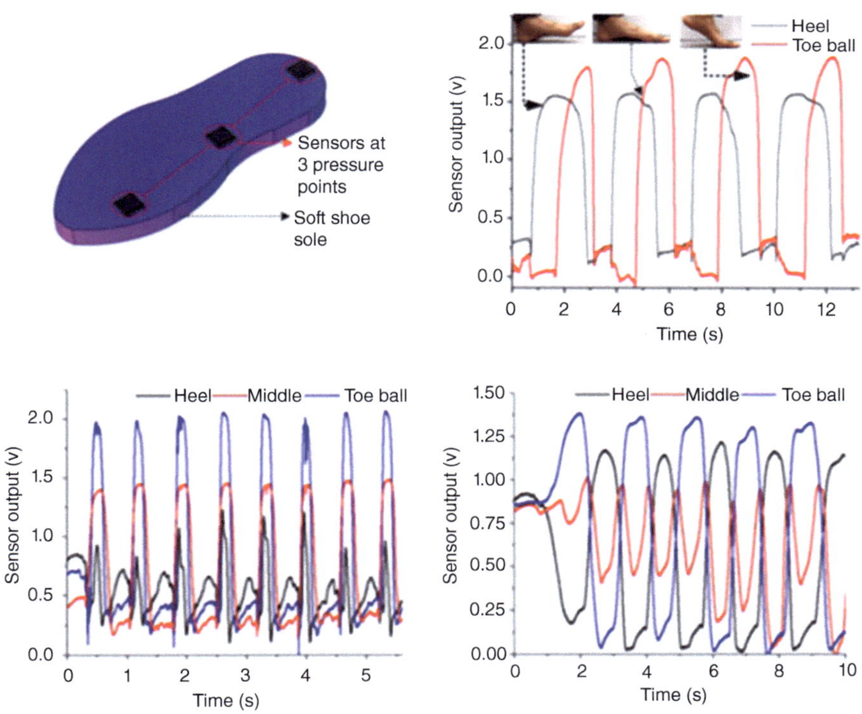

Figure 8.19 Application of graphene-PDMS foam sensors in human gait monitoring when assembled in a soft shoe sole. Source: Sengupta et al. [62]/with permission of American Chemical Society.

8.5.2 Tactile Perception

As the largest organ in the human body, the skin plays an important role in regulating the interaction between humans and the environment. In particular, the fingertip skin is more sensitive to external stimuli than other parts of the human skin, and the microstructure on the fingertip skin allows humans to discriminate texture roughness through friction signals between fingertips and objects, so the fingertip skin plays an important role in human perception of external stimuli. In the development of anthropomorphic robots and tactile-sensitive mechanical arms, tactile perception is a grand challenge. Highly sensitive piezoresistive pressure sensors offer a feasible approach to facilitate the development of tactile sensing units.

Inspired by the microstructures of fingertip skin, Xu and coworkers [63] designed a highly sensitive skin sensor with hierarchical microstructures on the surface (Figure 8.20a). By optimizing the porosity and graphite concentration, the fabricated skin sensor device exhibits a superior sensitivity of 245 kPa^{-1} over a wide linear pressure range from 5 Pa to 120 kPa and excellent response and recovery times (8 and 4 ms). Due to the high sensitivity of the tactile sensor, it can be used to distinguish the texture of different objects such as printing paper, and different sandpapers (Figure 8.20b). With the increase of object roughness, the current measured was increased obviously (Figure 8.20c). In addition, the sensor could be used to simulate the human figure tips when mounted on a robot arm, as shown in Figure 8.20d. It was able to measure the grasping force when taking up and release of different weights.

Huang et al. [64] proposed an auxetic PU foamed-based piezoresistive sensor, which was prepared by triaxial compression and surface coating of CB. The re-entrant structure of the foam rendered the sensor a GF of over 4.62 in a wide range of strain from 40% compressive strain to 80% tensile strain. The sensor demonstrated excellent stability of over 5000 repeating cycles and the ability to be used to detect the tactile pressure on a bionic hand (Figure 8.21). With perception data collected by the sensors mounted on the fingertips and palms, a generic method for real-time detection of unstable robotic grasping was established, which demonstrated the great value of the sensors and the detection approach in robotic perception and dexterous manipulation.

8.5.3 Temperature Monitoring

Changes in human body temperature can reflect human activity and health conditions and help maintain the thermal balance between the body and its environment. Temperature sensors have great potential to assist in disease diagnosis and medical health monitoring applications. The detection range of temperature sensors used to measure human body temperature is generally between 30 and 50 °C. Currently, temperature sensors commonly used in daily life include mercury temperature sensors and infrared temperature sensors, but these sensors are human unfriendly and inconvenient to use as wearable devices. Therefore, the development

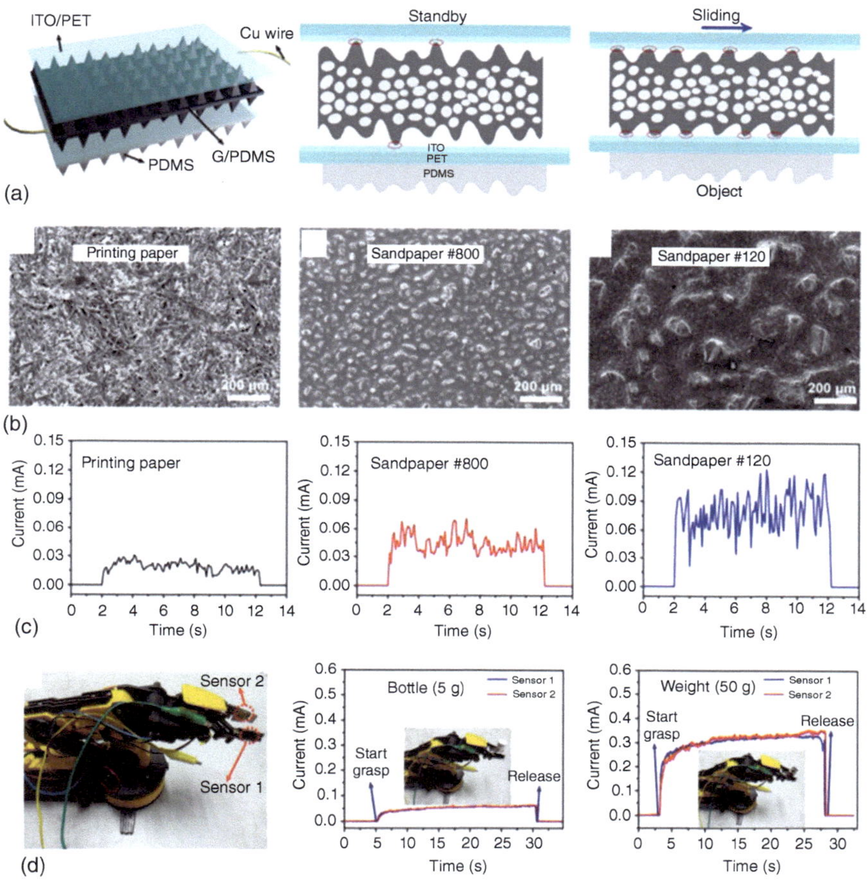

Figure 8.20 (a) Schematic illustration of the skin sensor for texture roughness recognition, (b) SEM images of the surface morphology of the test objects (printing paper, sandpaper #800, and sandpaper #120), (c) Current signals in response to various texture rough ness, and (d) two skin sensors attached to a robotic arm detecting the applied pressure in the process of grasping and releasing. Source: Sun et al. [63]/with permission of John Wiley & Sons.

of flexible temperature sensors with attributes such as comfort, lightweight, close fit on the human skin surface, and continuous monitoring of the human body temperature are highly desired. Piezoresistive sensors are also capable of detecting temperature change by incorporating thermal response materials, which could induce the change in resistivity of the sensor when temperature changes.

For example, Cui et al. [65] developed conductive CB/chlorinated poly (propylene carbonate) (CB/CPPC) foams with lightweight and cycle stability via the combination of facile melt blending and the subsequent chemical foaming process. The electrical percolation and the average cell size of the foam were 0.138 vol% and 0.26 ± 0.05 mm, respectively. As shown in Figure 8.22a, the CB/CPPC foams exhibited a nearly linear negative temperature coefficient effect of resistance due

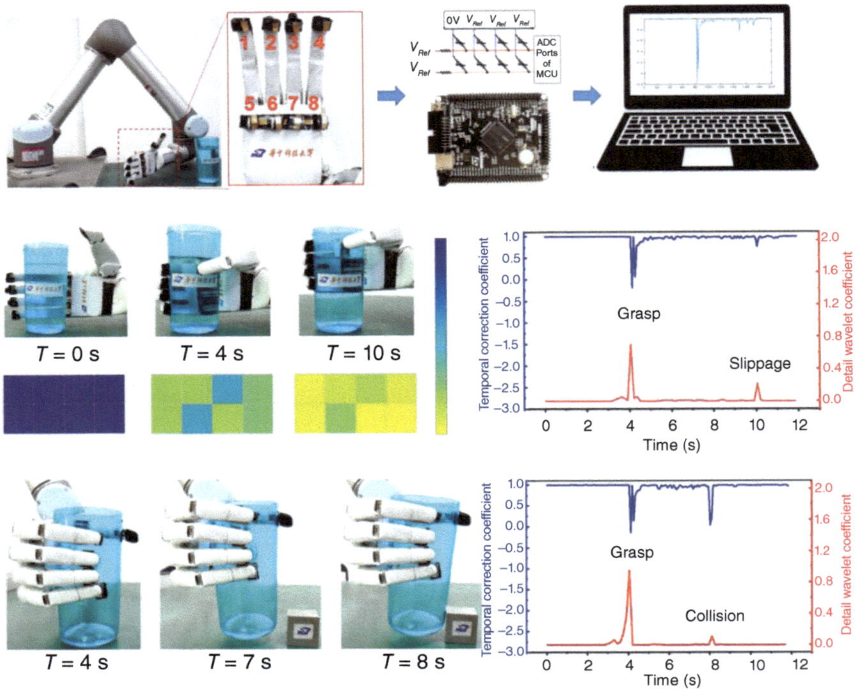

Figure 8.21 The application of the CB-coated auxetic PU-foam-based pressure sensors for detecting the grasping state of the bionic hand. Source: Huang et al. [64]/with permission of John Wiley & Sons.

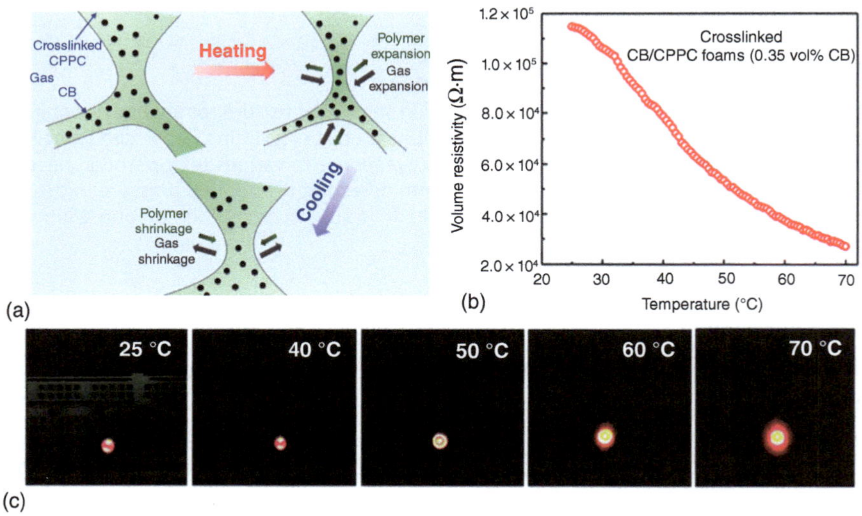

Figure 8.22 (a) Schematic illustration of microstructural changes in CB/CPPC foams during the heating process, (b) the volume resistivity of CB/CPPC foams as a function of temperature from 25 to 70 °C, (c) and the brightness of LED lamp changes as the foam (0.35 vol% CB) is heated. Source: Cui et al. [65]/with permission of Royal Society of Chemistry.

to the repetitive expansion of the polymer matrix and the gas in the cells. With increasing temperature from 25 to 70 °C, the volume resistivity of the CB/CPPC foam was reduced from $1.15 \times 10^5\ \Omega{\cdot}m$ to $2.73 \times 10^4\ \Omega\,m$ (Figure 8.22b). The lamp became brighter and brighter with the increase in temperature (Figure 8.22c).

In addition, various sensing properties can be integrated into a single sensor to achieve close mimicking of human skin. Zhu and coworkers [66] developed a skin-inspired quadruple tactile sensor integrated into a robot hand to enable object recognition. The quadruple tactile sensor consists of a skin-inspired multilayer microstructure containing a top sensing layer, a PDMS layer, a porous layer, and a bottom layer. The sensor is capable of perceiving the thermal conductivity of a material and measuring contact pressure at the same time. When integrated into a robot hand, the sensors could sense object temperature and environment temperature simultaneously and independently. With the help of machine learning, the smart hand could precisely recognize different shapes, sizes, and materials in a diverse set of objects.

References

1 Han, S.T., Peng, H., Sun, Q. et al. (2017). An overview of the development of flexible sensors. *Advanced Materials* 29: 1700375.

2 Meng, J. and Li, Z. (2020). Schottky-contacted nanowire sensors. *Advanced Materials* 32: 2000130.

3 Yang, J., Luo, S., Zhou, X. et al. (2019). Flexible, tunable, and ultrasensitive capacitive pressure sensor with microconformal graphene electrodes. *ACS Applied Materials & Interfaces* 11: 14997–15006.

4 Guan, X., Wang, Z., Zhao, W. et al. (2020). Flexible piezoresistive sensors with wide-range pressure measurements based on a graded nest-like architecture. *ACS Applied Materials & Interfaces* 12: 26137–26144.

5 Cao, X., Liu, H., Cai, J. et al. (2021). Chinese ink coated melamine foam with Joule heating and photothermal effect for strain sensor and seawater desalination. *Composites Part A: Applied Science and Manufacturing* 149: 106535.

6 Yang, H., Shang, J.-C., Wang, W.-F. et al. (2022). Polyurethane sponges-based ultrasensitive pressure sensor via bioinspired microstructure generated by pre-strain strategy. *Composites Science and Technology* 221: 109308.

7 Zhao, X., Wang, W., Wang, Z. et al. (2020). Flexible PEDOT:PSS/polyimide aerogels with linearly responsive and stable properties for piezoresistive sensor applications. *Chemical Engineering Journal* 395: 125115.

8 Gong, S., Schwalb, W., Wang, Y. et al. (2014). A wearable and highly sensitive pressure sensor with ultrathin gold nanowires. *Nature Communications* 5: 3132.

9 Liu, H., Dong, M., Huang, W. et al. (2017). Lightweight conductive graphene/thermoplastic polyurethane foams with ultrahigh compressibility for piezoresistive sensing. *Journal of Materials Chemistry C* 5: 73–83.

10 Wang, X., Gu, Y., Xiong, Z. et al. (2013). Silk-molded flexible, ultrasensitive, and highly stable electronic skin for monitoring human physiological signals. *Advanced Materials* 26: 1336–1342.

11 Yao, H.B., Ge, J., Wang, C.F. et al. (2013). A flexible and highly pressure-sensitive graphene–polyurethane sponge based on fractured microstructure design. *Advanced Materials* 25: 6692–6698.

12 Lu, N., Lu, C., Yang, S., and Rogers, J. (2012). Highly sensitive skin-mountable strain gauges based entirely on elastomers. *Advanced Functional Materials* 22: 4044–4050.

13 Wang, Y., Yue, Y., Cheng, F. et al. (2022). Ti3C2Tx MXene-based flexible piezoresistive physical sensors. *ACS Nano* 16: 1734–1758.

14 Hwang, J., Walczak, R., Oschatz, M. et al. (2019). Micro-blooming: hierarchically porous nitrogen-doped carbon flowers derived from metal-organic mesocrystals. *Small* 15: 1901986.

15 Choi, T.Y., Hwang, B.-U., Kim, B.-Y. et al. (2017). Stretchable, transparent, and stretch-unresponsive capacitive touch sensor array with selectively patterned silver nanowires/reduced graphene oxide electrodes. *ACS Applied Materials & Interfaces* 9: 18022–18030.

16 He, W., Li, G., Zhang, S. et al. (2015). Polypyrrole/silver coaxial nanowire aero-sponges for temperature-independent stress sensing and stress-triggered Joule heating. *ACS Nano* 9: 4244–4251.

17 Bhattacharjee, M., Soni, M., Escobedo, P., and Dahiya, R. (2020). PEDOT:PSS microchannel-based highly sensitive stretchable strain sensor. *Advanced Electronic Materials* 6: 2000445.

18 Zhou, Z., Zhang, X., Wu, X., and Lu, C. (2016). Self-stabilized polyaniline@ graphene aqueous colloids for the construction of assembled conductive network in rubber matrix and its chemical sensing application. *Composites Science and Technology* 125: 1–8.

19 Tewari, A., Gandla, S., Bohm, S. et al. (2018). Highly exfoliated MWNT-rGO ink-wrapped polyurethane foam for piezoresistive pressure sensor applications. *ACS Applied Materials & Interfaces* 10: 5185–5195.

20 Wang, H., Zhou, R., Li, D. et al. (2021). High-performance foam-shaped strain sensor based on carbon nanotubes and Ti3C2Tx MXene for the monitoring of human activities. *ACS Nano* 15: 9690–9700.

21 Chen, J., Jiang, W.-j., Zeng, Z. et al. (2023). Multifunctional shape memory foam composites integrated with tunable electromagnetic interference shielding and sensing. *Chemical Engineering Journal* 466: 143373.

22 Hammock, M.L., Chortos, A., Tee, B.C.K. et al. (2013). 25th anniversary article: the evolution of electronic skin (e-skin): a brief history, design considerations, and recent progress. *Advanced Materials* 25: 5997–6038.

23 Lipomi, D.J., Lee, J.A., Vosgueritchian, M. et al. (2012). Electronic properties of transparent conductive films of PEDOT:PSS on stretchable substrates. *Chemistry of Materials* 24: 373–382.

24 Choi, J., Kwon, D., Kim, K. et al. (2019). Synergetic effect of porous elastomer and percolation of carbon nanotube filler toward high performance capacitive pressure sensors. *ACS Applied Materials & Interfaces* 12: 1698–1706.

25 Wang, M., Gurunathan, R., Imasato, K. et al. (2018). A Percolation model for piezoresistivity in conductor–polymer composites. *Advanced Theory and Simulations* 2: 1800125.

26 Hu, N., Karube, Y., Yan, C. et al. (2008). Tunneling effect in a polymer/carbon nanotube nanocomposite strain sensor. *Acta Materialia* 56: 2929–2936.

27 Amjadi, M., Pichitpajongkit, A., Lee, S. et al. (2014). Highly stretchable and sensitive strain sensor based on silver nanowire–elastomer nanocomposite. *ACS Nano* 8: 5154–5163.

28 Yang, H., Yuan, L., Yao, X., and Fang, D. (2020). Piezoresistive response of graphene rubber composites considering the tunneling effect. *Journal of the Mechanics and Physics of Solids* 139: 103943.

29 Hamedi, M. and Atashparva, M. (2017). A review of electrical contact resistance modeling in resistance spot welding. *Welding in the World* 61: 269–290.

30 Gao, L., Zhu, C., Li, L. et al. (2019). All paper-based flexible and wearable piezoresistive pressure sensor. *ACS Applied Materials & Interfaces* 11: 25034–25042.

31 Doshi, S.M. and Thostenson, E.T. (2018). Thin and flexible carbon nanotube-based pressure sensors with ultrawide sensing range. *ACS Sensors* 3: 1276–1282.

32 Timsit, R.S. (1999). Electrical contact resistance: properties of stationary interfaces. *IEEE Transactions on Components and Packaging Technologies* 22: 85–98.

33 Zhang, S., Liu, H., Yang, S. et al. (2019). Ultrasensitive and highly compressible piezoresistive sensor based on polyurethane sponge coated with a cracked cellulose nanofibril/silver nanowire layer. *ACS Applied Materials & Interfaces* 11: 10922–10932.

34 Samad, Y.A., Li, Y., Schiffer, A. et al. (2015). Graphene foam developed with a novel two-step technique for low and high strains and pressure-sensing applications. *Small* 11: 2380–2385.

35 Tee, B.C.K., Chortos, A., Dunn, R.R. et al. (2014). Tunable flexible pressure sensors using microstructured elastomer geometries for intuitive electronics. *Advanced Functional Materials* 24: 5427–5434.

36 Yu, G., Hu, J., Tan, J. et al. (2018). A wearable pressure sensor based on ultraviolet/ozone microstructured carbon nanotube/polydimethylsiloxane arrays for electronic skins. *Nanotechnology* 29: 115502.

37 Peng, S., Blanloeuil, P., Wu, S., and Wang, C.H. (2018). Rational design of ultrasensitive pressure sensors by tailoring microscopic features. *Advanced Materials Interfaces* 5: 1800403.

38 Li, T., Li, Y., and Zhang, T. (2019). Materials, Structures, and functions for flexible and stretchable biomimetic sensors. *Accounts of Chemical Research* 52: 288–296.

39 Liu, H., Chen, X., Zheng, Y. et al. (2021). Lightweight, superelastic, and hydrophobic polyimide nanofiber /MXene composite aerogel for wearable

piezoresistive sensor and oil/water separation applications. *Advanced Functional Materials* 31: 2008006.

40 Luo, N., Dai, W., Li, C. et al. (2015). Flexible piezoresistive sensor patch enabling ultralow power cuffless blood pressure measurement. *Advanced Functional Materials* 26: 1178–1187.

41 Mannsfeld, S.C.B., Tee, B.C.K., Stoltenberg, R.M. et al. (2010). Highly sensitive flexible pressure sensors with microstructured rubber dielectric layers. *Nature Materials* 9: 859–864.

42 Frutiger, A., Muth, J.T., Vogt, D.M. et al. (2015). Capacitive soft strain sensors via multicore–shell fiber printing. *Advanced Materials* 27: 2440–2446.

43 Xia, Q., Wang, S., Zhai, W. et al. (2021). Highly linear and low hysteresis porous strain sensor for wearable electronic skins. *Composites Communications* 26: 100809.

44 Cohen, D.J., Mitra, D., Peterson, K., and Maharbiz, M.M. (2012). A highly elastic, capacitive strain gauge based on percolating nanotube networks. *Nano Letters* 12: 1821–1825.

45 Cao, C.-F., Zhang, G.-D., Zhao, L. et al. (2019). Design of mechanically stable, electrically conductive and highly hydrophobic three-dimensional graphene nanoribbon composites by modulating the interconnected network on polymer foam skeleton. *Composites Science and Technology* 171: 162–170.

46 Dargahi, J. and Najarian, S. (2004). Analysis of a membrane-type polymeric-based tactile sensor for biomedical and medical robotic applications. *Sensors and Materials* 16: 25–41.

47 Yao, D., Wu, L., Peng, S. et al. (2021). Use of surface penetration technology to fabricate superhydrophobic multifunctional strain sensors with an ultrawide sensing range. *ACS Applied Materials & Interfaces* 13: 11284–11295.

48 Ren, X., Tian, Q., Zhu, X. et al. (2023). Multi-scale closure piezoresistive sensor with high sensitivity derived from polyurethane foam and polypyrrole nanofibers. *Chemical Engineering Journal* 474: 145926.

49 Lee, J., Lim, M., Yoon, J. et al. (2017). Transparent, flexible strain sensor based on a solution-processed carbon nanotube network. *ACS Applied Materials & Interfaces* 9: 26279–26285.

50 Wang, X., Gao, X., Wang, Y. et al. (2022). Development of high-sensitivity piezoresistive sensors based on highly breathable spacer fabric with TPU/PPy/PDA coating. *Polymers* 14: 859.

51 Li, X.Y., Wu, M.H., Ma, W.Y. et al. (2023). Development of eco-friendly and high-strength foam sensors based on segregated elastomer composites with a large work range and high sensitivity. *ACS Applied Materials & Interfaces* 15: 57613–57625.

52 Yang, J.B., Ding, M.Q., Cai, W.R. et al. (2023). Lightweight and flexible sensors based on environmental- friendly poly(butylene adipate-co-terephthalate) composite foams with porous segregated conductive networks. *Journal of Applied Polymer Science* 140.

53 Wang, Y., Chen, Z., Mei, D. et al. (2022). Highly sensitive and flexible tactile sensor with truncated pyramid-shaped porous graphene/silicone rubber composites for human motion detection. *Composites Science and Technology* 217: 109078.

54 Wu, Y., Yan, Z.H., Wang, T. et al. (2023). Cellulose nanofibers/PEDOT:PSS conductive aerogel for pressure sensing prepared by a facile freeze-drying method. *ACS Applied Polymer Materials* 5: 3938–3948.

55 Wang, C., Pan, Z.Z., Lv, W. et al. (2019). A directional strain sensor based on anisotropic microhoneycomb cellulose nanofiber-carbon nanotube hybrid aerogels prepared by unidirectional freeze drying. *Small* 15: 1805363.

56 Jing, Z., Zhang, Q., Cheng, Y. et al. (2020). Highly sensitive, reliable and flexible piezoresistive pressure sensors based on graphene-PDMS @ sponge. *Journal of Micromechanics and Microengineering* 30: 085012.

57 Xu, S., Li, X., Sui, G. et al. (2020). Plasma modification of PU foam for piezoresistive sensor with high sensitivity, mechanical properties and long-term stability. *Chemical Engineering Journal* 381: 122666.

58 Samad, Y.A., Li, Y., Alhassan, S.M., and Liao, K. (2015). Novel graphene foam composite with adjustable sensitivity for sensor applications. *ACS Applied Materials & Interfaces* 7: 9195–9202.

59 Kaiqiang, W. and Xingyang, L. (2020). Wearable pressure sensor for athletes' full-range motion signal monitoring. *Materials Research Express* 7: 105003.

60 Wan, Y., Qin, N., Wang, Y. et al. (2020). Sugar-templated conductive polyurethane-polypyrrole sponges for wide-range force sensing. *Chemical Engineering Journal* 383: 123103.

61 Yu, T., Zhang, D., Wu, Y. et al. (2021). Graphene foam pressure sensor based on fractal electrode with high sensitivity and wide linear range. *Carbon* 182: 497–505.

62 Sengupta, D., Pei, Y., and Kottapalli, A.G.P. (2019). Ultralightweight and 3D squeezable graphene-polydimethylsiloxane composite foams as piezoresistive sensors. *ACS Applied Materials & Interfaces* 11: 35201–35211.

63 Sun, Q.J., Zhao, X.H., Zhou, Y. et al. (2019). Fingertip-skin-inspired highly sensitive and multifunctional sensor with hierarchically structured conductive graphite/polydimethylsiloxane foams. *Advanced Functional Materials* 29: 1808829.

64 Huang, X., Guo, W., Liu, S. et al. (2022). Flexible mechanical metamaterials enabled electronic skin for real-time detection of unstable grasping in robotic manipulation. *Advanced Functional Materials* 32: 2109109.

65 Cui, X., Chen, J., Zhu, Y., and Jiang, W. (2018). Lightweight and conductive carbon black/chlorinated poly(propylene carbonate) foams with a remarkable negative temperature coefficient effect of resistance for temperature sensor applications. *Journal of Materials Chemistry C* 6: 9354–9362.

66 Li, G.Z., Liu, S.Q., Wang, L.Q., and Zhu, R. (2020). Skin-inspired quadruple tactile sensors integrated on a robot hand enable object recognition. Science. *Robotics* 5: eabc8134.

9

Triboelectric Nanogenerators Based on Polymer Foams

9.1 Overview of TENG

9.1.1 The Development of Energy Harvesting Nanogenerators

Energy is an important material basis for human activities and supports scientific and technological development and economic growth. However, the excessive use of fossil energy, including coal, oil, natural gas, and other resources, had a serious impact on human society and economic development [1, 2]. Moreover, the overuse of fossil energy has brought serious environmental pollution, such as air pollution, the greenhouse effect, and acid rain, which also resulted in the crisis of energy exhaustion due to overexploitation. Electricity is the core of modern human life and human industries. Thermal power generation is the most prevalent way of generating electricity at present, which is causing serious climate change.

To promote the sustainable development of human society, it is an urgent need to find green renewable energy to replace traditional fossil energy. Since the twenty-first century, the world has been paying great attention to the development of renewable and sustainable energy sources. As shown in Figure 9.1, geothermal energy, solar energy, biochemical energy, wind energy, hydraulic energy, and nuclear energy are the most frequently used renewable energy sources so far. Although these renewable energy sources solve the energy crisis to some extent, there are still many problems associated with them. For example, wind power is easily affected by many factors, such as season, direction, and even geographical location. Solar photovoltaic power generation is greatly affected by the light conditions during the day, the weather, and the seasons. The nuclear power plant has a risk of radiation pollution and the leakage of radioactive elements [3].

Relatively speaking, mechanical energy or heat energy in the environment is not affected by day and night, climate, season, and place, so it is applicable to be used in a wider range. Due to the low accessibility of geothermal energy, the ubiquitous mechanical energy could be a more feasible green energy source.

With the arrival of the fourth industrial revolution based on the Internet of Things, wireless sensor networks, and artificial intelligence, mobile electronic devices or microelectronic devices tend to develop in the miniaturized and portable direction. Although the power consumption of wearable and portable devices has been greatly

Functional Polymer Foams: Green Fabrication Methods, Performance and Applications,
First Edition. Haoyang Mi.

Figure 9.1 Various new renewable energy sources have frequently been used so far. Source: Wang et al. [3]/with permission of John Wiley & Sons.

reduced to a few watts, even at the mW or μW level, the total energy consumption is still huge from a worldwide macro perspective.

In recent years, tremendous efforts have been made to harness the weeny mechanical energy from the environment or the activities of human beings by utilizing novel micro-nano systems or devices that are composed of advanced micro-nanomaterials. It has been aroused into a new research branch of nano energy. Various technologies have been developed to efficiently capture and store energy in the environment, providing long-lasting, maintenance-free, and self-driven energy for different applications. Micro-nano energy is based on availability, conversion efficiency, and stability as the basic performance indicators. Among them, availability is the emphasis on the availability of collected energy sources in the specific working environment. At the same time, to maintain the long-term continuous operation of micro-nano systems, the stability of energy sources is also crucial [4, 5].

9.1.2 TENG and Critical Factors

TENG was first proposed by Prof. Wang Zhonglin in 2012 as a new nanogenerator, which is based on the triboelectrification effect and can effectively convert tiny mechanical energy that is common in our daily life into electricity [6]. TENG is a brand new energy conversion device after electromagnetic induction, piezoelectric, and electrostatic mechanical energy devices, which brings new opportunities for the efficient use of mechanical energy [7]. TENG is considered to be a promising energy capture device due to its unique advantages of high efficiency, low cost, high output rate, easy manufacturing, and rich material selection. The triboelectric materials used in TENG and the macroscopic and microscopic structures of the friction materials and device configuration are the most important factors affecting the performance of TENGs.

9.1.2.1 Triboelectric Materials

Since the energy of TENG originates from the potential difference between two dissimilar materials with different triboelectric properties, the materials used as the friction layers have a great influence on the performance of TENGs. There are typically two major factors deciding the performance of friction materials in TENG, which are the tribopolarity of the material and the microstructure of the material.

The tribopolarity of the material describes the ability of the material to lose or gain electrons when in contact with another material. The tribopolarity of different materials is ranked in a triboelectric series, in which the material located on the top of the list tends to lose electrons and exhibit a positively charged surface, while the material located on the bottom of the list tends to gain electrons and present a negatively charged surface [8, 9]. In addition to the tribopolarity, other characteristics of the friction material are often considered to meet specific application requirements, such as easy degradation, pollution-free, flexibility, abrasion resistance, etc.

9.1.2.2 Macroscopic and Microscopic Structures

The TENG has a simple working mode and a variety of material options. With reasonable structural design, TENG can not only realize efficient conversion of mechanical energy in the environment but also can be used to collect almost all forms of mechanical energy. TENGs generate energy based on the repetitive contact and separation of two friction materials with opposite tribopolarities. Hence, it is important to design a rational micro-device that realizes the repetitive contact and separation of two friction materials under the trigger of external forces. So far, there are generally four kinds of TENGs developed based on different working modes, namely, contact-separation mode, sliding mode, single-electrode mode, and independent layer mode. The different working modes and the TENG device design are recognized as the macroscopic structure influencing the TENG performance.

In addition, the microscopic structure of TENG materials also has a great influence on the TENG performance. The surface morphology of friction materials directly influences the surface roughness and the specific surface area of the friction layer, which are crucial for the triboelectric output performance. Typically, the material with a rougher surface would have higher charge density due to its greater surface area.

9.1.3 TENGs Based on Porous Materials

Increasing the surface roughness and effective contact area of the friction layer in the TENG has been proven to be an effective strategy for improving the output performance of TENG. Thus, various methods, including template patterning, laser lithography, surface etching, chemical grafting, and particulate coating, have been used to produce a complex surface.

As a new type of friction material, porous foams have shown promising application prospects in TENG development due to their advantages of lightweight, high specific surface area, and structural stability. They also exhibit superior charge generation ability and higher charge density due to the additional surface provided by the internal pores [10, 11]. Compared with traditional surface-engineered solid triboelectric material-based TENGs, a porous polymer as a triboelectric material enables TENG to have the advantages of stability, high flexibility, low cost, and high durability.

There are many techniques that have been used to prepare porous materials that are capable of being used in TENGs, such as electrospinning, phase separation/freeze drying, solution pouring/particle leaching, 3D printing, pyrolysis, chemical foaming, physical foaming, and supercritical CO_2 drying.

9.2 Fundamentals of TENG

9.2.1 Charge and Current Generation Mechanism

Static electricity is ubiquitous in nature. Different kinds of objects may produce a great amount of static electricity due to the friction between different materials. Triboelectric effect in daily life usually causes an explosion, fire, static shock or signal interference, and other negative effects, but at the same time, this effect can also bring great help for our life production, such as electrostatic dust removal, electrostatic copying, air purification, etc. Therefore, the exploration of potential positive use of the triboelectric effect is an ongoing pursuit for human beings.

The basic principle of TENG can be attributed to the coupling of triboelectrification and electrostatic induction [12]. Triboelectrification provides electrically polarized charge, and electrostatic induction is the main mechanism for converting mechanical energy into electric energy. Generally, when two materials are in contact with each other, corresponding chemical bonds are formed at certain parts of the interface, and corresponding charges are transferred between the interfaces to balance the electrochemical potential between materials, thus generating friction charges [13]. This effect is the origin of the triboelectrification mechanism. When the oppositely charged friction materials are moving away from each other, the potential difference causes electron redistribution in the current collector or electrode attached to the back of the friction materials. At this time, if the two electrodes attached to the dissimilar charged materials are connected via an external circuit, an instant current would pass through the circuit to balance the potential difference. When the two materials are moving toward each other again, an instant current in the opposite direction would be generated due to the elimination of the potential difference between the two electrodes. This process can be repeated by reciprocal external force so that the TENG can generate periodic current signals during operation.

In 2017, Wang Zhonglin proposed the fundamental theory of TENG based on the Maxwell displacement current [14]. Maxwell displacement current is a theory proposed by British scientist Maxwell in 1861, which is derived from Maxwell's equations and closely links the magnetic field and electric field. Therefore, Maxwell's equations become one of the theoretical sources of nanogenerators. Maxwell's displacement current is defined as:

$$J_D = \frac{\partial D}{\partial t} = \varepsilon \frac{\partial E}{\partial t} + \frac{\partial P_S}{\partial t} \tag{9.1}$$

where D is the displacement field, ε is the permittivity of the medium, E is the electric field, and P_S is the polarization field of the surface polarized charge generated by piezoelectric or triboelectric effects. The time-varying electric field in the formula is related to the origin of electromagnetic waves. The surface polarization field is the origin of triboelectric energy for TENGs. Specifically, in piezoelectric nanogenerators, this polarization field originates from the piezoelectric polarization charge generated by the strain. In the TENG, the polarization field is derived from the surface polarization charge formed by the static charge changing with the displacement after the triboelectrification of two contact materials.

As an example, a contact-separation mode TENG is shown in Figure 9.2a. The basic model of the device in this mode has two back electrodes connected to an external load and two dielectric layers for contact electrification. The surfaces of the dielectric layers have opposite charges due to the contact charging effect, and the surface charge density is $\pm\sigma_{tribo}$, which is saturated after the initial contact and is independent of the gap distance z. An electrostatic field caused by friction charges drives free electrons through an external load and between electrodes. Figure 9.2b shows the equivalent circuit model of TENG.

There are some functional relationships between the amount of charge transferred on the electrodes and the gap distance z generated by displacement. The corresponding displacement current generated by the mechanical energy of displacement variable z can be calculated as:

$$J_D = \frac{\partial D_Z}{\partial t} = \frac{\partial \sigma_{tr}(z,t)}{\partial t} \tag{9.2}$$

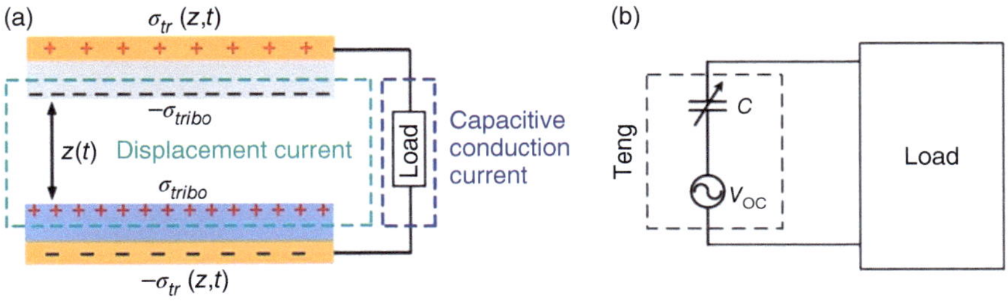

Figure 9.2 (a) Displacement current model of TENG in contact-separation mode, and (b) equivalent circuit model of triboelectric nanogenerator. Source: Wu et al. [15]/John Wiley & Sons.

The dielectric surface of TENG with opposite charge spacing variation can be regarded as a capacitor with capacitance variation. In this perspective, the capacitance model provides a more intuitive description for TENG. The current can be calculated by the following formula:

$$I = \frac{dQ}{dt} = A\frac{d\sigma_{tr}}{dt} \tag{9.3}$$

The result of the capacitance model is the same as Equation (9.2), which proves that Maxwell's displacement current is the basis of the capacitance model. Therefore, the corresponding TENG output voltage can be expressed as:

$$V = -\frac{1}{C(z)} \times Q + V_{oc}(z) \tag{9.4}$$

Hence, the capacitance model and Ohm's law are theoretical tools for the study, design, and optimization of the TENG.

9.2.2 Working Mode and Principle of TENG

According to the direction of polarity change and the contact modes of the TENG, four operating modes of TENG have been developed so far. According to the relative displacement can be divided into vertical contact mode (Figure 9.3a), horizontal sliding mode (Figure 9.3b), ingle-electrode mode (Figure 9.3c), and independent layer mode (Figure 9.3d) [16].

9.2.2.1 Vertical Contact-Separation Model

The vertical contact separation mode is the most frequently used operation mode for TENGs due to its simple setup design. The typical structure in vertical contact separation mode consists of two dielectric films with electrodes mounted on the back

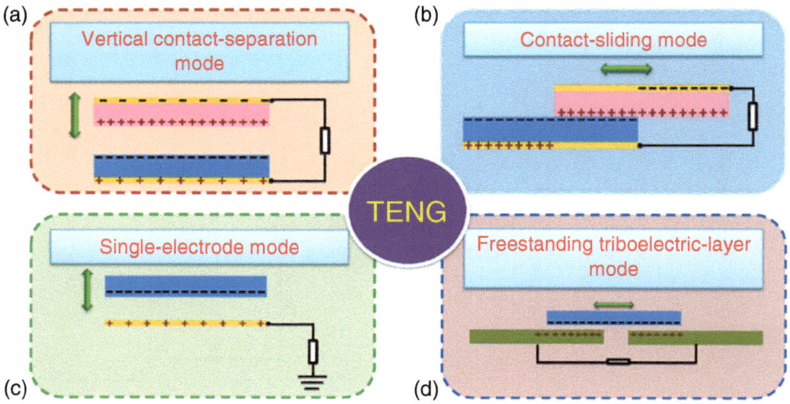

Figure 9.3 Four basic operating modes of triboelectric nanogenerators: (a) vertical contact-separation mode, (b) horizontal sliding mode, (c) single-electrode mode, and (d) independent layer mode. Source: Wang [16]/Royal Society of Chemistry.

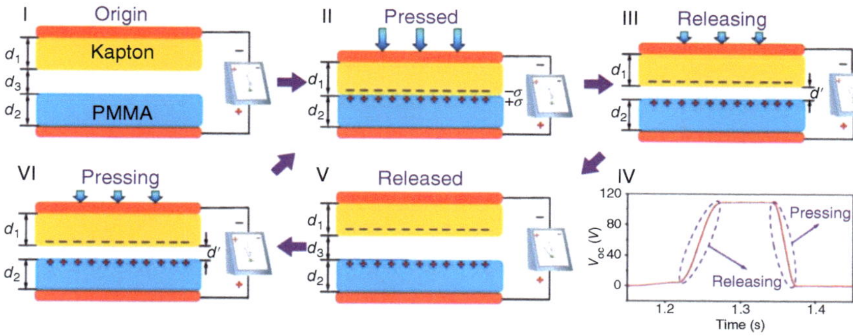

Figure 9.4 Working principle of triboelectric nanogenerator in vertical contact separation mode. Source: Zhu et al. [18]/American Chemical Society.

of them to collect the inducted charges. The two layers are separated via spacers or springs to create an air gap between them. When the external force is applied, the dielectric films of two different materials contact and friction each other, resulting in equal positive and negative charges generated on the surface of the dielectric film. When the surfaces of the two dielectric films are separated by the retraction of the flexible layers when the external force is removed, the generated charges on the positive and negative surfaces are also separated, which results in an induced potential difference between the two electrodes [17]. If the two electrodes are connected by an external circuit, the induced potential difference drives electrons between the electrodes, creating a reverse potential difference to balance the electrostatic field. When the two dielectric films come into contact again, the potential difference caused by the friction charge disappears and the electrons flow in the reverse direction causing a current flow in the opposite direction as illustrated in Figure 9.4 [18]. TENGs can produce periodic alternating current output under the action of periodic external force in the vertical contact separation mode. TENG with this working mode has the advantages of simple structure design, easy multilayer integration, and high instantaneous output power [19].

9.2.2.2 Horizontal Sliding Mode

The structure of the horizontal sliding mode is similar to that of the vertical contact-separation mode except for the elimination of the air gap since the contact and separation of the two friction layers were realized by the reciprocal sliding of the top material back and forth. As illustrated in Figure 9.5, in the initial state, the two friction layers contact each other, and electron migration occurs due to the contact electrification effect. When the external force is applied in the horizontal direction, the relative displacement of the two friction layers occurs, and more electrostatic charges are generated on the surfaces of the friction layers due to the relative displacement in the microscale, which causes enhanced contact force on the interface between the two materials [20]. At the same time, the charges accumulated on the friction layers induced the generation of electron redistribution on the electrodes in a similar way as the vertical contact-separation mode. Thus,

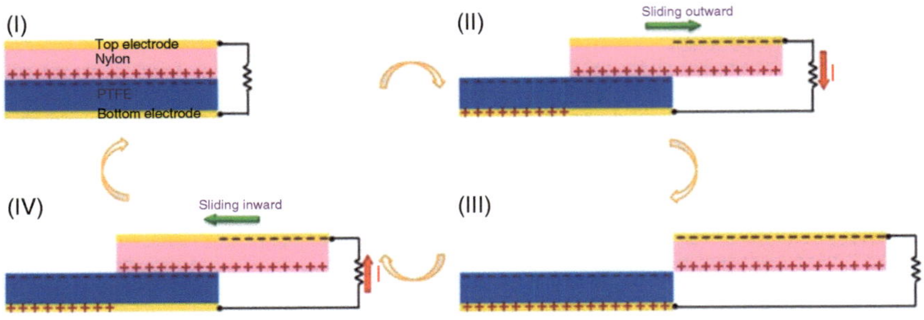

Figure 9.5 Working principle of the horizontal sliding mode TENG. Source: Wang et al. [20]/American Chemical Society.

the electrons can flow through the external circuit to form an instant current flow. When the two friction layers approach each other again, another instant current flow would be generated in the opposite direction due to the compensation of the potential difference between the two electrodes.

Therefore, the working principle of the horizontal sliding mode TENG is based on the coupling effect of sliding friction generation and electrostatic induction. Compared with TENGs in vertical contact-separation mode, the TENGs in horizontal sliding mode require no air gap but a relatively large moving distance in the horizontal direction. In addition, the TENG could normally recover to its initial state autonomously in vertical contact-separation mode, while it takes additional work to reset the horizontal sliding mode TENG to its initial state.

9.2.2.3 Single-Electrode Mode

One distinct characteristic of TENG is that it can generate energy using just one electrode. The working mechanism of the single electrode mode TENG is similar to that of the vertical contact-separation mode TENG. As illustrated in Figure 9.6, opposite charges are also generated on the surfaces of the friction layer and the free-moving friction material, and a potential difference still exists between the friction layer and the free-moving object. The charges on the surface of the friction layer would induce electrostatic induction charges on the electrode as well when the free-moving object is moved away from the friction layer, thus leading to the transfer of electrons between the single electrode and the ground. Different from the vertical contact-separation mode, the electrode in the single-electrode mode is grounded, thus the output performance would be about half of the conventional double electrodes TENG since the potential of the ground is zero. When the free-moving object contacts the friction layer again, the potential difference is compensated, and a current flow in the opposite direction appears in the external circuit.

Compared to conventional double-electrode TENGs, single-electrode TENGs have more flexibility in the material selection and device design. The overall setup is very simple and does not require redundant spacers, back supports, fasten tapes, and additional wires. However, it has the disadvantages of unstable output and relatively low power density. Based on these characteristics, single-electrode TENGs

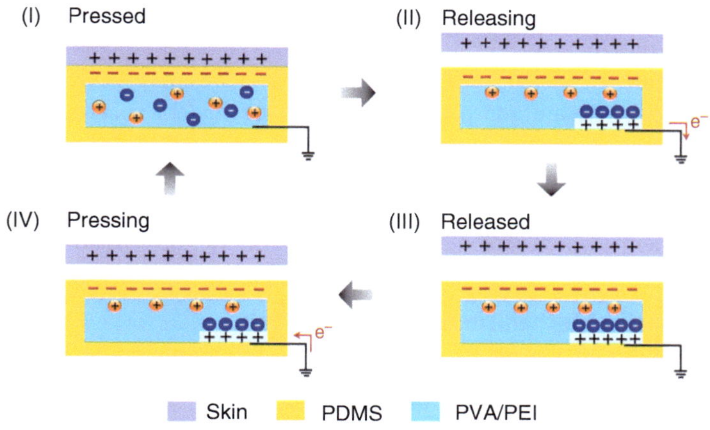

Figure 9.6 Working principle of TENG in single-electrode mode. Source: Wang and Daoud [21]/John Wiley & Sons.

can be used in specific applications that require thin and flexible devices, and touch detection situations that require less signal accuracy.

9.2.2.4 Independent Friction Layer Mode

The independent layer mode TENG is an ingenious design that usually contains an independent friction layer and a pair of fixed friction layers with electrodes attached to their back. Figure 9.7 illustrates the working mechanism of the independent layer mode TENG. The periodic movement of the friction layer between the two fixed friction layers causes the change of potential difference between the electrodes attached to them, which drive periodic electrons flows between the two electrodes through the external circuit [22]. When conductive materials like aluminum and copper sheets were used as the fixed friction layers and the electrodes at the same time in the independent layer mode, a potential difference would be generated

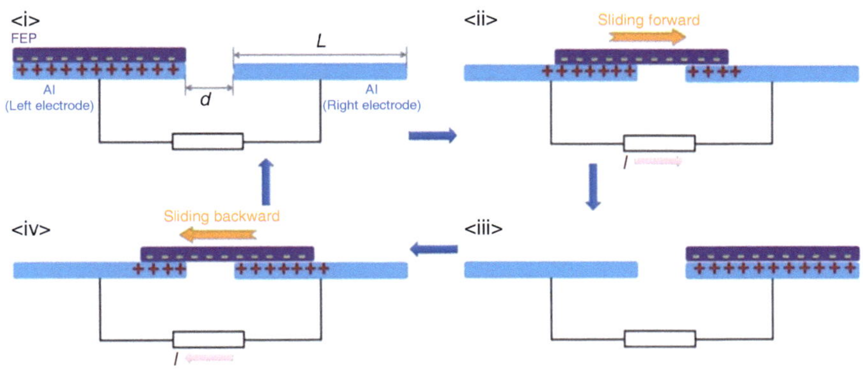

Figure 9.7 Working principle of triboelectric nanogenerator with the independent layer mode. Source: Wang et al. [22]/John Wiley & Sons.

between the two electrodes during the reciprocal movement of the negatively charged friction material.

Similar to the structure of the single-electrode mode TENG, the independent layer mode TENG is simple in design and production. The independent friction layer moves unrestricted periodically and is free from the influence of the electrostatic shielding effect. In addition, the friction layer in this mode does not need close contact, which greatly reduces the friction loss of the friction layer and the conversion of energy into heat energy, which prolongs the service life of the TENG device.

9.2.3 Power Generation Mechanism of Porous Materials

It is believed that the triboelectricity is produced by the coupling effect of contact charging and electrostatic induction for TENGs. Although the basic energy generation principles of TENGs with different modes have been revealed and the effect of surface microstructure on the triboelectric output performance has been elucidated, there are additional effects for the TENGs based on porous polymer foam, which would have a substantial influence on the performance of TENGs.

9.2.3.1 Charge Accumulation in the Porous Structure

Common friction materials generate charges on their micro-structured surface due to the contact electrification effect, and the surface with a more complicated hierarchical structure would possess a higher ability to generate more charges. In the case of friction layers made of porous materials, not only do the pores exposed on the surface affect the roughness of the material but the inner pores also assist the accumulation of charges due to the effect of contact charge transfer. As illustrated in Figure 9.8, the charges initially generated on the material surface would be transferred to the additional surfaces of the inner pores upon the compression of the porous material, which caused the transfer of charges to the inner cells. Thus, the

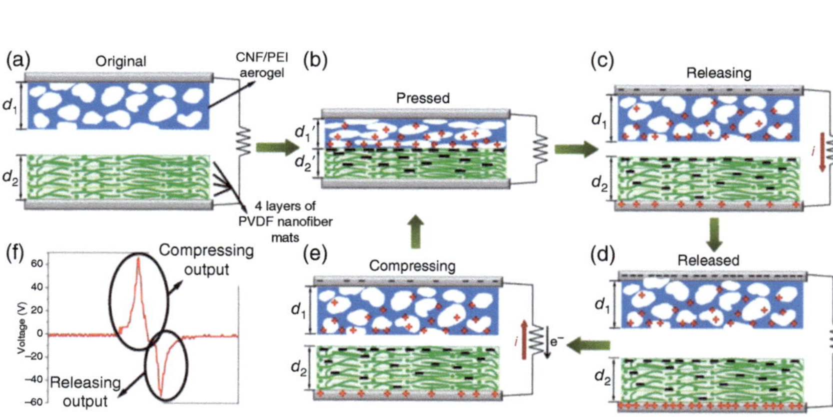

Figure 9.8 The tribological power generation schematic diagram of TENGs based on porous materials shows the charge accumulation mechanism during the cyclic compressing and releasing process. Source: Mi et al. [23]/with permission of Elsevier.

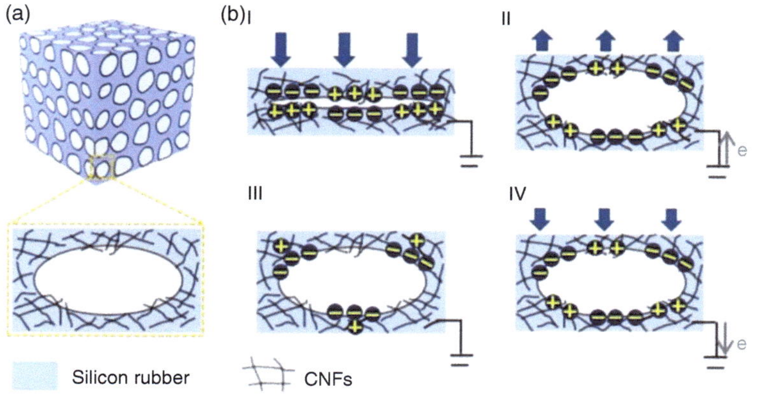

Figure 9.9 Schematic diagram of single-hole power generation process of conductive porous TENG showing the charge separation mechanism in a microcell. Source: Shao et al. [24]/with permission of Elsevier.

charges on the foam surface are reduced, and the surface could generate new charges in the next compression cycle. This charge accumulation phenomenon is frequently observed in TENGs, especially those composed of porous friction materials [23].

9.2.3.2 Charge Separation in the Micro Cells

When a conductive polymer foam is used as a friction material in a single electrode TENG, the micro-sized cells in the foam layer could influence the output performance of TENG significantly. As illustrated in Figure 9.9, the composite foam composed of silicon rubber and carbon nanofibers (CNFs) was used as a single-electrode TENG, and the entire sample consists of a large number of cells, each of which can be treated as a single TENG, so the output of the entire sample is the sum of all individual TENG. In the initial state, before the porous structure deforms, there is no charge transfer due to insufficient friction between the exposed CNF and the rubber. When the foam is compressed, the cells collapse, which results in the contact of the exposed CNF with the rubber cell wall. Then, both CNF and the rubber cell wall were charged oppositely. Due to the different electron affinity between CNF and silicone rubber, a micro potential difference is established on the two sides of a cell unit. Since all the cells were compressed in the same direction, the micro potential difference should be the same for all cell units, like electro dipoles. The sum of the micro potential difference caused the flow of electrons from the ground to the rubber/CNF foam TENG. When the foam is pressed again, the charges on the CNF and the rubber cell wall compensate each other and eliminate the micro potential difference, which results in the reverse flow of electrons [24]. Compared with traditional single-electrode TENG composed of solid friction material, the single-electrode TENG made of conductive polymer foam could utilize the charge separation mechanism on individual cells, which is beneficial for enhancing the output performance of the TENG.

9.3 Performance and Characterization of Polymer Foam-Based TENGs

9.3.1 Influencing Factors for Polymer Foam-Based TENGs

The output performance of TENG mainly depends on the tribopolarity, the microstructure of the friction layers, and the configuration of the TENG device. As porous materials are mainly used as the triboelectric materials of TENG, this section mainly focuses on the influencing factors for the triboelectric materials and structure factors.

9.3.1.1 Effect of Triboelectric Materials

Triboelectric Sequence (ability to gain and lose electrons) Since the charge density on the friction layer is a key factor that determines the output performance of TENGs, it is important to choose materials with distinctly different tribopolarity in TENG fabrication. The essence of triboelectrification is the behavior of materials to gain and lose electrons in the process of friction. In the TENG process, the material that loses electrons is positively charged, while the material that gains electrons is negatively charged. Therefore, triboelectric materials are divided into tribopositive material and tribonegative material according to their ability to gain and lose electrons. The triboelectric sequence of common materials was updated by Prof. Wang in 2013, as shown in Figure 9.10.

In practice, it is desired to choose tribopositive material from the top of the list and choose tribonegative material from the bottom of the list to achieve the largest tribopolarity difference which is expected to obtain the highest contact electrification charges.

Flexibility of Materials Flexible TENGs exhibit high sensitivity to impact force. For comfort use, flexible TENGs need to be able to generate electrical signals when subjected to palm press, stretch, and twist, which is not easily achieved by rigid triboelectric materials. When selecting materials for TENGs, it is important to consider their mechanical properties since the mechanical property of the friction layer not only decides the flexibility of the TENG device but also affects the contact tightness and the surface charge density during the TENG process.

Generally, soft material would achieve tighter contact with another material under compression force due to its elastic surface, which will usually lead to a greater charge density. In addition, flexible TENGs are often desired in the applications of wearable devices, portable power sources, and human motion sensors. Therefore, it is suggested to select soft triboelectric materials where it is possible.

In terms of flexibility, the porous structure provides a perfect approach to alter the flexibility of friction materials and the TENG device. A rigid material with high modulus will be flexible once it is processed into a thin film. For the same reason, many rigid materials can be flexible when they are prepared into highly porous membranes, which greatly broadens the applicable fields of rigid materials.

	Aniline-formol resin	Polyvinyl alcohol	
	Polyformaldehyde 1.3–1.4	Polyester (Dacron) (PET)	
	Etylcellulose	Polyisobutylene	
Positive	Polyamide 11	Polyuretane flexible sponge	
	Polyamide 6-6	Polyethylene terephthalate	
	Melanime formol	Polyvinyl butyral	
	Wool, knitted	Formo-phenolique, hardened	
	Silk, woven	Polychlorobutadiene	
	Polyethylene glycol succinate	Butadiene-acrylonitrile copolymer	
	Cellulose	Nature rubber	
	Cellulose acetate	Polyacrilonitrile	
	Polyethylene glycol adipate	Acrylonitrile-vinyl chloride	
	Polydiallyl phthalate	Polybisphenol carbonate	
	Cellulose (regenerated) sponge	Polychloroether	
	Cotton, woven	Polyvinylidine chloride (Saran)	
	Polyurethane elastomer	Poly(2,6-dimethyl polyphenyleneoxide)	
	Styrene-acrylonitrile copolymer	Polystyrene	
	Styrene-butadiene copolymer	Polyethylene	
	Wood	Polypropylene	
	Hard rubber	Polydiphenyl propane carbonate	
	Acetate, Rayon	Polyimide (Kapton)	
	Polymethyl methacrylate (Lucite)	Polyethylene terephthalate	Negative
	Polyvinyl alcohol	Polyvinyl chloride (PVC)	
	(continued)	Polytrifluorochloroethylene	
		Polytetrafluoroethylene (Teflon)	

Figure 9.10 The triboelectric sequence of common materials. Source: Wang [25]/American Chemical Society.

9.3.1.2 Effect of Microstructure

Surface Morphology and Roughness The surface morphology of porous material is another key factor influencing the output performance. Generally, the rougher surface with higher contact area the higher the charge density and the energy output for the same friction material. For porous materials, the surface morphology is an extension of their internal structures.

So far, there are typically three types of surface structures for porous materials, including concave cellular structure, convex humped structure, and fibrous web-like structure. Most polymer foams prepared by gas foaming, phase separation, and porogen sacrificial possess a concave cellular structured surface. Foams model using a negative template typically have convex humped structures on their surfaces. One example is replicating micro-nano features from other objects such as lotus leaves, fish skins, sandpapers, etched silicon wafers, etc. Porous materials with fibrous weblike surface structures are always fabricated via electrospinning, wet spinning, or melt spinning techniques. In addition, 3D printing as a versatile approach is capable of constructing porous materials with almost all types of surface morphologies. However, the size scale of the constructed features is mostly on a micro-scale. It is still difficult to produce nano-size features on a large scale via 3D printing so far.

Porous Structure and Surface Area to Volume Ratio The porous structure of the polymer foam would be influential for the TENG performance. Generally, porous materials with finner cellular structure and high porosity are more likely to generate higher charge density when paired with the same opposite triboelectric material. This is highly related to the specific surface area provided by the porous material per volume. Thus, the surface area to volume ratio is a characteristic parameter often investigated to evaluate the energy generation capability of a flexible porous material. By increasing the surface area to volume ratio, the material should have a greater effective contact area and be capable of accumulating more charges in the material.

In addition, compressive force and compression frequency are important operation factors that affect the performance of the TENGs. Normally, as the force increases, a higher electrical output would be achieved due to the increase of effective contact area under high-impact force.

9.3.2 Material Characterization for TENGs

9.3.2.1 Scanning Electronic Microscope (SEM)

Scanning electron microscope, using a high-speed focused beam of electrons to sample, to produce a sample surface morphology and the element information of the sample. The scanning image reflects the surface characteristics of the sample. For porous materials, the characteristic parameters such as pore size, pore diameter, fiber diameter, the aspect ratio of structures, and the combination of micro-nano hierarchical structure.

9.3.2.2 Fourier Transform Infrared Spectrometer (FT-IR)

Fourier-infrared spectrum analysis, the use of substances in the interference type infrared radiation, the interference pattern containing sample information, through data acquisition. For polymer materials, FT-IR spectra can reflect the chemical structure of the material and the chemical bonds present in the material molecular chains. This information can be used to evaluate the ability of the material to gain or lose electrons in the triboelectric process.

9.3.2.3 X-ray Diffraction (XRD) Analysis

XRD analysis is a technique used to analyze the structure of a material by using the diffraction effect of an X-ray in a crystalline substance based on the relationship between the intensity and direction of the diffraction X-ray. Therefore, the XRD characterization can realize the qualitative and quantitative analysis of the material

Figure 9.11 Schematic diagram illustrating the working principle of KPFM. Source: Bae et al. [26]/with permission of American Chemical Society.

phase, lattice constant, grain size, and residual stress measurement, which are influential for the charge generation property of TENGs.

9.3.2.4 Kelvin Probe Force Microscopy (KPFM)

Since the output characteristics of TENG are closely related to the surface friction charge density σ, the characterization and analysis of the surface potential of the friction materials will help to explore the correlation between material properties and device performance. The KPFM can be used to analyze the surface potential of different friction materials and qualitatively compare their abilities to gain or lose electrons.

KPFM is a technique that mainly utilizes microcantilever detection and amplified sharp probes to measure the interaction forces between the atoms of the tested sample, enabling surface detection and analysis of materials. KPFM is carried out on an AFM instrument, and an AC voltage is applied to the probe to generate electrostatic force between the probe and the sample. By balancing the electrostatic force at a specific frequency through a compensation voltage VDC (in amplitude modulation mode) or force gradient (in frequency modulation mode), as shown in Figure 9.11, the potential difference VCPD between the probe and the sample surface can be obtained, satisfying the following relationship:

$$eV_{CPD} = \varphi_{tip} - \varphi_{sample} \tag{9.5}$$

where φ_{tip} represents the work function of the probe, and φ_{sample} represents the work function of the sample. When using KPFM to characterize two different samples, sample1 and sample2, their work functions and contact potential differences satisfy Equation (9.6):

$$e(V_{CPD1} - V_{CPD2}) = -(\varphi_{sample1} - \varphi_{sample2})$$

$$e\Delta V_{CPD} = \Delta\varphi_{sample} \tag{9.6}$$

Samples with larger contact potential differences exhibit a stronger tendency toward positive frictional electrification, making them more likely to lose electrons and carry a positive charge during contact friction. Conversely, samples with smaller contact potential differences exhibit a stronger tendency toward negative frictional electrification, making them more likely to gain electrons and carry a negative charge during contact friction.

9.3.3 Performance Evaluation Criteria

9.3.3.1 Figure of Merit (FOM)

Despite numerous TENGs made of different materials with diverse configurations that have been developed, there is currently no unified system for evaluating TENG performance to facilitate comparison and assessment since different TENGs may have their unique advantages. In 2015, Zi et al. reported using the FOM to evaluate the overall quality of TENGs. The proposed FOM includes both the structural figure of merit (FOM_S) and the material figure of merit (FOM_M) [27]. This model focuses on the impact of structure and material on TENG performance. FOM_S is based on device design, while FOM_M is based on the surface charge density of materials used during operation.

Different TENG operating modes follow the trend of FOM_S in the order of perpendicular free friction layer mode, contact-separation mode, sliding free friction layer mode, planar contact sliding friction mode, and single-electrode mode. Figure 9.12a–d illustrates the FOMs of TENGs with four basic modes. From the simulation results, it can be concluded that double-electrode TENG performs better than single-electrode TENG under the same conditions. In addition, the triggering condition is also a very important factor because the contact-separation mode has better performance than the sliding mode when the surface charge density is

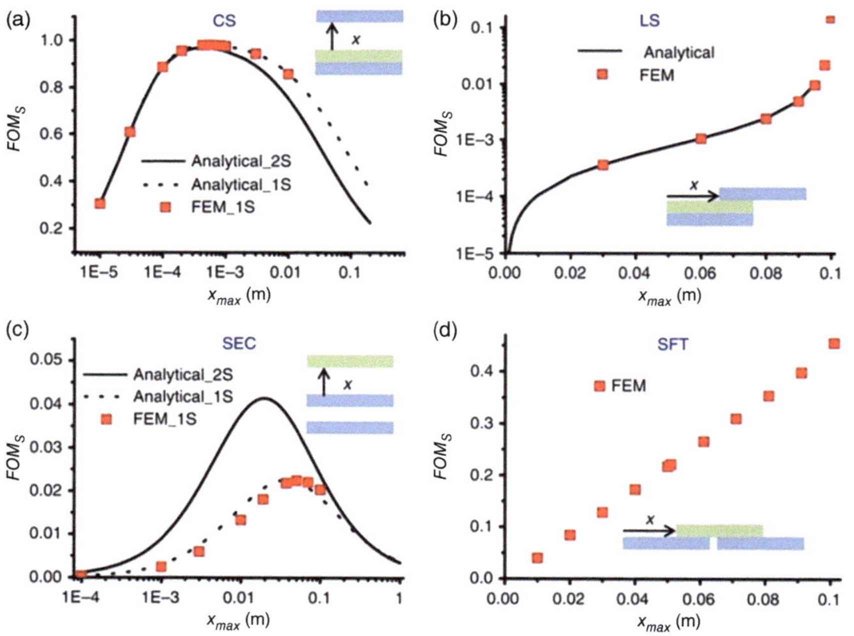

Figure 9.12 Standards and optimal values (FOMs) for evaluating the performance of TENG: (a–d) FOMs for vertical contact-separation mode, horizontal-sliding mode, single-electrode mode, and independent friction layer mode. Source: Zi et al. [27]/ Springer Nature / CC BY 4.0.

the same. The single-electrode mode and independent friction layer mode have relatively lower FOM_S since they did not make full use of the charge generated on the surface of the counterpart friction material.

9.3.3.2 Output Current, Voltage, and Charge Measurement

To evaluate the energy output of TENG, short circuit current, open circuit voltage, and open circuit charges are typically measured and compared among TENGs prepared with different materials and designs. Generally, the electrical signals are measured using an electrometer such as Keithley 6514, which provides high resolution, current sensitivity, and charge measurement capability. Another instrument used to measure the instant open circuit voltage is an oscilloscope, and the short circuit current can also be measured using a low-noise current measurement system (e.g. Stanford Research SR570) or a picoammeter with a high-sampling rate and sensitivity.

9.3.3.3 Measurement of Output Power Density

To assess the actual power of a TENG, it is usually connected to external loads to measure the power that the TENG supplies to the external load. Typically, a series of resistors with resistance ranging from a few Ohms to thousands of megaohms are connected to the TENG, and the current or the voltage of the resistor is measured. The output power of the TENG is then calculated using Equation (9.7)

$$P = UI = I^2R = U^2/R \tag{9.7}$$

where P is the output power, U is the voltage across the external load, I is the current through the external load, and R is the electrical resistance of the external load.

Then, the power density of the TENG on a specific resistor can be calculated by dividing the power value by the area or volume of the TENG device. When plotting the power density against the resistance of the external load, a maximum power density can be obtained. This power density is recognized as the optimum power density of the device and the corresponding resistance is recognized as the internal impedance of the TENG device.

9.4 Fabrication of Polymer Foams for TENGs

9.4.1 Supercritical CO_2 Foaming

Supercritical CO_2 ($scCO_2$) foaming is an approach that produces polymer foams with micro-sized cells by using CO_2 as the physical blowing agent. The process typically involves four stages, including gas dissolving, cell nucleation, cell growth, and foam solidification. $ScCO_2$ foaming is an environmentally friendly, safe, and inexpensive method for producing porous polymer foams on a large scale. On the one hand, this method does not release any toxic gas, making the foam products suitable for recycling. On the other hand, the pore structure can be controlled in the range of several micrometers to several hundred micrometers by controlling the foaming parameters

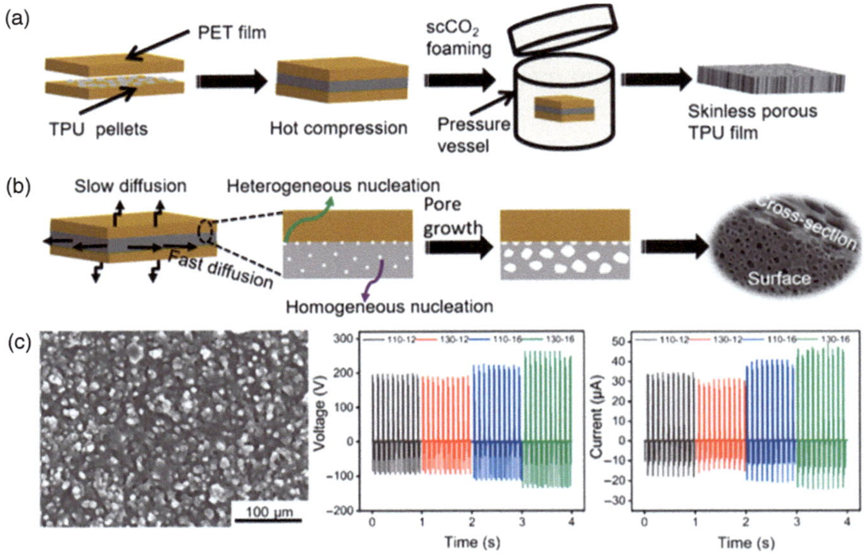

Figure 9.13 (a) Schematic illustration of the fabrication process of skinless TPU films, (b) foaming mechanism of the surface-constrained foaming process, and (c) output voltage and current of TENGs comprised of convex structured PDMS film and skinless TPU films foamed at different conditions. Source: Ni et al. [28]/with permission of Elsevier.

or adding fillers. Therefore, using the $scCO_2$ foaming method may be a reasonable method for preparing fully shape-adaptive flexible TENG with a porous structure.

Gas foaming techniques encounter challenges in generating porous thin films due to the rapid escape of CO_2 molecules when pressure drops suddenly. Addressing these challenges is pivotal in expanding the utility of gas-foaming technology for crafting high-performance TENGs. Mi and coworkers adopted a surface-constrained $scCO_2$ foaming approach to fabricate a skinless porous TPU film as a novel friction-positive material. As illustrated in Figure 9.13a–b, a sandwich-structured PET/TPU/PET film was created through vacuum-assisted hot pressing, followed by $scCO_2$ foaming within a high-pressure vessel. The inclusion of a PET layer effectively curbed $scCO_2$ diffusion, preventing CO_2 molecules from escaping during foaming. The resulting skinless porous structure led to a substantial enhancement in output voltage and current, with increases of 340% and 460%, respectively (Figure 9.13c). Furthermore, pairing the concave porous TPU film with a convex polydimethylsiloxane (PDMS) film further boosted the TENG performance by over 40% and achieved a high power density of 4.6 W m^{-2} when connected to an external load of $3.3 \times 10^6\ \Omega$ [28].

Mi and coworkers developed a novel internally constrained foaming process, which involved the creation of a systematically patterned porous composite TPU foam by integrating a stainless steel mesh (SSM) through $scCO_2$ foaming techniques [29]. By encapsulating the composite foam between two silicone rubber membranes, simultaneous utilization of both sides of the composite foam could be achieved. Notably, this design effectively mitigated SSM electrode corrosion,

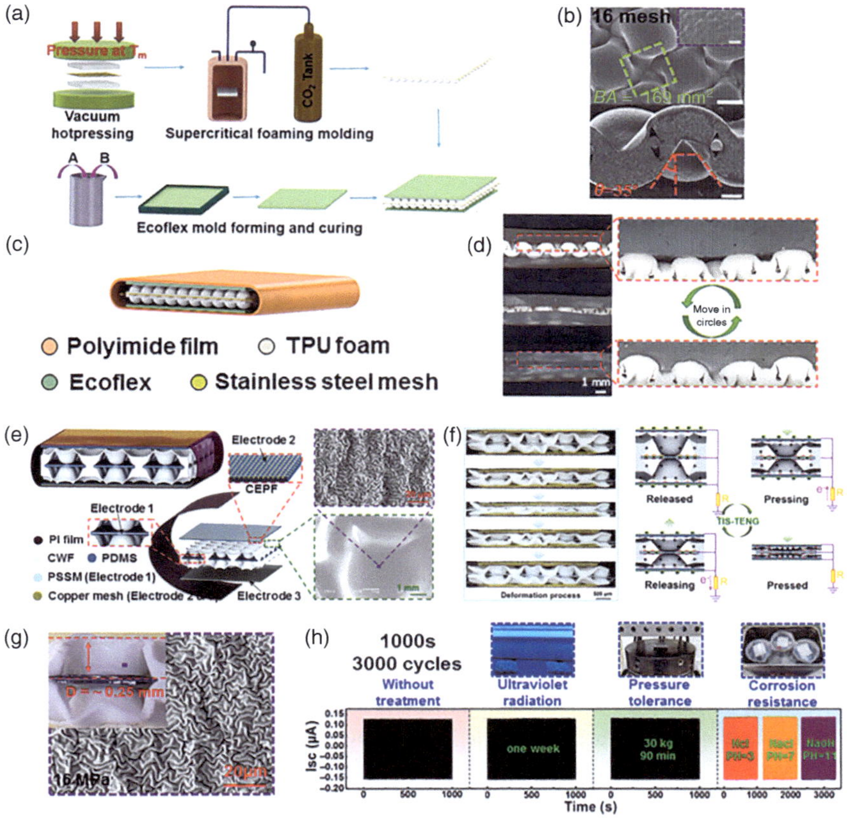

Figure 9.14 (a) Fabrication process and (b) SEM images of the composite TPU foam prepared via internally constrained foaming process, (c) illustration of flexible double side-TENG assembled using PI packaging, and (d) its compressing and releasing states. (e) Structure design of the TIS-TENG, (f) photographs showing the process of compressing and releasing for the TIS-TENG and its working mechanism, (g) macroscopic and microscopic morphologies of the inner surface of the TIS–TENG, and (h) durability of the TENG when subjected to ultraviolet irradiation, high-intensity extrusion, and corrosion treatment (acid, alkali, and salt). Source: Adapted from [29, 30]/with permission of Elsevier.

leading to a substantial enhancement in output performance and durability. They also invented a triple-electrode-integrated self-supporting (TIS-TENG) by combining the dynamic $scCO_2$ foaming and the internally constrained foaming method. As shown in Figure 9.14a–e, the fabricated internal electrode structured foam has a wrinkled porous surface. The foamed sample could be directly used for triboelectric energy generation. By combining the foam with two additional silicon rubber layers, a TIS-TENG device with three electrodes and eight friction layers was prepared. The output performance of the device could be enhanced about four times with this ingenious design. In addition, the output could be further boosted by adding tea powders and patterning micro domes on the PDMS negative friction

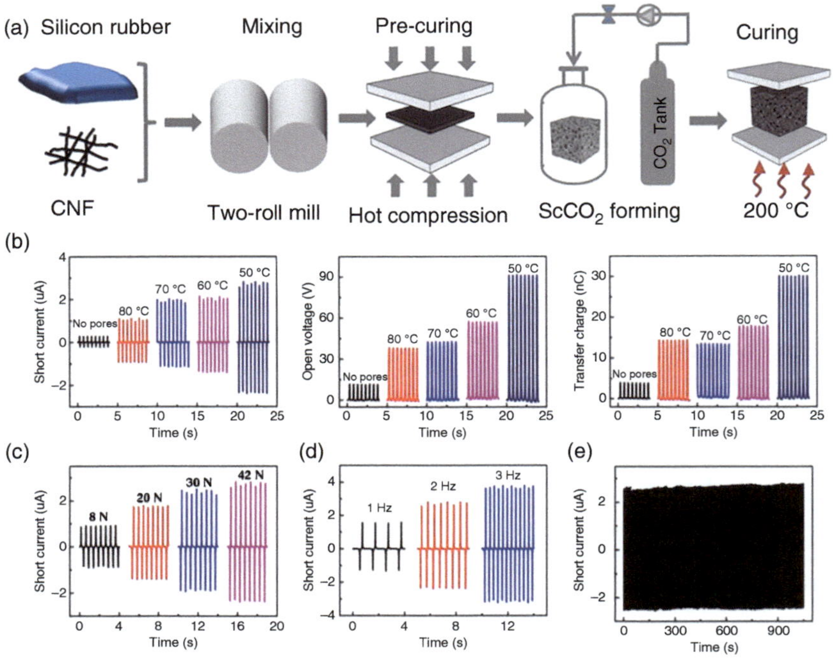

Figure 9.15 (a) Schematic diagram of the preparation process for silicon/CNF nanocomposite foams, (b) the output performance of TENGs based on the foams fabricated at a saturation temperature of 50–80 °C, (c) Isc of TENG under different compressive forces, (d) Isc of TENG under different frequencies, and (e) durability of the silicon/CNF foam based TENG in the long term. Source: Shao et al. [24]/with permission of Elsevier.

layers, which resulted in a high Voc of 106.3 V and a power density of 32.5 mW m^{-2} (Figure 9.14f–h) [30].

Shao et al. engineered an integrated frictional TENG utilizing a composite material comprising rubber and carbon nanofibers (CNFs). The procedure commences with the placement of the rubber/CNFs composite within a high-pressure foaming vessel, followed by the controlled introduction of $scCO_2$ and rapid foaming. Then, the foamed component undergoes complete curing at 200 °C (Figure 9.15a). Under optimized conditions, this TENG yields a remarkable open-circuit voltage of 91 V, a short-circuit current of 2.87 μA, and a transferred charge of 40 nC. The TENG possesses the capability to harness intricate mechanical energies encompassing stretching, twisting, bending, and compression (Figure 9.15b–e) [24]. Furthermore, Li et al. devised a single-electrode TENG, leveraging $scCO_2$-foamed TPU in its construction. The configuration yielded an output voltage of approximately 26 V when subjected to pressure via a rubber airbag [31].

9.4.2 Chemical Foaming

Chemical foaming constitutes a plastic foaming process facilitated by the generation of gas through chemical reactions that are realized by their ability to decompose

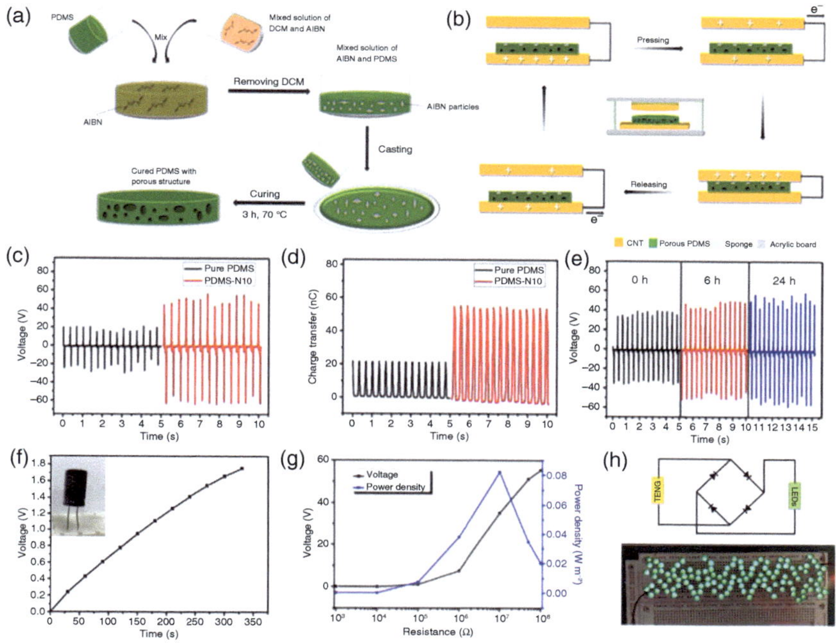

Figure 9.16 (a) Schematic illustration of the manufacturing process for the porous PDMS film, (b) the working mechanism of TENG, (c) the output voltage, (d) the charge transfer of pure PDMS and 12 h PDMS-N10 film, (e) the output voltage of TENGs based on PDMS porous films with different pre-curing times, (f) the voltage-time curve charging 22 μF capacitor, (g) the relationship between resistance and voltage or power density, and (h) digital photographs of green LEDs lighted by the TENG. Source: Yang et al. [32]/John Wiley & Sons.

and emit one or more types of gases within specific temperature ranges. Yang et al. utilized azobisisobutyronitrile (AIBN) as a self-reactive foaming agent to modify PDMS, yielding a self-assembled porous structure. During the pre-curing phase, AIBN decomposition remains negligible at room temperature. Subsequently, as the PDMS/AIBN mixture undergoes full curing, cells were generated by the nitrogen gas emitted during the AIBN decomposition, which gave rise to the formation of a porous structure in PDMS (Figure 9.16a). Leveraging this porous PDMS membrane, the electrical performance of the TENG experienced a remarkable improvement, boasting a 2.5-fold increase in voltage and a 2.7-fold boost in current compared to pure PDMS (Figure 9.16b–e). The developed TENG also demonstrated excellent output stability and power supply ability due to the robustness of the PDMS matrix (Figure 9.16f–h) [32].

9.4.3 Phase Separation and Freeze Drying

Thermal-induced phase separation (TIPS) and freeze-drying methods are frequently used to fabricate porous polymer materials. In TIPS and freeze drying, the pores were generated by the removal of the solvent phase, and the residual solid phase would form a porous or fibrous morphology. Zheng et al. [33] first investigated the capability of using polymer-based porous aerogel films as friction materials in the TENG. Positive friction materials used were porous cellulose nanofiber (CNF) or chitosan (CTS) aerogel films, and negative friction materials used were porous PDMS and polyimide (PI) aerogel films. In the fabrication process, the aerogels were obtained by frozen-induced phase separation and freeze drying. The aerogel films were obtained by compressing the aerogels using high pressure. Mi et al. investigated the effect of introducing polyethyleneimine (PEI) into the CNF aerogel on the triboelectric performance of the TENG composed of CNF aerogel and polyvinylidene fluoride (PVDF) nanofiber membrane [23]. It was found that the addition of PEI increases the number of amino bonds, which are beneficial for positive charge generation while reducing the fibrous structure of the aerogel film. Therefore, a PEI content of 10% was recognized as the optimum parameter for enhancing the TENG output (Figure 9.17).

In addition to freeze drying, the solvent phase after phase separation can be removed by supercritical drying, which utilizes the ability of $scCO_2$ to dissolve the solvent and vaporize it when the $scCO_2$ is turning from the supercritical state to the

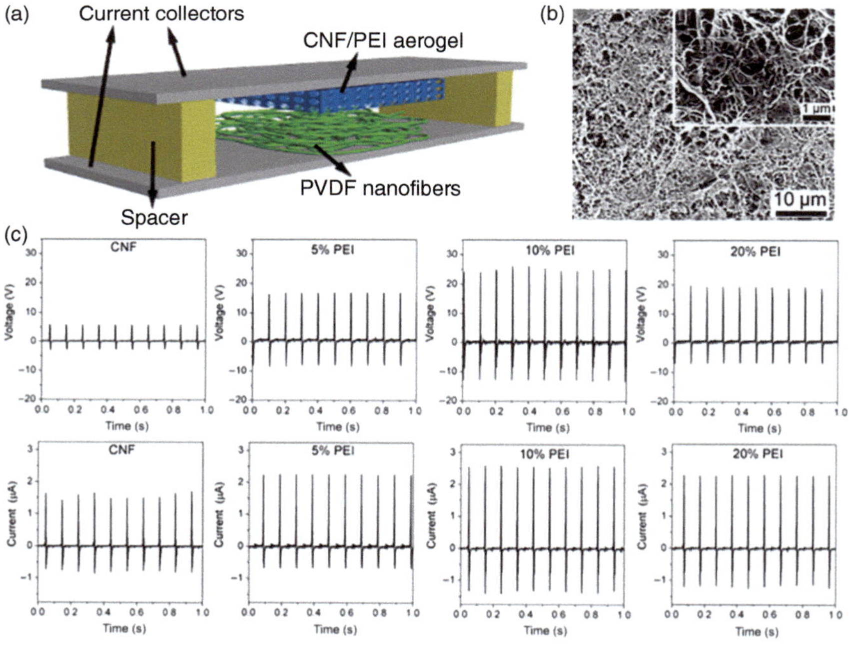

Figure 9.17 (a) Schematic illustration of the TENG made of CNF/PEI aerogel film paired with PVDF nanofiber mats, (b) morphology of CNF/PEI aerogel with 5% PEI, and (c) effect of PEI content on the triboelectric output voltage and current generated by different aerogels paired with one layer of PVDF nanofiber mat. Source: Mi et al. [23]/with permission of Elsevier.

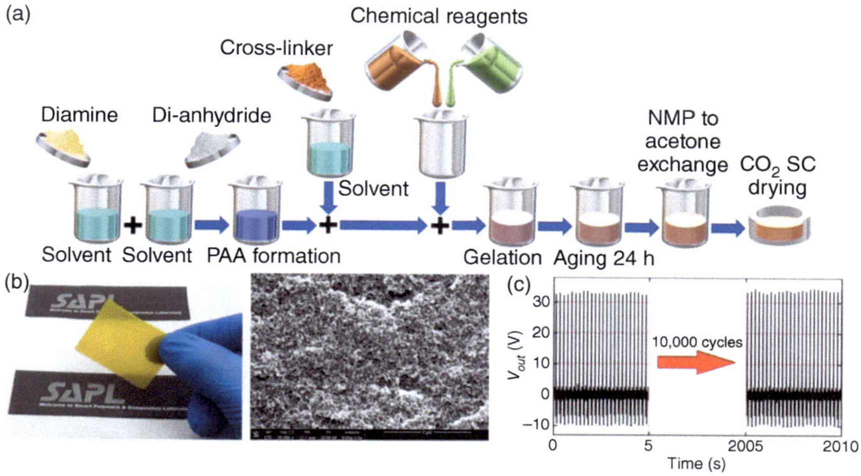

Figure 9.18 (a) The schematic preparation process of the polyimide (PI) aerogel flm, (b) photo and SEM images of the fabricated PI aerogel, and (c) mechanical robustness of the aerogel-based TENG under continuous harmonic loads. Source: Saadatnia et al. [34]/Springer Nature/CC BY 4.0.

gas state. Saadatnia et al. [34] fabricated PI aerogel films with different open-cell content by polycondensation reaction and $scCO_2$ drying (Figure 9.18a). When used as TENG friction materials, the PI aerogel films showed remarkable enhancement compared with fully dense PI-based TENG due to the increase in the effective surface area, charge generation inside the open cells, and increase in the relative capacitance (Figure 9.18b). The PI aerogel film with 50% open-cell content achieved the highest output and excellent long-term operation performance (Figure 9.18c).

9.4.4 Other Methods to Fabricate Porous Films

In addition to the above-mentioned approaches, many other approaches have been used to develop porous polymer films. Wu et al. [35] fabricated sponge-structured PDMS foam through particle filtration and water-leaching techniques. The PDMS foam was fabricated by blade-coating a PDMS/NaCl mixture and soaking it in hot water, and the foam was then spray-coated with an AgNW layer as the conduction component in the TENG. The porous PDMS foam-based TENG could be used for thin skin-integrated sensors and energy harvesters.

Electrospinning is another widely used technique to produce porous membranes. Electrospinning produced nanofibers from a polymer solution with the existence of a high electric field. The collected membrane has a fibrous structure and can typically achieve a porosity of 80%. Mi et al. [36] fabricated polyamide (PA) nanofibrous membrane by electrospinning and assembled TENGs by pairing the PA porous membrane with PI aerogel (Figure 9.19). It was found that the output performance increased gradually with the increase of PA membrane thickness,

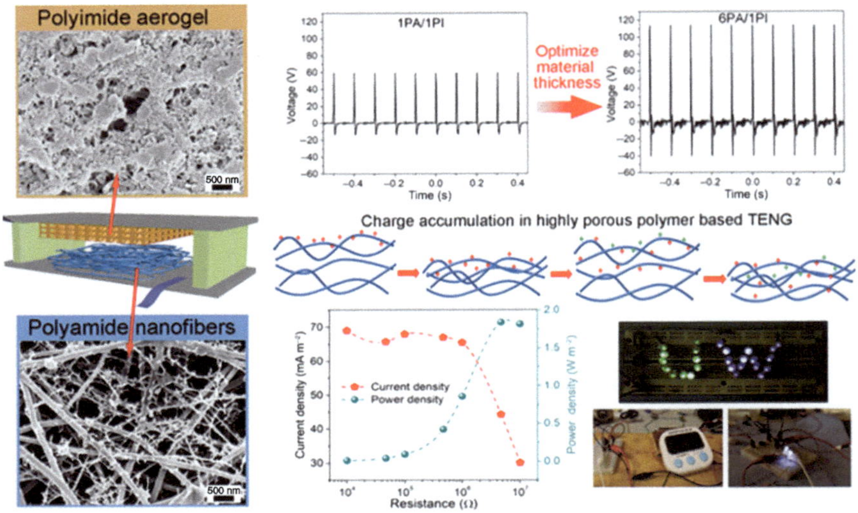

Figure 9.19 The schematic diagram and electrical output performance of TENG made of PA nanofiber mat and PI aerogel film. Source: Mi et al. [36]/with permission of American Chemical Society.

which testified that the porous structures of the inner pores also contribute to the charge generation. They thus proposed the charge accumulation mechanism in highly porous materials.

9.5 Advanced Application of TENG

Owing to the excellent energy conversion performance, wide material species, and flexible device configuration, TENGs have shown tremendous applicable fields so far. This section will introduce several potential application areas of advanced TENGs that have been heavily investigated in recent years, such as sustainable micro/nano energy for self-powered systems, self-powered sensors, large-scale blue energy collection systems for harvesting low-frequency water energy, and high-voltage sources for specific instruments. Figure 9.20 shows a roadmap for the development of TENGs and related technologies proposed by the TENG inventor Prof. Wang [15]. The invention of TENGs has opened up a new research field for material scientists, chemists, and engineers to efficiently convert mechanical energy using organic nanogenerators and proposed a new industry discipline for green energy generation.

9.5.1 Micro and Nano Energy

TENGs have been used as micro and nano energy sources in wearable electronics, implantable electronics, and self-powered systems. In recent years, with the increasing popularity of personal electronic devices and their integration into various

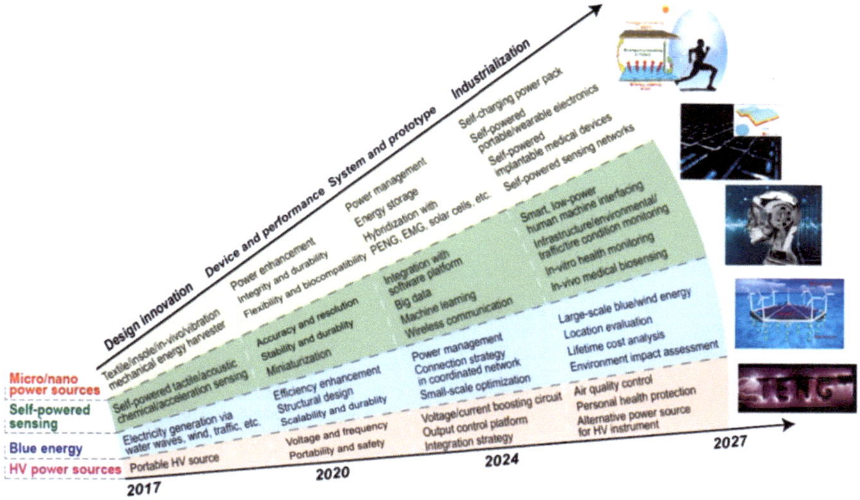

Figure 9.20 Roadmap for the development of TENGs from 2017 to 2027. Source: Wu et al. [15]/with permission of John Wiley & Sons.

aspects of daily life, such as medical monitoring, motion sensing, and security control, people have become eager to collect energy from their body movements to power numerous microelectronic devices. Therefore, various configurations of TENGs, such as multilayered assembled TENGs, have been developed to effectively collect energy from human movements.

Using TENG, energy management circuits, and low leakage energy storage devices, Niu et al. [37] built the first self-driving system that truly meets the milliwatt-level requirements of personal electronic devices, as shown in Figure 9.21a. The multilayer additional electrode structure TENG effectively collects the biological energy of the human body and converts it into alternating current, and then converts it into direct current that conforms to the input of the energy storage unit through the energy management circuit, and finally realizes the power supply for the whole functional circuit of sensors, data processors, displays, and wireless transmitters. The total energy conversion efficiency of the system is 59.8%, and it is demonstrated to be used to drive a variety of commercial electronic devices, such as temperature sensors, electrocardiogram systems, and pedometers.

In addition to *in vitro* energy collection for implantable electronic devices, such as cardiac pacemakers [39], laser therapy [40], etc., implantable TENG has also aroused widespread research interest. Figure 9.21b demonstrates a fully implanted symbiotic pacemaker developed by Wang and coworkers [38] based on an implantable TENG using Al and PTFE as the friction layers, which achieves energy harvesting and storage as well as cardiac pacing on a large-animal scale simultaneously. Implantable TENGs for implantable medical devices offer the advantages of excellent output performance, high power density, and good durability.

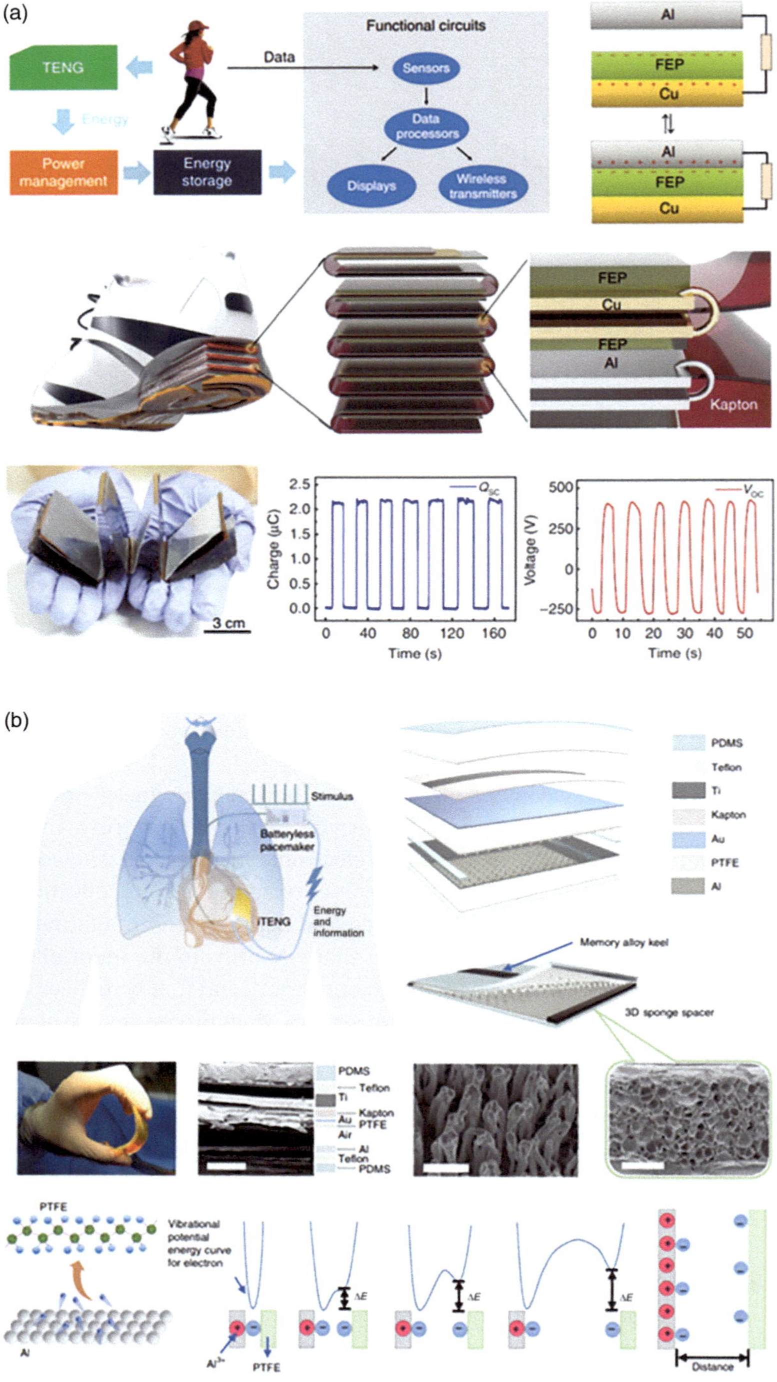

Figure 9.21 Applications of TENG micro and nano energy: (a) a coexistent pacemaker system based on TENG and its *in vivo* energy harvesting, (b) a fully self-powered symbiotic cardiac pacemaker driven by human biomechanical energy. Source: (a) Niu et al. [37]/Springer Nature/CC BY 4.0. (b) Ouyang et al. [38]/Springer Nature/CC BY 4.0.

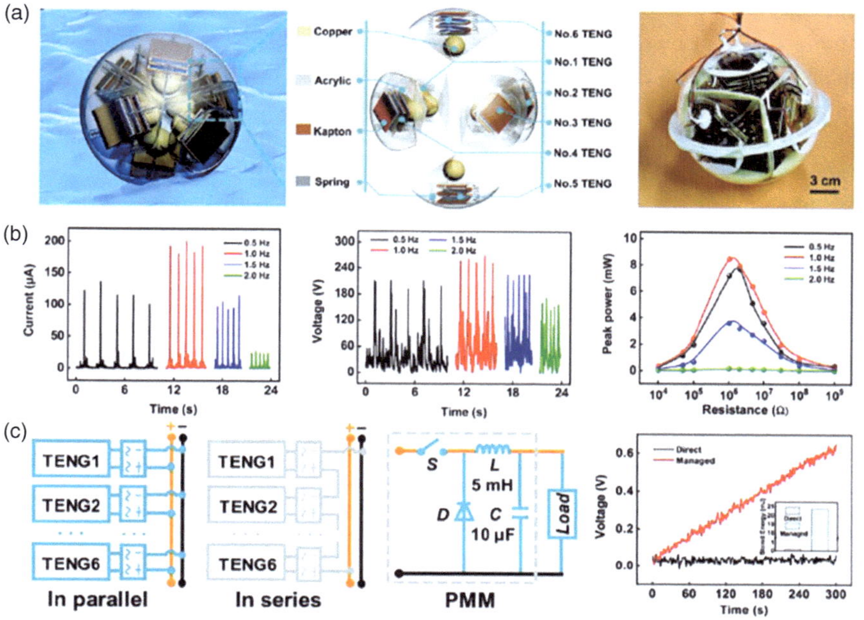

Figure 9.22 TENG spherical structure for collecting multidirectional water wave energy: (a) schematic and actual images of the structure, (b) output performance of TENG underwater waves of different frequencies, and (c) power management module circuit and charging curve of 0.1 F supercapacitor. Source: Liang et al. [43]/with permission of Royal Society of Chemistry.

9.5.2 Blue Energy

About 70.8% of the earth's surface is covered by water, which contains abundant energy in its waves, making it one of the most important clean and renewable energy sources. However, the low frequency, wide distribution, and random peak characteristics of water waves pose many challenges for traditional electromagnetic induction-based wave energy collection technologies. In this specific case, TENG can convert low-frequency, multidirectional, and random water wave vibrations into electricity, and a network of multiple TENG units can be used to achieve large-scale ocean energy collection, known as blue energy [41].

With the concept of wave energy, various structured TENGs have been designed and constructed, such as rolling ball structures, swing structures, impeller structures, spring-assisted structures, and wave structures [42], to efficiently collect ocean energy. As shown in Figure 9.22, Liang et al. designed and fabricated a spherical structure device integrated with six spring-assisted multilayer TENG units. Under the action of 1.0 Hz water waves, the maximum output voltage is 250 V, the current is 80 μA, and the power reaches 8.5 mW. Based on the structural design of the device, it can realize the conversion of water wave energy in any triggering direction [43].

9.5.3 Self-Powered Sensors

Apart from their role in converting ambient mechanical energy into electrical power, TENGs possess the capability to directly transduce external mechanical, chemical, and biochemical stimuli into electrical signals, which makes them suitable for self-powered sensing applications. These TENG-based self-powered sensors encompass a broad spectrum, including tactile sensors, motion sensors, pressure sensors, strain sensors, chemical sensors, biomechanical sensors, vibration sensors, acceleration sensors, and acoustic sensors. Mi and coworkers [29] developed porous composite TPU foams with regular bumps and valleys on both sides of the foam via $scCO_2$ foaming. After assembling with two layers of PDMS films and wrapping with PI tape, the obtained TENG works in a single-electrode mode (Figure 9.23a) and demonstrates excellent stability and tolerance. The device is flexible and can be used as a self-powered sensor to detect tapping and bending, and can be directly used in the monitoring of human activities such as walking, sitting, and assisting in boxing training (Figure 9.23b,c). In another work, they invented a triple-electrode integrated self-supporting TENG, which is composed of a surface wrinkled TPU foam with regular hump structures. The TENG has eight friction layers in a single device, which renders the device outstanding output performance and stability. They successfully employed the TENG in self-powered sensing applications including an intelligent parking system that can monitor the parking and departure of vehicles, and a safety monitoring and alarming system for construction workers that is capable of making alarm signals when the helmet and the vamp are subjected to strikes. It can also be used to monitor the walking pace and working intensity of the worker [30].

Moreover, self-powered sensors based on TENGs have shown great potential in safety monitoring, environmental monitoring, biomedical applications, smart homes, smart sports, human–machine interface (HMI), smart machines, and autonomous driving. In recent years, researchers have been focused on structural design, performance optimization, and scenario applications. Based on previous explorations, the construction of TENG self-powered sensor systems with technology platforms such as big data, machine learning, and wireless communication will truly achieve their application potential in multiple fields and scenarios in the future.

9.5.4 High Voltage Power Supply

In contrast to conventional electromagnetic generators, TENGs exhibit distinctive output characteristics, characterized by high open-circuit voltage and low short-circuit current, which position TENGs as influential elements within the realm of high-voltage power supply. TENG-based high-voltage power supplies have garnered considerable attention and engendered proactive exploration of their potential applications across diverse domains. Li et al. [44] harnessed the high output voltage of TENGs to construct a nano-electrospray ionization (Nano ESI) system for high-sensitivity mass spectrometry analysis, which leverages the AC pulse output derived from a sliding-independent layer TENG to induce electrospray ionization (Figure 9.24a). The inherent high voltage imbued this system with remarkable mass spectrometry analysis sensitivity, even at exceedingly low substance concentrations. In applications necessitating high-voltage power

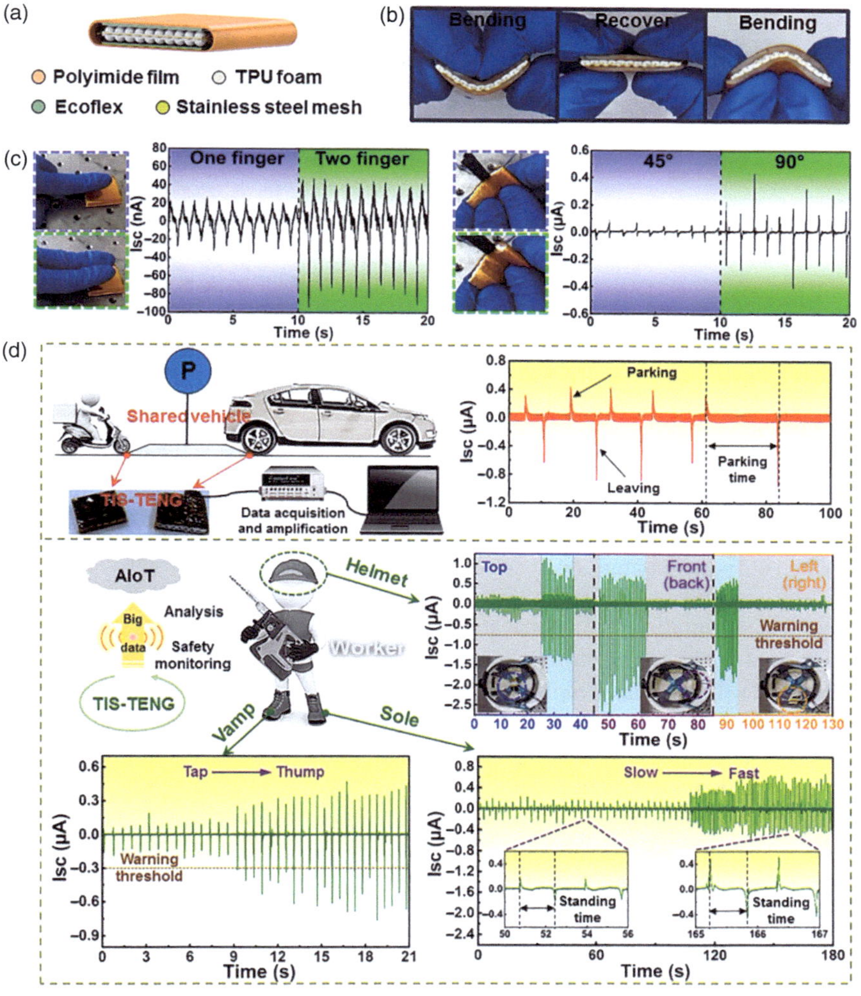

Figure 9.23 (a) Illustration of flexible TENG, (b) digital photographs of a TENG being bent and recovered, (c) Isc signals of a TENG pressed using different fingers and repetitively bent to ~45° and 90° [29], and (d) safety monitoring and alarming system built using TENGs including the Isc signals detected on the top, front, and left surface of a helmet, Isc signals detected on the vamp of a worker when the foot is subjected to tap or thump, and Isc signals detecting the walking pace of a worker. Source: (a) Xie et al. [30]/With permission of Elsevier. (b) Xie et al. [30]/With permission of Elsevier.

supply, such as micro-plasma generation, electrostatic spinning, and electrostatic adsorption, the output voltage generated by a simplistic TENG device often falls short of the requisite levels. Cheng et al. [45] successfully integrated an optimized independently rotating TENG with a plasma source. Through the series connection of multiple TENGs, they pioneered the development of the inaugural frictional electric-micro-plasma source (Figure 9.24b). This innovative device harnesses

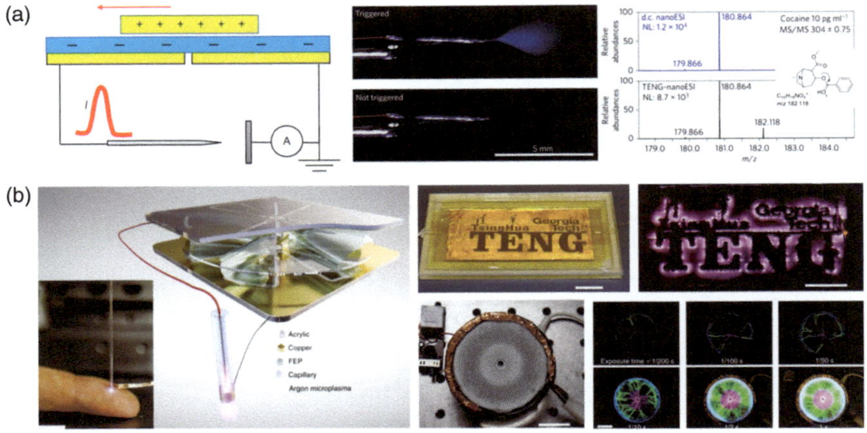

Figure 9.24 Applications of TENG-based high voltage power sources: (a) schematic diagram and application of TENG-based nanoelectrospray ionization for high sensitivity mass spectrometry, and (b) TENG-based frictional-electric micro-plasma source and its surface treatment and patterned luminescence application. Source: (a) Li et al. [44]/with permission of Springer Nature. (b) Cheng et al. [45]/Springer Nature/CC BY 4.0.

mechanical stimulation to produce atmospheric pressure plasma, manifesting four distinctive atmospheric pressure plasma types.

In addition, TENG-based high voltage power supplies offer notable advantages in terms of safety, portability, structural simplicity, and cost-effectiveness when juxtaposed with their traditional counterparts, which typically entail high output current and intricate power conversion mechanisms. Up to date, polymer foams used in these advanced fields are very limited, especially those fabricated via scalable physical foaming methods. Therefore, there is a great chance to expand the application of polymer foams in the development of high-performance and durable TENGs for advanced applications.

References

1 Shahsavari, A. and Akbari, M. (2018). Potential of solar energy in developing countries for reducing energy-related emissions. *Renewable and Sustainable Energy Reviews* 90: 275–291.

2 Musa, S.D., Zhonghua, T., Ibrahim, A.O., and Habib, M. (2018). China's energy status: a critical look at fossils and renewable options. *Renewable and Sustainable Energy Reviews* 81: 2281–2290.

3 Wang, A.C., Wu, C., Pisignano, D. et al. (2018). Polymer nanogenerators: opportunities and challenges for large-scale applications. *Journal of Applied Polymer Science* 135: 45674.

4 Lu, X. (2022). Research on the construction system of computer big data security platform based on artificial intelligence. 2022 4th International Conference on Artificial Intelligence and Advanced Manufacturing (AIAM), 865–869.

5 Cheng, T., Shao, J., and Wang, Z.L. (2023). Triboelectric nanogenerators. *Nature Reviews Methods Primers* 3: 39.

6 Wang, K., Weng, Y., Chen, G. et al. (2021). Coupling electrostatic induction and global electron circulation for constant-current triboelectric nanogenerators. *Nano Energy* 85: 105929.

7 Siddique, A.R.M., Mahmud, S., and Heyst, B.V. (2015). A comprehensive review on vibration based micro power generators using electromagnetic and piezoelectric transducer mechanisms. *Energy Conversion and Management* 106: 728–747.

8 Zhang, R. and Olin, H. (2022). Advances in inorganic nanomaterials for triboelectric nanogenerators. *ACS Nanoscience Au* 2: 12–31.

9 He, X., Zhang, H., Jiang, J., and Liu, X. (2023). Output characteristics of series-parallel triboelectric nanogenerators. *Nanotechnology* 34: 155403.

10 Tantraviwat, D., Buarin, P., Suntalelat, S. et al. (2020). Highly dispersed porous polydimethylsiloxane for boosting power-generating performance of triboelectric nanogenerators. *Nano Energy* 67: 104214.

11 Zhang, L., Jin, L., Zhang, B. et al. (2015). Multifunctional triboelectric nanogenerator based on porous micro-nickel foam to harvest mechanical energy. *Nano Energy* 16: 516–523.

12 Fan, F.-R., Tian, Z.-Q., and Lin Wang, Z. (2012). Flexible triboelectric generator. *Nano Energy* 1: 328–334.

13 Galembeck, F., Burgo, T.A.L., Balestrin, L.B.S. et al. (2014). Friction, tribochemistry and triboelectricity: recent progress and perspectives. *RSC Advances* 4: 64280–64298.

14 Wang, Z.L. (2017). On Maxwell's displacement current for energy and sensors: the origin of nanogenerators. *Materials Today* 20: 74–82.

15 Wu, C., Wang, A.C., Ding, W. et al. (2019). Triboelectric nanogenerator: a foundation of the energy for the new era. *Advanced Energy Materials* 9: 1802906.

16 Wang, Z.L. (2014). Triboelectric nanogenerators as new energy technology and self-powered sensors – principles, problems and perspectives. *Faraday Discussions* 176: 447–458.

17 Zhu, G., Lin, Z.-H., Jing, Q. et al. (2013). Toward large-scale energy harvesting by a nanoparticle-enhanced triboelectric nanogenerator. *Nano Letters* 13: 847–853.

18 Zhu, G., Pan, C., Guo, W. et al. (2012). Triboelectric-generator-driven pulse electrodeposition for micropatterning. *Nano Letters* 12: 4960–4965.

19 Yin, X., Liu, D., Zhou, L. et al. (2019). Structure and dimension effects on the performance of layered triboelectric nanogenerators in contact-separation mode. *ACS Nano* 13: 698–705.

20 Wang, S., Lin, L., Xie, Y. et al. (2013). Sliding-triboelectric nanogenerators based on in-plane charge-separation mechanism. *Nano Letters* 13: 2226–2233.

21 Wang, L. and Daoud, W.A. (2019). Highly flexible and transparent polyionic-skin triboelectric nanogenerator for biomechanical motion harvesting. *Advanced Energy Materials* 9: 1803183.

22 Wang, S., Xie, Y., Niu, S. et al. (2014). Freestanding triboelectric-layer-based nanogenerators for harvesting energy from a moving object or human motion in contact and non-contact modes. *Advanced Materials* 26: 2818–2824.

23 Mi, H.-Y., Jing, X., Zheng, Q. et al. (2018). High-performance flexible triboelectric nanogenerator based on porous aerogels and electrospun nanofibers for energy harvesting and sensitive self-powered sensing. *Nano Energy* 48: 327–336.

24 Shao, Y., Luo, C., Deng, B.-w. et al. (2020). Flexible porous silicone rubber-nanofiber nanocomposites generated by supercritical carbon dioxide foaming for harvesting mechanical energy. *Nano Energy* 67: 104290.

25 Wang, Z.L. (2013). Triboelectric nanogenerators as new energy technology for self-powered systems and as active mechanical and chemical sensors. *ACS Nano* 7: 9533–9557.

26 Bae, C., Kim, D., Moon, S. et al. (2010). Aging dynamics of solution-processed amorphous oxide semiconductor field effect transistors (vol 2, pg 626, 2010). *ACS Applied Materials & Interfaces* 2: 1273.

27 Zi, Y., Niu, S., Wang, J. et al. (2015). Standards and figure-of-merits for quantifying the performance of triboelectric nanogenerators. *Nature Communications* 6: 8376.

28 Ni, G.-L., Zhu, X., Mi, H.-Y. et al. (2021). Skinless porous films generated by supercritical CO_2 foaming for high-performance complementary shaped triboelectric nanogenerators and self-powered sensors. *Nano Energy* 87: 106148.

29 Xie, Y., Hu, J., Li, H. et al. (2022). Green fabrication of double-sided self-supporting triboelectric nanogenerator with high durability for energy harvesting and self-powered sensing. *Nano Energy* 93: 106827.

30 Xie, Y., Shan, T., Chen, R. et al. (2023). Triple-electrode integrated self-supporting triboelectric nanogenerators with high output and durability based on dynamic supercritical carbon dioxide foaming. *Nano Energy* 116: 108786.

31 Li, H., Sinha, T.K., Oh, J.S., and Kim, J.K. (2018). Soft and flexible bilayer thermoplastic polyurethane foam for development of bioinspired artificial skin. *ACS Applied Materials & Interfaces* 10: 14008–14016.

32 Yang, Y., Jing, T., and Xu, B. (2020). Self-assembly of porous microstructured polydimethylsiloxane films for wearable triboelectric nanogenerators. *Macromolecular Materials and Engineering* 305: 2000276.

33 Zheng, Q., Fang, L., Guo, H. et al. (2018). Highly porous polymer aerogel film-based triboelectric nanogenerators. *Advanced Functional Materials* 28: 1706365.

34 Saadatnia, Z., Mosanenzadeh, S.G., Esmailzadeh, E., and Naguib, H.E. (2019). A high performance triboelectric nanogenerator using porous polyimide aerogel film. *Scientific Reports* 9: 1370.

35 Wu, M., Gao, Z., Yao, K. et al. (2021). Thin, soft, skin-integrated foam-based triboelectric nanogenerators for tactile sensing and energy harvesting. *Materials Today Energy* 20: 100657.

36 Mi, H.-Y., Jing, X., Meador, M.A.B. et al. (2018). Triboelectric nanogenerators made of porous polyamide nanofiber mats and polyimide aerogel film: output

optimization and performance in circuits. *ACS Applied Materials & Interfaces* 10: 30596–30606.

37 Niu, S., Wang, X., Yi, F. et al. (2015). A universal self-charging system driven by random biomechanical energy for sustainable operation of mobile electronics. *Nature Communications* 6: 8975.

38 Ouyang, H., Liu, Z., Li, N. et al. (2019). Symbiotic cardiac pacemaker. *Nature Communications* 10: 1821.

39 Zheng, Q., Shi, B., Fan, F. et al. (2014). In vivo powering of pacemaker by breathing-driven implanted triboelectric nanogenerator. *Advanced Materials* 26: 5851–5856.

40 Tang, W., Tian, J., Zheng, Q. et al. (2015). Implantable self-powered low-level laser cure system for mouse embryonic osteoblasts' proliferation and differentiation. *ACS Nano* 9: 7867–7873.

41 Wang, Z.L. (2017). Catch wave power in floating nets. *Nature* 542: 159–160.

42 Wen, X., Yang, W., Jing, Q., and Wang, Z.L. (2014). Harvesting broadband kinetic impact energy from mechanical triggering/vibration and water waves. *ACS Nano* 8: 7405–7412.

43 Liang, X., Jiang, T., Liu, G. et al. (2020). Spherical triboelectric nanogenerator integrated with power management module for harvesting multidirectional water wave energy. *Energy & Environmental Science* 13: 277–285.

44 Li, A., Zi, Y., Guo, H. et al. (2017). Triboelectric nanogenerators for sensitive nano-coulomb molecular mass spectrometry. *Nature Nanotechnology* 12: 481–487.

45 Cheng, J., Ding, W., Zi, Y. et al. (2018). Triboelectric microplasma powered by mechanical stimuli. *Nature Communications* 9: 3733.

10

Porous Polymers for Solar Steam Generation

10.1 Overview of SSG

Billions of people worldwide face freshwater scarcity, while oceans cover 70% of our planet's surface [1]. Developing scalable, eco-friendly technologies is imperative to meet the growing demand for freshwater [2]. Traditional methods like distillation [3], electrodialysis [4], and reverse osmosis [5] have been extensively used for water purification. However, these techniques often lead to significant environmental issues and high energy consumption [6]. Sunlight, a sustainable and eco-friendly energy source, has been vital in various fields from ancient times to the present [7]. SSG is emerging as a promising approach for freshwater production due to its low energy requirements, high evaporation efficiency, environmental friendliness, and sustainability [8]. SSG devices are simple, portable, and suitable for off-grid use, facilitating field implementation [9]. In this process, solar irradiation (approximately 1 kW m^{-2}) is absorbed by a solar absorber, which is usually a polymer-based porous composite material that is capable of converting photon energy into thermal energy, thus heating the underlying water and accelerating the water evaporation [10].

10.1.1 Solar Evaporation-Based Desalination Methods

The application of solar energy in desalination has a long history. In a conventional single-slope basin solar still, saline water is heated and evaporated by direct solar exposure (Figure 10.1a) [11]. In the process, a black absorber in the seawater absorbs and converts solar energy into heat. However, this design limits the absorber's exposure to incoming solar radiation due to the poor light penetration through the thick water layer. Moreover, the generated heat easily dissipates into the bulk water and is lost to the surroundings through radiation, convection, and conduction from the water surface. Consequently, solar-to-steam conversion efficiency is relatively low [11]. The use of nanoparticle-dispersed fluids has been shown to harness solar energy more effectively for direct water vapor generation [12]. Solar conversion efficiencies of up to ≈70% have been achieved by reducing heat loss to bulk water (Figure 10.1b) [13]. The concept of interfacial evaporation was introduced in 2014 [14]. As shown in Figure 10.1c, the solar absorber is placed at the interface between

Functional Polymer Foams: Green Fabrication Methods, Performance and Applications, First Edition. Haoyang Mi.

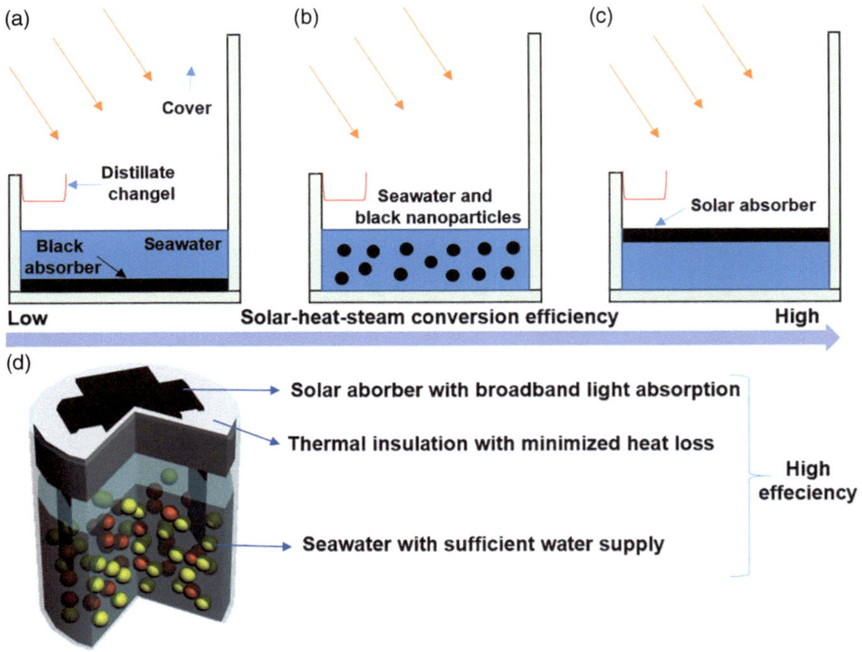

Figure 10.1 Structural comparisons of conventional solar evaporation-based desalination configurations and solar interfacial evaporation: (a) conventional solar distiller based on bottom heating pattern with low conversion efficiency, (b) solar volumetric evaporation using nanoparticles with mediate conversion efficiency, (c) solar interfacial evaporation by heat localization with high conversion efficiency, and (d) schematic illustrating the key components of a solar interfacial evaporation system.

saline water and air, which minimizes heat loss to the bulk water and increases the surface area for vapor release. In this method, the solar radiation is photothermally concentrated at the absorber's surface, resulting in high surface temperatures for rapid heat transfer to water [15]. This setup allows immediate heating and evaporation of water transported from a reservoir, enhancing steam generation rates [16].

A solar interfacial evaporation system comprises three key components, including photothermal materials, heat management, and water supply, each significantly contributing to conversion efficiency [17, 18]. The optical properties of photothermal materials govern light absorption and subsequent conversion, with a large surface area enhancing steam escape. Broadband light-absorbing solar absorbers are highly sought after for their superior efficiency in evaporator material preparation [19]. Heat management often involves engineering solutions, such as thermal insulation, to reduce heat losses [20]. Using a heat insulator placed between the solar evaporator and bulk water, the SSG system can effectively block downward heat conduction, thus minimizing heat loss and improving photothermal conversion (Figure 10.1d). Water supply is crucial to maintain evaporation, as maximizing solar evaporators' potential requires a steady water flow. The transportation of water from the reservoir to the evaporation site is realized by the capillary effect, offering additional benefits like heat loss reduction and salt rejection [21].

10.1.2 Key Elements in an SSG System

To engineer an efficient SSG system for desalination, it must meet six key criteria: (i) The solar absorber should efficiently absorb light across the entire solar spectrum (300–2500 nm); (ii) It must exhibit high light-to-heat conversion efficiency, facilitating the transformation of absorbed solar energy into thermal energy; (iii) The absorber should have low thermal conductivity to minimize heat loss from conduction, convection, and radiation; (iv) It needs good wettability and a porous structure for effective and continuous water transport and evaporation; (v) The absorber should be mechanically robust and chemically stable, suitable for practical applications in harsh hypersaline seawater environments and under prolonged exposure to continuous irradiation; and (vi) The absorber or the solar desalination system must incorporate salt-rejection properties to prevent salt accumulation from long-term evaporation [22].

In recent years, various materials have been explored for photothermal evaporators, including biomass-derived carbon materials [23], carbon-based materials [24], natural materials [25], synthetic polymers [22], plasmonic nanomaterials [26], and black semiconductor materials [27]. However, many of these materials are costly and require complex manufacturing processes, posing challenges for large-scale commercialization. In addition, salt buildup on the evaporating surface is a critical issue that limits the long-term effectiveness of solar evaporators, especially at high evaporation rates. Thus, the development of new, efficient, low-cost, and scalable photothermal materials is crucial for advancing SSG [28].

10.1.3 Porous Polymers for SSG

Porous polymers, with solar energy conversion efficiencies ranging from 50% to 90% due to multiple light reflections, offer benefits such as abundant resources, physicochemical stability, and salt resistance [29]. Moreover, their structural composition, including porosity and rich microchannels, makes them ideal substrates for integration with various nanostructures in highly efficient SSG devices.

Figure 10.2 demonstrates the basic approaches of using porous polymers to enhance the SSG performance. To enhance the light absorption ability, the porous polymer absorber needs to possess a chemical composition and microstructure that enable multiple light capture and scattering within the porous structures. The SSG system needs to facilitate water transportation so as to provide a continuous and moderate water supply. Hence, the polymer absorber needs to be hydrophilic enough to form numerous capillary channels and prevent excess water on the material surface in the meantime to maximize solar absorption. To achieve ideal heat management during the SSG process, the porous polymer needs to have high solar absorption and low thermal conductivity to minimize heat loss. It is preferred to have air passing through the water layer to facilitate water evaporation during the

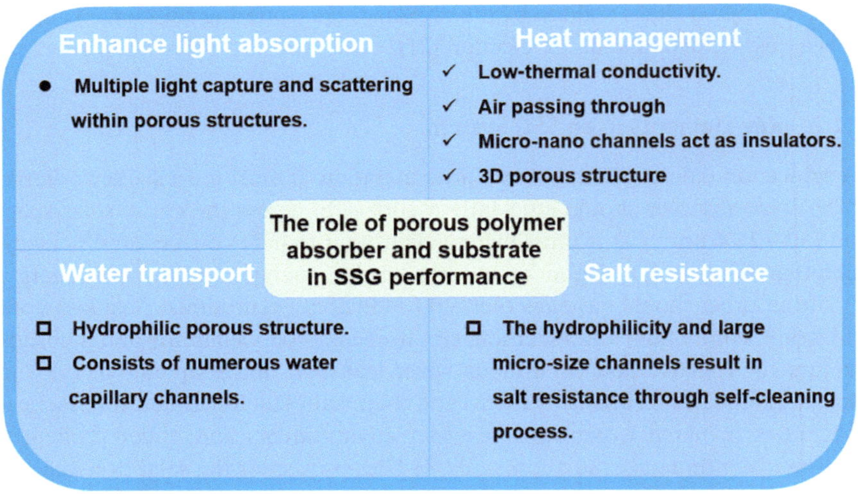

Figure 10.2 The role of porous polymer absorber and substrate in the development of high-performance SSG.

SSG process. To prevent salt particle accumulation on the polymer absorber foam surface, the salt precipitate needs to be dissolved and transferred back into the bulk water. Thus, the foam is desired to have hydrophilic relatively large micro-sized channels to allow the back diffusion of salt ions through the ion gradient to realize a salt precipitate self-cleaning process.

10.2 Energy Conversion in SSG

Evaporation is an energy-intensive process characterized by intricate heat transfer and exchange. A thorough understanding of the energy balance in a solar evaporation system is crucial for optimizing its performance and efficiency [30]. As mentioned above, photothermal porous materials convert part of the absorbed light into heat to drive water evaporation, while some part of the energy is lost through reflection. Additional heat losses during evaporation include conduction to the cooler bulk water, radiation, and convection to the atmosphere, as illustrated in Figure 10.3.

In the process of SSG, heat transfer and loss are important factors affecting evaporation performance. The heat loss during photothermal conversion includes three aspects: heat convection, heat radiation, and heat conduction. The heat radiation loss can be calculated using the Stefan–Boltzmann equation,

$$\Phi = \varepsilon A \sigma \left(T_1^4 - T_2^4\right) \tag{10.1}$$

where Φ is the radiant heat, A is the surface or lateral area of the device, σ is the Stefan–Boltzmann constant ($5.67 \times 10^{-8}\ \mathrm{W\,m^{-2}\,K^{-4}}$), ε is the thermal radiation coefficient of the material, T_1 is the average surface or lateral temperature of the evaporation surface under one solar irradiation, and T_2 is the ambient temperature.

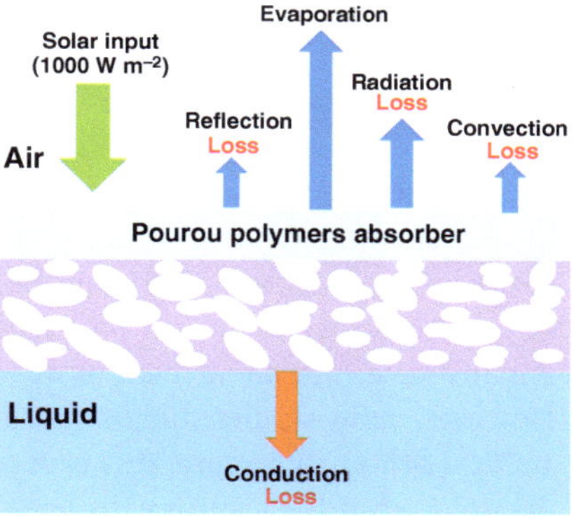

Figure 10.3 The energy balance of an SSG system includes solar input, conduction loss, reflection loss, evaporation, radiation loss, and convection loss.

The heat convection losses can be calculated by Newton's formula,

$$Q_2 = hA\Delta T \tag{10.2}$$

where Q_2 is the convective heat, h is the convective heat transfer coefficient, and ΔT is the difference between the evaporation surface and the ambient temperature under one 1 sun.

The heat conduction loss can be calculated using the following equation [31],

$$Q_3 = Cm\Delta T \tag{10.3}$$

where, Q_3 is the heat conduction loss, m is the mass of water, and ΔT is the temperature difference before and one hour after 1 sun irradiation. According to the above equations, it is beneficial to improve the performance by increasing the surface temperature, reducing heat loss, and obtaining more heat from the environment during water evaporation.

In addition, downward conduction loss, driven by the temperature gradient and water convection, can be quantified by monitoring the water reservoir's temperature change throughout the evaporation process. However, determining surface radiative and convective losses, caused by heat transfer from solar absorbers to the environment due to temperature differences, is currently a subject of debate [32]. While room temperature is commonly used as a reference in most studies, some argue that considering the temperature of the hot water vapor zone near the evaporation surface yields results that are more in line with theoretical energy balance and experimental observations. Nonetheless, the exact measurement strategies and boundary conditions for these calculations are not yet definitively established. The solar-to-steam conversion efficiency is defined by the proportion of light energy used for steam production, which is calculated using the formula (10.4) [33],

$$\eta = \frac{mh_{lv}}{C_{opt}P_0} \tag{10.4}$$

where m represents the evaporation rate (kg m^{-2} h^{-1}), h_{lv} is the total enthalpy required to transform liquid water into vapor, and C_{opt} P_0 is the total solar energy input with $P_0 = 1$ kW m^{-2} and C_{opt} as the optical concentration.

Although this formula is well-established, its accuracy in efficiency calculation requires careful consideration. First, natural evaporation under dark conditions should be deducted from the measured evaporation flux under illumination to ensure m represents evaporation due to light alone. h_{lv} encompasses both the sensible heat of water and the latent heat of liquid–vapor phase change. Sensible heat calculations depend on the water's specific heat and the temperature difference between the solar absorbers and bulk water. Notably, the latent heat varies and should be calculated based on the surface temperature of the solar evaporators. For instance, the latent heat at 40 °C is 2406 kJ kg^{-1} and 2256 kJ kg^{-1} at 100 °C. However, many studies still use the latter value for temperatures significantly below 100 °C [34]. Furthermore, it is crucial to confine the exposure area strictly within the solar absorber's surface to prevent additional heat input from solar light or heated surroundings.

Moreover, the thermal path diagram also suggests potential modifications of the SSG system in terms of maximizing energy usage. For instance, the energy stored in steam could be recycled for electricity generation or multistage steam generation. In addition, the natural temperature difference between the solar absorber and bulk water could be harnessed using thermoelectric modules. The migration of salt ions across the SSG porous material can induce an ion concentration gradient, which can be used to generate constant current flow. All these modifications to the SSG system are ongoing research hot spots to expand the advanced applications of porous polymer-based solar absorbers.

10.3 Characterization of Solar Steam Generator

To evaluate the SSG performance of porous polymer-based SSG systems, it is necessary to investigate the microstructure, chemical composition, optical property, heat generation capability, water evaporation rate, salt resistance, and so on. The following sections introduce the specialized experiments for the evaluation of SSG material and system.

10.3.1 SSG Experiments

The experiment of SSG is generally tested by a lab-built online test system consisting of a solar simulator with a solar filter (AM 1.5), an infrared camera, an electronic balance, an optical power meter, and a wet and dry thermometer [35]. The room temperature is generally maintained at ~20 °C, and the relative humidity balance is about 50%.

For experiments on fabric like photothermal material, thermal insulation foam is typically used as a support body to reduce heat loss. As shown in Figure 10.4, two grooves are generally opened on the thermal insulation foam, and the photothermal material is fixed to the surface of the thermal insulation layer through the two

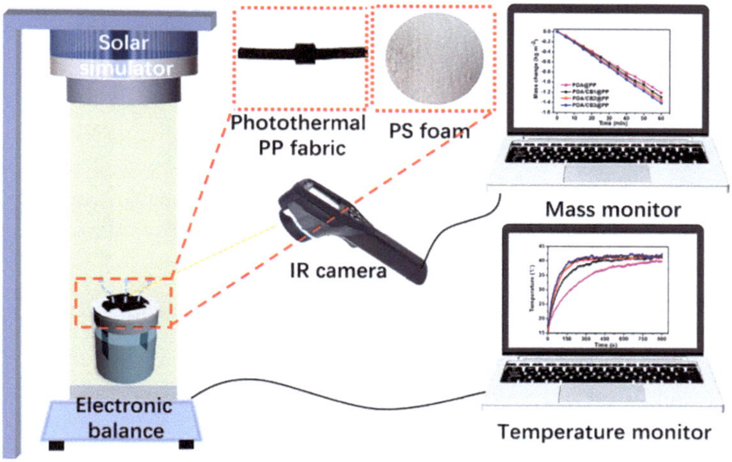

Figure 10.4 Schematic illustration of the solar steam generation test system. Source: Sun et al. [35]/with permission of Elsevier.

grooves to assemble the solar evaporator for testing. In the case of porous foam-based photothermal materials, the materials containing microchannels are often directly placed on the water surface, and the saline water can be transported to the material surface through the capillary effect.

10.3.2 Preparation of Simulated Wastewater and Artificial Seawater

The SSG system is capable of purifying saline water as well as wastewater as long as the contaminated components are nonvolatile under heat, such as heavy metal ion-contaminated wastewater. In the purification performance assessment experiments, simulated wastewater and artificial seawater are often need to be prepared.

Normally, dye-contaminated wastewater is prepared using methyl blue (MB) or methyl orange (MO).The decoloration efficiency of organic dyes is measured by UV–vis–NIR spectrometer (Figure 10.5a). Heavy-metal-contaminated wastewater is prepared by dissolving $ZnCl_2$, $SrCl_2$, $Cu(NO_3)_2$, and $FeCl_3{\cdot}6H_2O$ in DI water. Artificial seawater is prepared by dissolving NaCl, $MgCl_2$, KCl, and $CaCl_2$ into DI water. The ion concentrations of wastewater, seawater, and the corresponding purified water are generally measured by inductively coupled plasma optical emission spectrometer (ICP-OES) and inductively coupled plasma mass spectrometer (ICP-MS) as shown in Figure 10.5b and c [36].

10.3.3 Water Evaporation Rate

The water evaporation rate is an important parameter for SSG systems, which directly decides the effectiveness of the system in water purification. In a typical

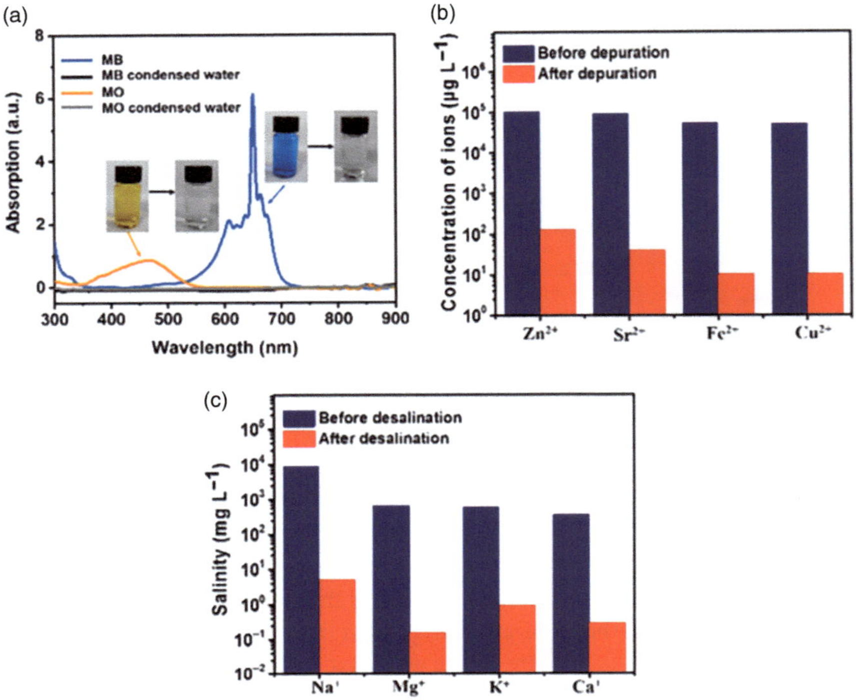

Figure 10.5 (a) UV–vis absorption spectra of MB and MO solutions before evaporation, (b) concentrations of heavy-metal ions in wastewater before and after purification, and (c) salinity values of the metal ions before and after purification. Source: Sun et al. [36]/with permission of American Chemical Society.

measurement, the water evaporation rate can be tested using a system illustrated in Figure 10.4, and the weight of the bulk water is measured before and after the SSG experiment. The solar energy is usually set to 1.0 kW m^{-2} light illuminations, and the water evaporation rate can be calculated by the formula (10.5):

$$m_0 - m = kt \tag{10.5}$$

where m_0 and m are the mass of water before and after evaporation, respectively. t is the irradiation time and k is the water evaporation rate.

10.3.4 Light Absorption

In an SSG system, given that solar radiation is the sole energy source, when light beams are falling on photothermal materials, they are absorbed and subsequently converted into heat that will be confined on the material surface. Therefore, light absorption is an important parameter, and it is generally measured using a UV–vis–NIR spectrophotometer (ranging from 300 to 2500 nm) equipped with an integrating sphere.

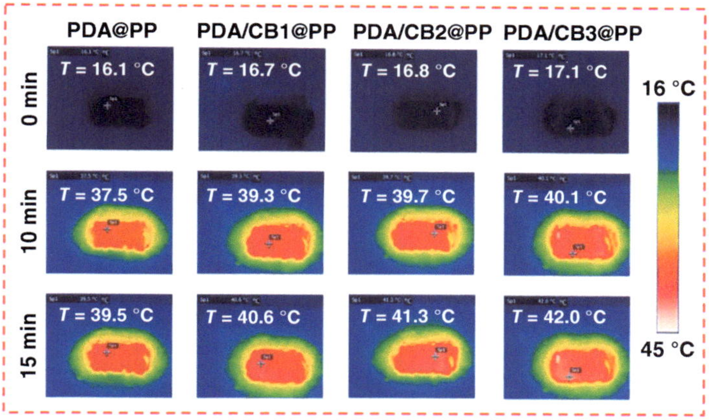

Figure 10.6 Infrared images show the surface temperature change of different photothermal materials when they are used to draw water. Source: Sun et al. [35]/with permission of Elsevier.

10.3.5 Surface Temperature

The surface temperature change of the photothermal material under sunlight directly affects the evaporation rate of water on the material surface. The surface temperature and the temperature distribution are generally evaluated in real-time using an infrared thermal imaging camera. As shown in Figure 10.6, the temperature generally will quickly rise when the photothermal material is placed under sunlight illumination, and the surface temperature changes are often evaluated on both dry and wet samples to determine the photothermal conversion ability [35].

10.3.6 Salt Resistance

As mentioned above, salt resistance is an important characteristic for an ideal SSG system to realize long-term continuous operation with high efficiency. To evaluate the salt resistance of an SSG system, solid NaCl particles are sometimes placed on the top of the photothermal sample during the continuous evaporation of seawater under 1 sun irradiation. If NaCl particles can be completely dissolved and transported into the bulk water, the SSG system is testified to have high salt resistance. As shown in Figure 10.7, salt particles deposited on the surface of a photothermal material are gradually removed within 3.5 hours while the system is generating freshwater vias steam generation [35].

10.4 Structural Design for Enhancing SSG Performance

10.4.1 Boosting the Light Absorption Characteristics

The light absorption characteristics of porous polymers are pivotal for photothermal conversion efficiency in SSG. An ideal solar absorber should efficiently absorb

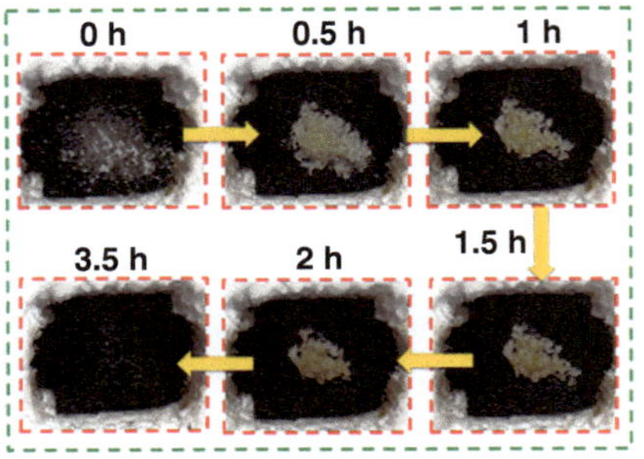

Figure 10.7 The ability of an SSG system to eliminate salt granules during operation. Source: Sun et al. [35]/with permission of Elsevier.

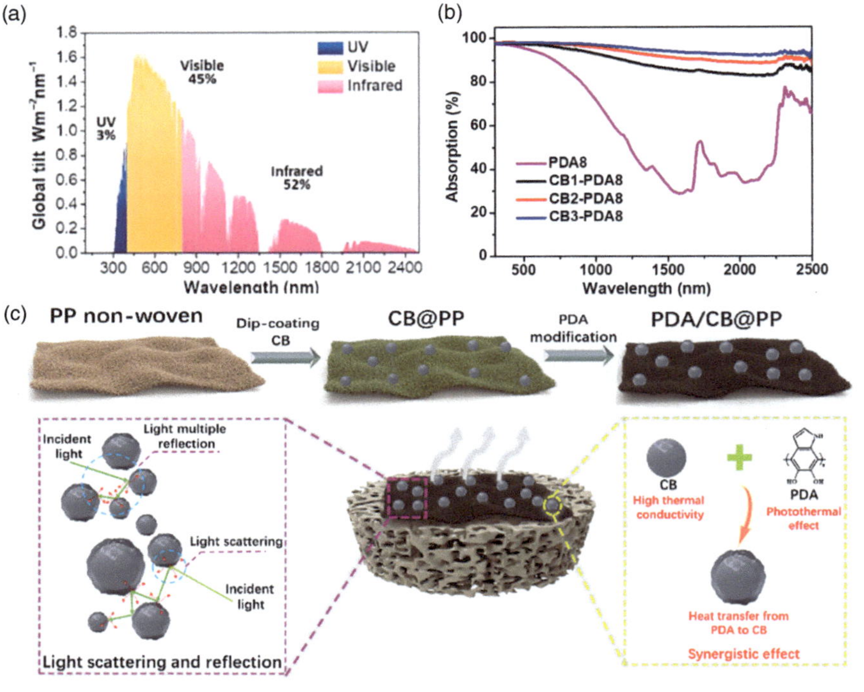

Figure 10.8 (a) Solar spectral irradiance, (b) the light absorption of carbon black and PDA-coated PP fabric, and (c) the fabrication process of the carbon black and PDA-coated PP fabric, and the light absorption mechanism of the composite porous fabric. Source: (a) Gao et al. [37]/Royal Society of Chemistry. (b,c) Sun et al. [35]/with permission of Elsevier.

sunlight across the entire solar spectrum, spanning wavelengths from 250 to 2500 nm, which includes the ultraviolet (UV), visible (vis), and near-infrared (NIR) regions (Figure 10.8a) [37]. Porous polymers, characterized by high porosity and open microchannels, serve as excellent substrates for enhancing light absorption [38]. Their intrinsic micropores and three-dimensional (3D) microstructures facilitate multiple sunlight scattering and trapping within pore vessels, thereby extending sunlight absorption across the solar spectrum.

The integration of nanoparticles into porous polymers preserves the substrate's microporosity while aiding in internal light trapping and reflection, thus broadening the range of light absorption. For instance, in the context of plasmonic metals, several strategies have been proposed to enhance light absorption by manipulating their size and morphology or by combining them with porous substrates such as electrospun poly/vinylidene fluoride (PVDF) [39], phase inversion PVDF [40], etc. Despite these advancements, the efficiency of light absorption predominantly remained confined to the UV–vis region, with extending it to the NIR region posing a significant challenge. This limitation can be addressed using porous polymer substrates, owing to their abundant meso-/micro-pores that enable multiple light reflections and scattering within the structure [26]. The resulting composite material of nanoparticle-coated porous polymer leads to more comprehensive light absorption across the entire solar spectrum. On these bases, Mi and coworkers [35] decorated carbon black on a polydopamine (PDA)-coated polypropylene (PP) fabric through a simple coating process and demonstrated exceptional light absorption (~95%) across the UV–vis–NIR spectrum. The fabrication process and vapor transport mechanism of the PP fabric are depicted in Figure 10.8c.

It was found that employing different nanoparticles offers similar enhancements. For instance, polypyrrole (PPy) nanoparticles exhibited a light absorption of 90.8% across the solar spectrum. After integration into porous microchannels, light absorption was increased to 97.5% owing to the synergistic effect of light absorption through PPy and light scattering via the porous structure [41]. Moreover, employing polymers as substrates for solar absorbers, integrated systems have been developed where the top surface of the porous structure is carbonized to act as the solar absorber, while the noncarbonized part serves as the support and wicking channels. This carbonization process transforms the color of the biomass to black and increases the ratio of p-conjugated bonds, thereby achieving excellent light absorption characteristics [42].

10.4.2 Suppressing Heat Loss in Porous Polymers Systems

Thermal management is a critical aspect that significantly impacts the photothermal conversion efficiency in SSG. The efficiency of SSG devices is closely linked to the light-to-heat conversion proficiency of the absorber surface. The substrate must have excellent heat localization capabilities while restricting heat flow to the bulk water. By minimizing heat transference to the bulk water and surrounding areas, a substantial amount of heat can be localized, allowing the absorbed sunlight to effectively heat the water layer. The heat converted to energy must be fully utilized for optimal water evaporation performance. In SSG systems utilizing porous polymers, heat loss suppression, and efficient utilization are achievable through three strategies, including: (i) reducing downward heat loss to the bulk water, (ii) preventing heat loss to the surroundings, and (iii) maintaining a lower temperature at the solar absorber compared to the surroundings to harness environmental energy, mainly achieved via 3D structural design [43]. A common approach to promote heat localization and reduce conductive heat transfer to the bulk water is by incorporating thermal insulating polymer foams, such as polystyrene (PS) foam and expandable polyethylene (EPE) foam, between the solar absorber and the bulk water. Additionally, the integration of water capillary channels is essential due to the hydrophobic nature of these insulators.

Considerable research has been focused on constructing porous absorbers that effectively minimize heat loss. For instance, the potential of materials like graphene foams in localizing heat during water evaporation has been explored [44]. However, their complex and costly fabrication processes limit practical applications. Consequently, research efforts are shifting toward finding eco-friendly, cost-effective materials with low thermal conductivity and open multichannel structures for water transport. Hydrophilic porous polymer materials have emerged as one of the most promising candidates. Various types of porous polymers have been proposed for their efficacy in heat localization and utilization. This is largely attributed to their pore microchannels, which allow external air to act as a thermal insulator, aiding in the transport and management of water vapor. For example, Liao et al. fabricated porphyrin-/aniline-based conjugated microporous polymers (PACMPs) by coating the commercial PU sponges with a hydrophilic black polymer layer, which not only facilitated solar energy absorption but also enhanced the water transportation in the porous foam (Figure 10.9a). The PACMP-coated foams showed a hierarchical structure that can induce light scattering (Figure 10.9b) and demonstrated excellent photothermal heating behavior under both dry and wet states. The PACMP modified using 120 mg polymer coating achieved the highest evaporation rate and the optimum solar-thermal conversion efficiency of 1.31 $kg\,m^{-2}\,h^{-1}$ and 86.3%, respectively (Figure 10.9c) [45].

10.4.3 Fast Water Transport of Porous Polymers

The water supply from the bulk to the upper region of the solar absorber, facilitated by capillary action, is a fundamental element for high-efficiency interfacial SSG systems. Porous polymers, composed of hydrophilic cellulose with a plethora of micro/nanochannels, are exceptionally conducive to efficient water transport, thereby enhancing water evaporation efficiency. An additional intrinsic feature of these materials is their self-floating capability, eliminating the need for external support. The water transport mechanism in porous polymers can be categorized into three pathways, namely, 1D, 2D, and 3D, each defined by the structural and compositional characteristics of the polymer.

The 1D water channel strategy has proven to be effective for continuous water transport from bulk to the evaporation region, offering salt resistance. For example, Mi and coworkers [22] developed a facile approach to construct a one-way fluidic SSG that combines photothermal and electro-thermal effects for all-weather-efficient SSG (Figure 10.10a and b). The SSG features a photothermal layer created through chemical vapor deposition polymerization of PPy on microfiber non-woven fabric, complemented by an electro-thermal layer made by coating silver nanoparticles (Ag NPs) and silicone resin on polypropylene (PP) fabric. A separating porous polymer layer, serving as insulation, is sandwiched

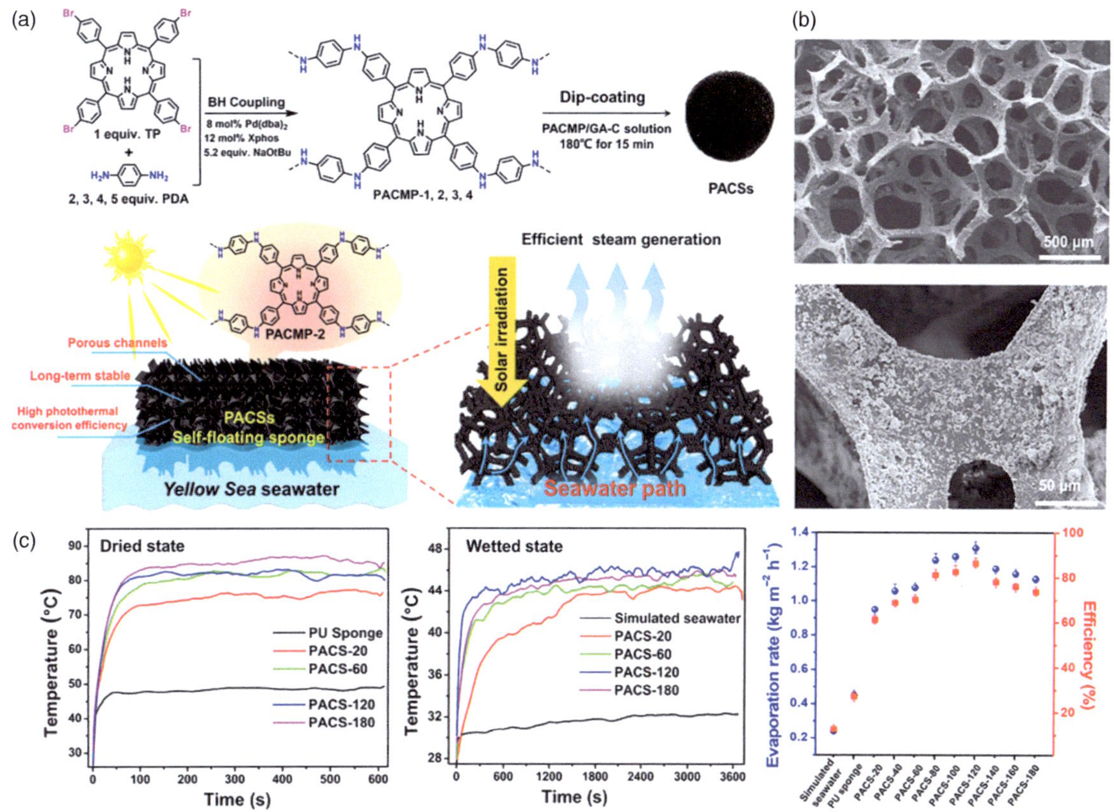

Figure 10.9 (a) Preparation process of PACMPs and the structure design of PACSs as a solar steam evaporator, (b) SEM images of modified PU sponge using 120 mg PACMP, and (c) surface temperature changes of dried and wetted foams by seawater under simulated sunlight illumination, and the evaporation rate and solar-thermal conversion efficiency results of foams coated with different amount of PACMPs. Source: Shi et al. [45]/with permission of American Chemical Society.

between these two layers to avoid direct contact of the electro-thermal layer with saline water. The optimized device achieved a remarkable evaporation rate of 2.61 $kg\,m^{-2}\,h^{-1}$ under 1 sun, with an input voltage of 1 V (Figure 10.10c). In addition, the developed SSG system exhibited excellent salt resistance for highly concentrated saline solutions (10–15 wt%) and the capability to purify wastewater and dye-contaminated water with high evaporation rates and efficiency (Figures 10.10d and e).

The 2D channel represents another innovative approach in water channel design, achieved by enveloping a thermal insulator with hydrophilic porous polymer materials. This design effectively facilitates water transport to the top solar absorber layer through a 2D water path within the porous structure, driven by capillary force. For instance, Mi and coworkers [36] enabled 2D water transport from the bulk water by wrapping a PS foam with PPy and CNT-coated porous polymer fabric (Figure 10.11a). This strategically designed architecture features a hierarchical microstructure and Janus wetting properties, which are instrumental in enhancing solar absorption and heat generation on the evaporation surface while effectively preventing salt crystallization. The superhydrophilic/underwater superoleophobic bottom surface of this layer forms a water layer that repels oil droplets, creating a channel to redirect concentrated salt back into the bulk water. The special design enabled efficient separation of oil-in-water emulsions and continuous desalination of seawater without diminishing the evaporation rate, as shown in Figure 10.11b.

3D channels also offer efficient water transport in SSG systems, such as those using hydrophilic porous polymers with a 3D mesoporous structure featuring abundant and well-aligned open micro/nanochannels. Zhang et al. [46] fabricated a degradable SSG device using a polybutylene adipate terephthalate (PBAT) monolith with a hierarchical porous morphology, which was achieved by *in situ* polymerization of PDA and subsequent dip coating with MXene as shown in Figure 10.11c. Owing to the synergistic effects of MXene and PDA, the resulting SSG device exhibited promising solar absorbance (up to 97.07%), superhydrophilicity, and high evaporation efficiency.

10.4.4 Salt Resistance of Porous Polymers

Salt crystallization poses a significant challenge in water desalination for SSG. Mitigating or preventing this issue is crucial for ensuring the reusability and stability of the SSG devices in the long term. Salt crystallization occurs when water travels toward the absorber and its supporting structures, leading to salt ions migrating with vapor and crystallizing on the solar absorber surface. The build-up of salts on the absorber can severely impede water evaporation, particularly when high concentrations of salts crystallize, resulting in substantially reduced photothermal conversion and evaporation efficiency. Consequently, a variety of methods have been developed to impart anti-salt crystallization properties to solar absorber systems, achieving notable improvements [22].

For example, Zhu et al. [47] demonstrated a flexible Janus porous polymer absorber that is fabricated through sequential electrospinning of polyacrylonitrile

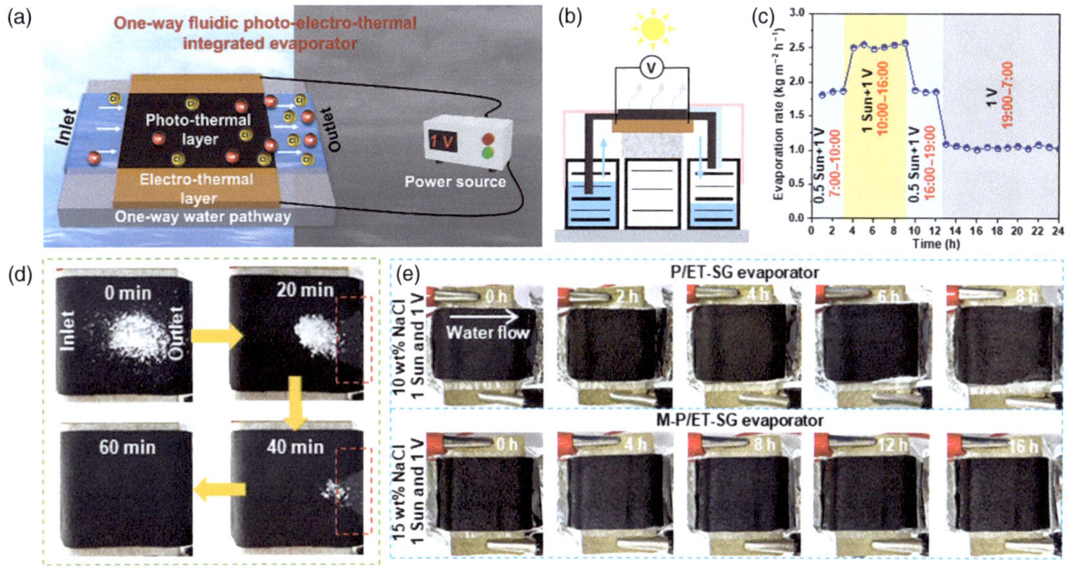

Figure 10.10 (a) The one-way fluidic photo-electro-thermal evaporator for all-weather desalination, (b) schematic of the one-way fluid SSG system, (c) the evaporation rates of the SSG when drawing seawater in the simulated different light environments of the day and night, (d) the ability of the SSG to eliminate salt granules during operation, and (e) salt precipitation on the surface of the SSG when continuously drawing saline water. Source: Sun et al. [22]/with permission of Springer Nature.

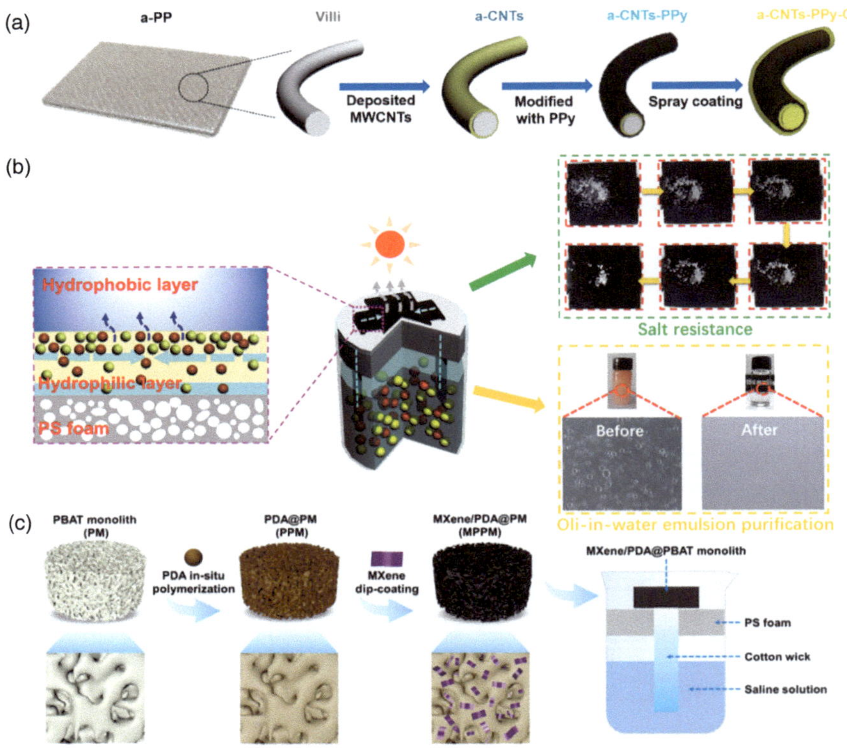

Figure 10.11 (a) Schematic illustrations of the fabrication process of PP nonwoven fabric-based photothermal absorber, (b) schematics of the transport path of water and salt ions salt resistance, and the oil-in-water emulsions purification performance of the SSG, and (c) schematic illustration of the preparation process of the PDA and MXene-coated PBAT monolith. Source: (a,b) Sun et al. [36]/with permission of American Chemical Society. (c) Zhang et al. [46]/with permission of Elsevier.

(PAN) and polymethylmethacrylate (PMMA) layers followed by spray coating of a carbon black nanoparticle layer (Figure 10.12a). The Janus property is realized by the hydrophobic carbon black coating and the hydrophilic PAN layer, which limits the salt deposition only in the hydrophilic PAN layer, thus preventing NaCl accumulation. Owing to the Janus effect, the absorber showed a high efficiency of 72% and consistent water output of 1.3 kg m^{-2} h^{-1} over 16 days under 1 sun (Figure 10.12b).

In addition, the salt crystallization on porous photothermal materials can be addressed by controlling the water transport direction and utilizing edge-preferential crystallization methods (Figure 10.12c). This design spatially isolates the salt crystallization sites from the water evaporation surface, enabling continuous steam generation and salt harvesting over 600 hours of nonstop operation [48]. Zhang et al. [49] developed a fluidic photothermal structure that successfully prevented salt formation during durable steam generation and simultaneously enabled electricity generation during one-way fluid transport. As shown in Figure 10.12d,

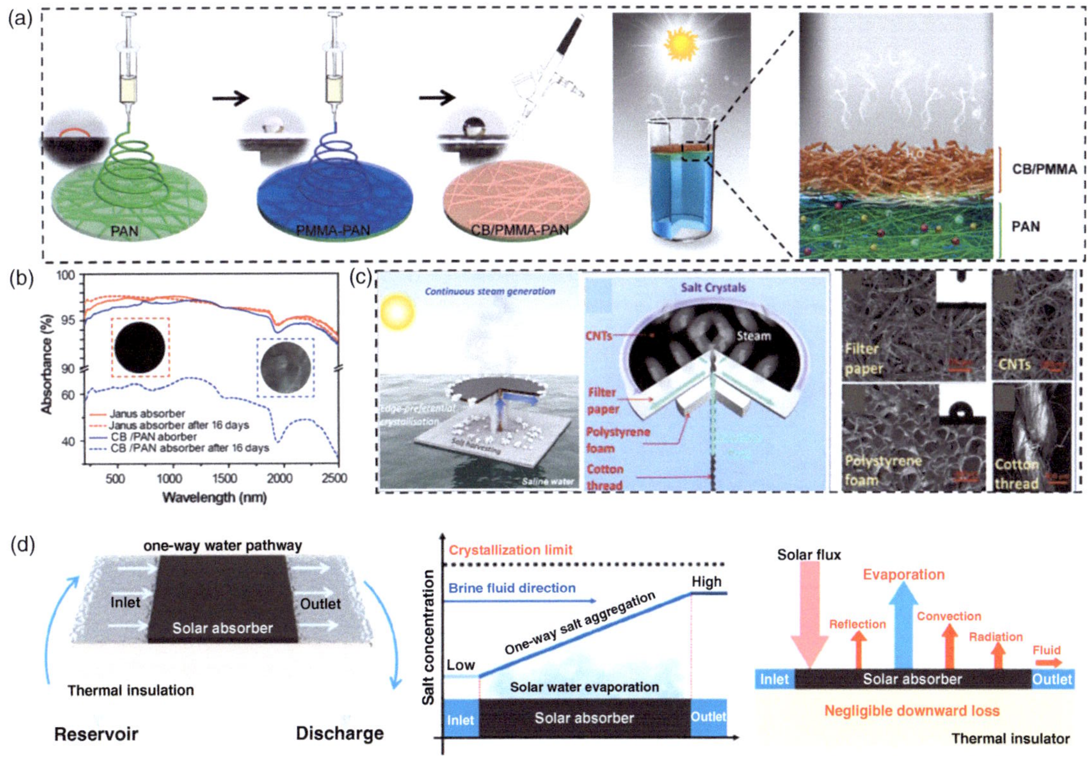

Figure 10.12 (a) Flow chart for the fabrication of Janus absorber and schematic of solar steam generation and desalination, (b) the absorption spectra of the Janus absorber and CB/PAN membrane before and after the cycle test, (c) schematic illustration of the novel design for continuous solar steam generation and salt harvesting, and (d) schematic illustration of the fluidic photothermal structure for completely salt-rejecting solar water evaporation from one-way saline fluid. Source: (a,b) Xu et al. [47]/with permission of John Wiley & Sons. (c) Adapted from Xu et al. [48]/with permission of Royal Society of Chemistry. (d) Zhang et al. [49]/American Chemical Society.

the saline fluid was continuously directed through the solar absorber composed of a porous polymer-based polyaniline (PANi)/cellulose bilayer, and the salt rejection was achieved even during intense steam generation.

10.5 Opportunity and Challenge of Porous Polymer for SSG

The utilization of porous polymers in SSG has shown promising results in a variety of water purification applications. The exceptional performance of porous polymers in SSG can be attributed to several factors: (i) their high porosity and microchannels facilitate multiple internal light reflections and trapping; (ii) their relatively low thermal conductivity ensures localized heat at the air–water interfaces while minimizing heat losses; (iii) their surface wettability and interconnected micro-channels lead to efficient water supply channels; and (iv) their microsize channels, biomass composition, and hydrophilicity contribute to extraordinary anti-salt crystallization and self-cleaning properties. Despite these advancements, several challenges remain that need to be addressed: (i) Exploring new classes of photothermal materials integrated with porous polymer substrates could further improve solar thermal conversion efficiency; (ii) The structural stability and long-term reusability of these systems are critical for practical applications; (iii) The techniques of large-scale production of SSG is another challenge for the developed and practical usage of porous-polymer-based SSG.

In addition, current researches on porous polymer-based SSG are mostly focused on the modification of commercially available polymer foams, porous nature materials, or porous aerogels. Scalable techniques such as $scCO_2$ foaming are rarely used in the fabrication of SSG, mainly due to the closed cellular structure of most $scCO_2$-foamed materials. However, achievements in the precisely controlled fabrication of open-cellular foams were made, and it is expected that the large-scale production of $scCO_2$-foamed porous polymer foams would be a promising material for the application of SSGs.

References

1 Zhou, X., Zhao, F., Guo, Y. et al. (2018). A hydrogel-based antifouling solar evaporator for highly efficient water desalination. *Energy & Environmental Science* 11: 1985–1992.

2 Elsheikh, A.H., Sharshir, S.W., Ahmed Ali, M.K. et al. (2019). Thin film technology for solar steam generation: a new dawn. *Solar Energy* 177: 561–575.

3 Cabassud, C. and Wirth, D. (2003). Membrane distillation for water desalination: how to choose an appropriate membrane? *Desalination* 157: 307–314.

4 Campione, A., Gurreri, L., Ciofalo, M. et al. (2018). Electrodialysis for water desalination: a critical assessment of recent developments on process fundamentals, models and applications. *Desalination* 434: 121–160.

5 Greenlee, L.F., Lawler, D.F., Freeman, B.D. et al. (2009). Reverse osmosis desalination: water sources, technology, and today's challenges. *Water Research* 43: 2317–2348.

6 Yang, J., Chen, Y., Jia, X. et al. (2020). Wood-based solar interface evaporation device with self-desalting and high antibacterial activity for efficient solar steam generation. *ACS Applied Materials & Interfaces* 12: 47029–47037.

7 Ibrahim, I., Bhoopal, V., Seo, D.H. et al. (2021). Biomass-based photothermal materials for interfacial solar steam generation: a review. *Materials Today Energy* 21: 100716.

8 Peng, F., Xu, J., Bai, X. et al. (2021). A janus solar evaporator with 2D water path for highly efficient salt-resisting solar steam generation. *Solar Energy Materials and Solar Cells* 221: 110910.

9 Ibrahim, I., Seo, D.H., McDonagh, A.M. et al. (2021). Semiconductor photothermal materials enabling efficient solar steam generation toward desalination and wastewater treatment. *Desalination* 500: 114853.

10 Hu, Y., Ma, H., Wu, M. et al. (2022). A reconfigurable and magnetically responsive assembly for dynamic solar steam generation. *Nature Communications* 13: 4335.

11 Arunkumar, T., Raj, K., Dsilva Winfred Rufuss, D. et al. (2019). A review of efficient high productivity solar stills. *Renewable and Sustainable Energy Reviews* 101: 197–220.

12 Neumann, O., Urban, A.S., Day, J. et al. (2013). Solar vapor generation enabled by nanoparticles. *ACS Nano* 7: 42–49.

13 Wu, J., Mahajan, A., Riekehr, L. et al. (2018). Perovskite $Sr_x(Bi_{1-x}Na_{0.97-x}Li_{0.03})0.5TiO_3$ ceramics with polar nano regions for high power energy storage. *Nano Energy* 50: 723–732.

14 Wang, Z., Liu, Y., Tao, P. et al. (2014). Bio-inspired evaporation through plasmonic film of nanoparticles at the air–water interface. *Small* 10: 3234–3239.

15 Fu, J., Li, Z., Li, X. et al. (2023). Hierarchical porous metallic glass with strong broadband absorption and photothermal conversion performance for solar steam generation. *Nano Energy* 106: 108019.

16 Jia, S., Hao, L., Liu, Y. et al. (2023). Freestanding hydrophilic/hydrophobic janus covalent organic framework membranes for highly efficient solar steam generation. *ACS Materials Letters* 5: 458–465.

17 Yu, B., Wang, Y., Zhang, Y., and Zhang, Z. (2023). Nanoporous black silver film with high porosity for efficient solar steam generation. *Nano Research* 16: 5610–5618.

18 Li, Y., Zhang, S., Xia, Z. et al. (2023). Micro-macro-capillaries fabric-based evaporator for eliminating salt accumulation and highly efficient solar steam generation. *Separation and Purification Technology* 308: 122852.

19 Bae, K., Kang, G., Cho, S.K. et al. (2015). Flexible thin-film black gold membranes with ultrabroadband plasmonic nanofocusing for efficient solar vapour generation. *Nature Communications* 6: 10103.

20 Ni, G., Li, G., Boriskina Svetlana, V. et al. (2016). Steam generation under one sun enabled by a floating structure with thermal concentration. *Nature Energy* 1: 16126.

21 Huang, Q., Liang, X., Yan, C., and Liu, Y. (2021). Review of interface solar-driven steam generation systems: high-efficiency strategies, applications and challenges. *Applied Energy* 283: 116361.

22 Sun, S., Tian, Q., Mi, H.-Y. et al. (2022). Fabric-based all-weather-available photo-electro-thermal steam generator with high evaporation rate and salt resistance. *Science China Materials* 65: 2479–2490.

23 Wilson, H.M., Ahirrao, D.J., Raheman Ar, S., and Jha, N. (2020). Biomass-derived porous carbon for excellent low intensity solar steam generation and seawater desalination. *Solar Energy Materials and Solar Cells* 215: 110604.

24 Ivan, M.N.A.S., Saha, S., Saleque, A.M. et al. (2024). Progress in interfacial solar steam generation using low-dimensional and biomass-derived materials. *Nano Energy* 120: 109176.

25 Wang, Y., Zhang, M., Shen, X. et al. (2021). Biomass-derived carbon materials: controllable preparation and versatile applications. *Small* 17: 2008079.

26 Sheng, C., Yang, N., Yan, Y. et al. (2020). Bamboo decorated with plasmonic nanoparticles for efficient solar steam generation. *Applied Thermal Engineering* 167: 114712.

27 Lin, Y., Xu, H., Shan, X. et al. (2019). Solar steam generation based on the photothermal effect: from designs to applications, and beyond. *Journal of Materials Chemistry A* 7: 19203–19227.

28 Li, J., Zhou, X., Mu, P. et al. (2020). Ultralight biomass porous foam with aligned hierarchical channels as salt-resistant solar steam generators. *ACS Applied Materials & Interfaces* 12: 798–806.

29 Meng, S., Gong, T., Zhao, X. et al. (2021). Boosting solar steam generation in dynamically tunable polymer porous architectures. *Polymer* 226: 123811.

30 Shi, L., Wang, X., Hu, Y. et al. (2020). Solar-thermal conversion and steam generation: a review. *Applied Thermal Engineering* 179: 115691.

31 Zhou, W., Zhou, C., Deng, C. et al. (2022). High-performance freshwater harvesting system by coupling solar desalination and fog collection with hierarchical porous microneedle arrays. *Advanced Functional Materials* 32: 2113264.

32 Zhang, P., Liao, Q., Yao, H. et al. (2019). Direct solar steam generation system for clean water production. *Energy Storage Materials* 18: 429–446.

33 Saleque, A.M., Nowshin, N., Ivan, M.N.A.S. et al. (2022). Natural porous materials for interfacial solar steam generation toward clean water production. *Solar RRL* 6: 2100986.

34 Zhang, Y., Xiong, T., Nandakumar, D.K., and Tan, S.C. (2020). Structure architecting for salt-rejecting solar interfacial desalination to achieve high-performance evaporation with in situ energy generation. *Advanced Science* 7: 1903478.

35 Sun, S., Sun, B., Wang, Y. et al. (2021). Carbon black and polydopamine modified non-woven fabric enabling efficient solar steam generation towards seawater desalination and wastewater purification. *Separation and Purification Technology* 278: 119621.

36 Sun, S., Wang, Y., Sun, B. et al. (2021). Versatile janus composite nonwoven solar absorbers with salt resistance for efficient wastewater purification and desalination. *ACS Applied Materials & Interfaces* 13: 24945–24956.

37 Gao, M., Zhu, L., Peh, C.K., and Ho, G.W. (2019). Solar absorber material and system designs for photothermal water vaporization towards clean water and energy production. *Energy & Environmental Science* 12: 841–864.

38 Mu, P., Zhang, Z., Bai, W. et al. (2019). Superwetting monolithic hollow-carbon-nanotubes aerogels with hierarchically nanoporous structure for efficient solar steam generation. *Advanced Energy Materials* 9: 1802158.

39 Wu, T., Li, H., Xie, M. et al. (2019). Incorporation of gold nanocages into electrospun nanofibers for efficient water evaporation through photothermal heating. *Materials Today Energy* 12: 129–135.

40 Li, T., Fang, Q., Lin, H., and Liu, F. (2019). Enhancing solar steam generation through manipulating the heterostructure of PVDF membranes with reduced reflection and conduction. *Journal of Materials Chemistry A* 7: 17505–17515.

41 Huang, W., Hu, G., Tian, C. et al. (2019). Nature-inspired salt resistant polypyrrole–wood for highly efficient solar steam generation. *Sustainable Energy & Fuels* 3: 3000–3008.

42 Wang, H., Du, A., Ji, X. et al. (2019). Enhanced photothermal conversion by hot-electron effect in ultrablack carbon aerogel for solar steam generation. *ACS Applied Materials & Interfaces* 11: 42057–42065.

43 Wang, Y., Wu, X., Yang, X. et al. (2020). Reversing heat conduction loss: extracting energy from bulk water to enhance solar steam generation. *Nano Energy* 78: 105269.

44 Qiu, P., Liu, F., Xu, C. et al. (2019). Porous three-dimensional carbon foams with interconnected microchannels for high-efficiency solar-to-vapor conversion and desalination. *Journal of Materials Chemistry A* 7: 13036–13042.

45 Shi, Y., Meng, N., Wang, Y. et al. (2022). Scalable fabrication of conjugated microporous polymer sponges for efficient solar steam generation. *ACS Applied Materials & Interfaces* 14: 4522–4531.

46 Zhang, R., Han, W., Jiang, H. et al. (2022). PBAT/MXene monolithic solar vapor generator with high efficiency on seawater evaporation and swage purification. *Desalination* 541: 116015.

47 Xu, W.C., Hu, X.Z., Zhuang, S.D. et al. (2018). Flexible and salt resistant janus absorbers by electrospinning for stable and efficient solar desalination. *Advanced Energy Materials* 8.

48 Xu, Z., Zhang, L., Zhao, L. et al. (2020). Ultrahigh-efficiency desalination via a thermally-localized multistage solar still. *Energy & Environmental Science* 13: 830–839.

49 Zhang, Y., Xiong, T., Suresh, L. et al. (2020). Guaranteeing complete salt rejection by channeling saline water through fluidic photothermal structure toward synergistic zero energy clean water production and in situ energy generation. *ACS Energy Letters* 5: 3397–3404.

Index

Functional Polymer Foams: Green Fabrication Methods, Performance and Applications, First Edition. Haoyang Mi.

q

Functional Polymer Foams